AF770507

METAL POISONING IN FISH

ELSA M. B. SORENSEN
Environmental and Life Sciences Associates
Austin, Texas 78720

CRC Press
Boca Raton Ann Arbor Boston

Library of Congress Cataloging-in-Publication Data

Sorensen, Elsa M. B., 1948—
Metal poisoning in fish / Elsa M. B. Sorensen.
p. cm.
Includes bibliographical references and index.
ISBN 0-8493-4268-6
1. Heavy metals—Toxicology. 2. Fishes—Diseases. 3. Heavy metals—Physiological effect. 4. Fishes—Physiology. I. Title.
SH177.H4S67 1991
597'.02'4—dc20 91-7175
CIP

This book represents information obtained from authentic and highly regarded sources. Reprinted material is quoted with permission, and sources are indicated. A wide variety of references are listed. Every reasonable effort has been made to give reliable data and information, but the author and the publisher cannot assume responsibility for the validity of all materials or for the consequences of their use.

All rights reserved. This book, or any parts thereof, may not be reproduced in any form without written consent from the publisher.

Direct all inquiries to CRC Press, Inc., 2000 Corporate Blvd., N.W., Boca Raton, Florida, 33431.

© 1991 by CRC Press, Inc.

International Standard Book Number 0-8493-4268-6

Library of Congress Card Number 91-7175
Printed in the United States

THE AUTHOR

Elsa M. B. Sorensen, Ph.D., is founder of Environmental and Life Sciences Associates, a private consulting firm in Austin, Texas. Since 1979, she has been expert witness and consultant to the Environmental Protection Division of Office of the Texas Attorney General, Texas Department of Water Resources, Texas Parks and Wildlife, Duke Power Company, and other state agencies and utilities in determining the impact of toxic elements on reservoirs associated with electric generating stations and manufacturing companies. Dr. Sorensen obtained her doctorate in 1974 from the Zoology Department at the Univerisity of Texas in Austin, Texas. From 1976-1978, she was a postdoctoral fellow at the Division of Biological and Medical Research at Argonne National Laboratory. Other faculty and research appointments include those at the University of Texas and the University of Tennessee Health Sciences Center at Memphis.

Dr. Sorensen has presented over 50 invited and contributed research papers at national meetings. She has published over 100 research papers, and a portion of her work is now part of a permanent collection at the Smithsonian Institution, Washington D.C. Dr. Sorensen appears to be first to publish the bizarre hepatic abnormalities occurring at high hepatic As levels in sunfish (i.e. As-inclusions, necrotic bodies, and fibrous bodies). She quantitates intracellular elemental damage, localizes intracellular elemental deposits (to determine high-risk organelles and mechanism of action of the element under test), and evaluates plasma membrane metal binding and influx. Dr. Sorensen's research has furthered understanding of morphological and functional changes occurring during elemental challenge—changes which occur in a predictable fashion, allowing assessment of the rate and degree of tissue damage in fish.

To my husband and dedictated co-worker Dr. Alan L. Sorensen, who gave unselfishly of his time in discussions, in formating the final document, and in literature searches. Also, to our son David, who helped obtain the references upon which this volume is based.

PREFACE

This volume focuses on the teleost response to excess levels of seven elements (i.e., Se, As, Pb, Zn, Cd, Cu, and Hg) singly and in combination. Regulatory personnel, enforcement personnel, graduate students, and experienced scientists should find this information useful in understanding the impact of specific elements on fish. Although, this volume goes beyond a simple condensation of EPA and FDA data, a compilation of the U.S. EPA lowest observed effect levels for 24 h (L.O.E.L.) is presented in Appendix 1 for each of the seven elements for comparative purposes.

An up-to-date critique of literature concerns many aspects of metal poisoning in euryhaline and stenohaline fish. Emphasis is placed on the use of quantitative data gathered from a variety of sources to pinpoint a specific elemental agent as the causal factor in morbidity and mortality of fish. Topics include mechanism of action, toxicity, biological effects, accumulation, tissue distribution, concentration factors, maximum acceptable toxicant concentrations (MATC), application factors, biological half-lives, uptake kinetics, depuration kinetics, and detoxification mechanisms. Physiological, behavioral, histopathological, and biochemical effects are summarized, where possible. Despite the wealth of information currently available in this area, chapter content varies with the availability of published research. Great indebtedness goes to Harvey M. Kane, Life Science Editor, and CRC Press, Inc., Boca Raton, FL, who have made this publication possible.

CONTENTS

CHAPTER I
INTRODUCTION

Introduction

Loading of aquatic habitats with nondegradable, nonnutritive, cumulative pollutants such as As, Pb, Cd, and Hg can result in undeniably complex alterations of numerous teleost trophic levels. These effects can last for centuries, and studies spanning the globe document the ubiquity of such contamination. Industrial and agricultural discharges are considered the primary source of metal-poisoning of fish in Egypt (El Nabawi, *et al.*, 1987), Poland (Zamojski, *et al.*, 1986), England (Badsha and Goldspink, 1982), the Texas Outer Continental Shelf (Horowitz and Presley, 1977), and Canada (Cappon, 1987). Industrial and/or sewage effluents are a major source of metal poisoning of fish in Africa (Greichus, *et al.*, 1978), Wisconsin and Michigan (Kleinert, *et al.*, 1974), Canada (Johnson, 1987), and Mexico (Villarreal-Trevino, *et al.*, 1986). Highway or motorboat traffic are main sources of metal poisoning in Virginia (Pagenkopf and Neuman, 1974; Van Hassel, *et al.*, 1979; Van Hassel, *et al.*, 1980; Ney and Van Hassel, 1983), Arizona (Bussey, *et al.*, 1976), and Illinois (Rolfe and Jennett, 1979). Mine and smelting operations are said to be responsible for contamination of fish in Britain (Carpenter, 1924; Carpenter, 1925; Carpenter, 1926; Carpenter, 1927; Carpenter, 1930; Jones, 1940), Missouri (Gale, *et al.*, 1982; Schmitt and Finger, 1987), and Illinois (Rolfe and Jennett, 1979). Feces from waterfowl and airborne particulate are thought to be the primary sources of metals in fish in some places in Michigan (Mathis and Kevern, 1975). Even studies of fish from areas as remote as the Canadian High Arctic (Bohn and Fallis, 1978) and tropical islands (Taylor and Bright, 1973), or fish from worldwide sources (Zook, *et al.*, 1976), or fish from pristine Canadian lakes (Uthe and

Bligh, 1971) contain sufficiently elevated metal levels to be cause for concern (Chow, *et al.*, 1974; Settle and Patterson, 1980).

Chemical and Physical Properties of Elements Considered

A brief glance at the periodic table shows that light metals are located to the far left, transition elements or metals at the center, nonmetals to the right, and inner transition elements below. The dividing line between metals and nonmetals is not sharp; in general, metalloids lie along the stairstep-line separating the metals from the nonmetals. Amphoteric in nature, metalloids have some characteristics of metals and some characteristics of nonmetals.

Regardless of the position in the periodic table, most elements are toxic to fish if ionized (i.e. present in the unbound or free form). And if ionized, metals tend to loose electrons; whereas, nonmetals tend to gain electrons. Because electron loss in metals results in a positive charge, remaining electrons are pulled closer to the nucleus, resulting in an ionic radius smaller than the atomic radius (half the internuclear distance) (Sanderson, 1960). Nonmetals gain electrons, which offset the charge balance in the opposite direction, pulling electrons away from the nucleus to give a larger ionic radius. For example, ionized Se^{-2} (considered either a metalloid or a nonmetal) has an ionic radius almost twice the atomic radius. The reverse is generally true of metals (e.g. the heavy metal ion Pb^{+4} has an ionic radius approximately half that of the atomic radius).

Different valence states of As provide an example of the amphoteric nature of metalloids. Arsenic has either pentavalent (arsenate) or trivalent (arsenite) valence forms. When ionized, arsenate tends to loose electrons (as do metals), such that the resulting ionic radius is smaller than the atomic radius. However, arsenite tends to gain electrons (as do nonmetals), resulting in an ionic radius larger than the atomic radius. This difference in the valence state of As leads to a five-fold difference in ionic radius. The amphoteric nature of As and other parameters (e.g. steric hindrance) result in lower toxicity of arsenates than arsenites (Harvey, 1975).

Transition elements (e.g. Cu and Fe) are classified as “heavy metals” or simply “metals”. Some of these belonging to Group IB, have unfilled or partially filled *d* subshells in the stable state. Many (e.g. Cd, Cu, Zn) are known for their ability to induce metallothionein (Olafson, 1981) and to precipitate epithelial mucus of fish (e.g. Cu, Zn, Hg) (Coombs, *et al.*, 1972).

Metals donate *d* orbital electrons in coordinate compounds and stabilize several more proximate elements than the metal actually bonds (e.g. ferrous groups in heme).

Underlying orbitals participate in metallic bonding in Cu, for example, by promotion of at least one *d* electron into a higher energy orbital of the outermost principal quantum level (Sanderson, 1960). Here the electron is available for participation in electrical and thermal conduction. In fact, one of the most conspicuous features in passing from Group IB elements (Cu, Ag, and Au) to Group IIB elements (Zn, Cd, and Hg) is the abrupt and considerable reduction in metallic nature from Group IB to Group IIB. Group IIB elements have lower melting points, heats of fusion, boiling points, heats of atomization, and electrical and thermal conductivity than Group IB elements. The availability of *d* orbital electrons of IB, but not IIB, elements accounts for the difference. Incidentally, no evidence is available that the *d* orbital electrons of Group IIB metals are ever able to become involved in bonding; whereas, Group IB metals definitely participate in bonding.

Such differences in ionization and bonding ultimately determine how individual elements interact with biological ligands to form complexes. The elements selected for this review (Se, As, Pb, Zn, Cd, Cu, and Hg) are very toxic to fish (Jones, 1939; Brown, 1968). Current water quality criteria are provided in Appendix 1. Evaluation of the general complexing abilities of each shows that As, Se, and Pb (two metalloids and a metal) react in less complicated ways than the heavy metals Zn, Cd, Cu, and Hg. Multiple interactions are observed for some elements; whereas, a paucity of interactions is found for others (Table 1).

For example, some ions (e.g. As^{+5}) simply substitute for a second ion (P^{+5}), while others (e.g. Cd^{+2}) react with almost every cellular ligand. In substituting for phosphates, arsenates uncouple oxidative phosphorylation and inhibit dark repair (Petres and Berger, 1972). In contrast, Cd uncouples oxidative phosphorylation (Jacobs and Jacob, 1956; Vallee and Ulmer, 1972), inactivates some enzymes while activating others (Vallee and Ulmer, 1972; Prasada-Rao, *et al.*, 1983), displaces Zn bound to high molecular weight proteins in the cytoplasm (Bryan and Hidalgo, 1976; Frazier and Kingsley, 1976; Frazier and Puglese, 1978), alters ribonucleic acid synthesis (Bryan and Hidalgo, 1976), probably destabilizes deoxyribonucleic acid (Eichhorn, 1962; Stoll, *et al.*, 1976), activates metallothionein synthesis (Bryan and Hidalgo,

1976; Failla, *et al.*, 1979; Stacey, *et al.*, 1980), and alters the plasma membrane allowing release of ions such as potassium (Stacey, *et al.*, 1980) and/or cytoplasmic enzymes such as aspartate aminotransferase and lactate dehydrogenase (Stacey, *et al.*, 1980; Sorensen, 1985). Diagrammatic representations of such interactions between Cd and intracellular components (Sorensen, 1985) show these to be rather complicated. Certainly, in comparison to As^{+5}, Cd^{+2} is obviously more reactive intracellularly based on current scientific knowledge (Table 1).

Table 1. Complexes Formed by Elements Considered in This Volume (Based on Sanderson, 1960).

		1	2			3		4
	Group*	**O**	**S**	**P**	**C**	**F**	**Cl**	**N**
Selenium	VIA	•						
Arsenic	VA	•	•					
Lead	IVA	•	•					
Zinc	IIB		•	•	•	•	•	
Cadmium	IIB		•	•	•	•	•	•
Copper	IB		•	•	•	•	•	•
Mercury	IIB		•	•	•	•	•	

* Group in periodic table.

[1] Known chiefly as complexes with donor oxygen.

[2] Forms stable complexes in which the donor is halogen, sulfur, phosphorus, or carbon.

[3] Has a tendency to form complexes with other nonmetals, fluorine, or chlorine.

[4] Known chiefly as complexes with donor nitrogen.

Terminology of Metal Poisoning

Thusfar, we see that metal poisoning in fish is derived from a multitude of sources and that each element interacts in unique ways intracellularly. At this point, some of the terminology of metal poisoning will provide a basis for discussion of individual elements for those less familiar with the field.

Despite differences among various fields of toxicology, terminology used to evaluate metal poisoning in fish is akin to that of other fields. The term "bioassay", for example, is used in reference to toxicity testing of whole fish or preparations thereof (Warner, 1967; Sprague, 1969; Sprague, 1970; Sprague, 1971; Sprague, 1973), as well as in reference to the testing of phamacological agents. In either case, the active power of an agent (metal or drug) is determined by noting its effect on a live animal or an isolated preparation thereof compared with the effect on a standard or control preparation. Often scientists unfamiliar with this usage begrudge ichthyological toxicologists for use of a term sometimes immutably associated with pharmacology.

Even though this volume extends beyond simple toxicity testing of fish, toxicity testing (bioassays) form a good basis for beginning a discussion of terminology because accumulation, fate, and effects studies often are either based on toxicity tests or form an integral part of a teleost study. Also, such a discussion is essential since marked differences exist between toxicity testing of fish and that of air-breathing vertebrates. The very nature of the tests of fish are different, requiring knowledge of ionic form, chemical form, water hardness, pH, alkalinity, temperature, oxygen level, and a number of other interrelated parameters not considered as essential (or not considered at all) in testing of rodents and other vertebrates. Therefore, the terminology used in acute and chronic toxicity tests of fish will be presented briefly to facilitate a better understanding of the effects of metals on fish.

In acute and chronic toxicity testing, fish are administered agents by intravenous, subcutaneous, intragastric, dietary, or aqueous routes. Regardless of mode of administration, groups of fish are exposed to specific concentrations of an element using procedures which are fairly standardized to obtain a curve relating survival (or mortality) to time (Fig. 1) or to concentration of metal (Fig. 2). Stringent controls are placed on acclimation time, feeding, water quality, concentration range, numbers of fish, randomization of specimens, observation times, dosing apparatuses, water replacement in continuous-flow studies, and other factors (Sprague, 1969; Sprague, 1970; Sprague, 1971; Sprague, 1973; Alabaster and Lloyd, 1980).

In Figure 1, cumulative percent survival is plotted against time since the initiation of the exposure to aqueous solutions of Zn. While all controls survive, all bluegills exposed to 32 ppm Zn at 21°C die in about 7–8 h. In Figure 2, cumulative percent mortality is plotted against As exposure concentration for green sunfish exposed at 20°C. The absence of linearity in either Figure 1 or Figure 2 reduces the accuracy of estimation of the survival time or concentration for 50% of fish being tested; so, transformation of data is considered advisable.

Use of a logarithmic scale for metal concentration and a probability (or probit) scale for mortality generally produces a linear relationship between percent mortality (or, if desired, survival) and metal exposure level. Since

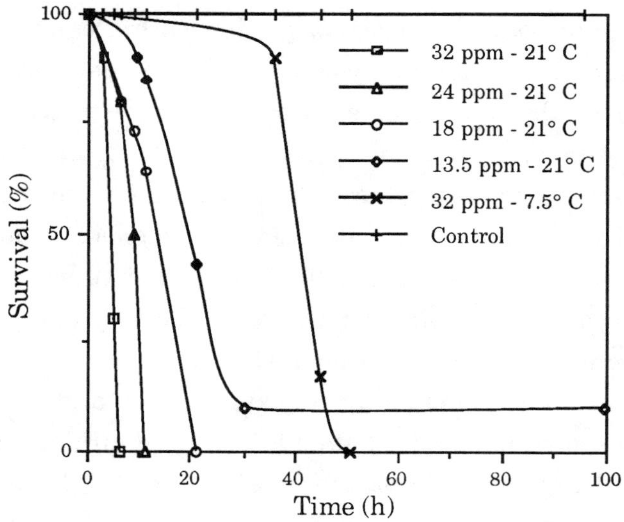

Figure 1. Relationship between survival and time in sunfish exposed to Zn, as well as the effect of temperature on results. Percent survival for warm- and cold-acclimated bluegill sunfish (*Lepomis macrochirus*) exposed to soluble Zn sulfate at pH 7.3 and 46 ppm average total hardness. From left to right, fish are exposed to 32, 24, 18, and 13.5 ppm Zn at 21°C. The fifth curve represents data from 32 ppm Zn at 7.5°C, and control values are shown as 100% survival at the top. For comparison, no fish succumbed within 96-h if insoluble Zn phosphate is used in comparable toxicity tests. (Modified and redrawn from Cairns *et al.*, 1971.)

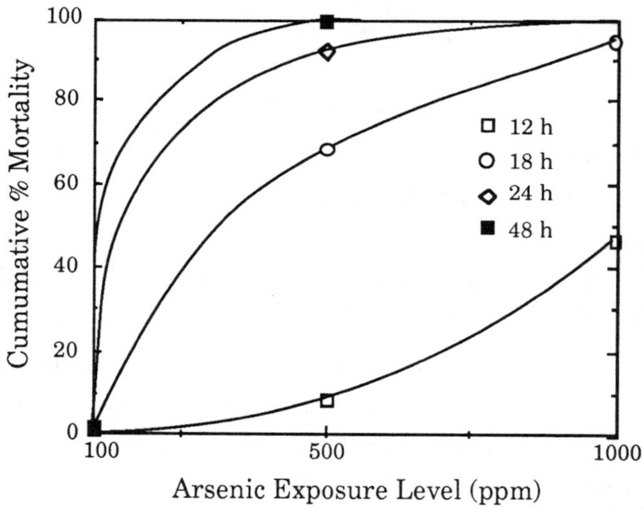

Figure 2. Relationship between mortality and concentration in sunfish exposed to As, as well as the effect of time on results. Percent survival for green sunfish (*Lepomis cyanellus*) is given for fish exposed to arsenate at 20°C, pH 8.4, and 92 ppm average total hardness. The LC50 values are 1000, 350, 175, and 150 ppm As for 12, 18, 24, and 48 h, respectively. (Modified and redrawn from Sorensen, 1976.)

the probability scale never reaches either 100% or 0%, the position of these terminal data points generally are plotted using an arrow for each data point. This method is considered theoretically sound to reduce possible inaccuracies associated with derivation of the LC50 value using a curved rather than linear plot. Nonetheless, use of raw data for LC50 values, rather than extrapolated values from the linear plot is preferred by many workers in the field.

The shape of the toxicity curve provides clues for mode of action of toxic elements. For example, when fish are exposed to Cu or to Zn and data are presented using probit and logarithm scales, a linear curve is produced over the majority of the concentration range (Sprague, 1964). However, an abrupt transition in the curve (referred to as the incipient lethal limit, ILL; the median threshold concentration; the lethal threshold concentration; or the ultimate median threshold limit) occurs at 420 ppb for Zn and 32 ppb for Cu (Sprague and Ramsay, 1965). In logarithm-probit plots, the toxicity curve parallels the ordinate at a specific concentration; in untransformed plots, the curve becomes asymptotic with the time axis at the same concentration. At this metal concentration, extending time of exposure theoretically does not produce a change in observed mortality. The abrupt change in the toxicity curve suggests that at extremely elevated levels of either metal, a sudden destruction of ionoregulatory mechanisms occurs (see Chapters V and VII).

In toxicity testing of teleosts, the level of metal killing half of fish in the group being tested is indicated as median lethal concentration (LC50 or LD50) for a specified time. Much of the early testing of fish involved estimation of the concentration killing half of a group of fish in 48 h. To distinguish the value of this concentration from the point at which the curve becomes asymptotic, this concentration is designated by various terms including the 48-h median threshold concentration, 48-h LC50 (lethal concentration killing 50% of the group), 48-h TC50 (toxic concentration for 50% of the group), 48-h LCm (median lethal concentration), 48-h ILC50 (incipient lethal concentration for 50% of the group), or the 48-h TLm (median toxic level) (Brown, 1968). Obviously, circular terminology is common. However, regardless of the existence of multiple terms for the same parameter, reasonable estimates of the effects of elements on fish can be derived from these measurements.

Today, 96-h tests commonly provide a LC50 value for toxic elements. In fact, the 96-h LC50 is commonly referred to as the incipient lethal level (ILL) (Wilson, 1972). Also, 96-h LC50 values are used in the calculation and use of application levels, safety factors, and other relationships. However, in thermal relationships, LC50 measurements can be less sensitive than survival time, incipient lethal levels, and median tolerance limits.

An example of the reduced sensitivity of LC50 measurements is observed in exposure of fish to both elevated temperature and excess Zn. Thermal increases reduce survival time (Fig. 1) (Lloyd, 1960; Sprague, 1964; Cairns, *et al.*, 1971; Burton, *et al.*, 1972) and incipient lethal level (Sprague, 1964) for fish exposed to Zn. However, temperature does not alter 96-h LC50 values for bluegills (Cairns and Scheier, 1957; Cairns, *et al.*, 1971). Lack of a thermal effect on LC50 in this case is probably due to lack of sensitivity of the LC50 measurement and a reduced thermal effect at the higher temperatures. Therefore, either elevated temperature or elevated Zn alone can produce the death of fish; moreover, the rate of temperature change can markedly alter Zn toxicity. The complexity of interactions between temperature and metal toxicity in fish, as well as the insensitivity of common measures such as LC50 values, show that much remains in elucidating interactions between temperature and metal-induced toxicity (see Chapter V).

Another term, "application factor", provides useful predictive information to the fisheries toxicologist (Mount, 1968; Mount and Stephen, 1969; McKim and Benoit, 1971). This factor is obtained by dividing the maximum acceptable toxicant concentration (MATC) by the 96-h LC50 value. The MATC is the highest concentration which has no adverse effect on the survival, growth, and/or reproduction of fish. The status of survival, growth, and reproduction can be combined using laboratory fish production indices; therefore, at best, application factors equal to MATC (based on the laboratory production index) divided by the 96-h LC50 value for the species under consideration. Inherent in the use of application factors is the idea that "no-effect" concentrations or "safe levels" can be derived from available toxicity tests of fish. The safe level is a presumably harmless concentration, empirically derived from one or many effects (Sprague, 1971). Therefore, safe levels are no more than statistics based on available laboratory data.

While application factors have been alternately referred to as immutable constants and oversimplifications, they do provide useful information. Application factors can be used to predict the no-effect level of an agent on a fish species for which no information is available. The application factor of a known species is multiplied by the 96-h LC50 value for an unknown species to obtain an estimated no-effect level for the unknown species. For example, use of Mount and Stephen's (Mount and Stephen, 1969) application factor for fathead minnows in soft water (0.13–0.22) and O'Hara's (O'Hara, 1971) value of 2.4 ppm Cu as the 96-h LC50 for bluegills in comparable aqueous solutions, produces a no-effect level for bluegills at 0.31–0.53 ppm Cu, a level only slightly higher than the experimentally determined no-effect level. Therefore, despite some lack of sensitivity in the LC50 value upon which it is based, application factors may be one of the better parameters in approximating the limit of harmfulness for toxic elements.

To provide a better understanding of experimental parameters which alter results of toxicity tests, response surfaces are used (see Chapters V and VII). Such three-dimensional figures can be used to illustrate the effect of water quality parameters on LC50 and other values. For example, pH, total water hardness, and/or alkalinity alter 96-h LC50 values for Cu (Fig. 3) and incipient lethal concentrations which cause 50% mortality (ILC50) for Cu (Fig. 4) (Howarth and Sprague, 1978; Miller and Mackay, 1980). At a water hardness of 300 ppm, increasing pH significantly increases Cu toxicity (Fig. 3), and Cu is about four-hundred times as toxic at pH 9 as at pH 5. Copper toxicity at pH 9 may be due to CO_2-induced pH reduction at the gill surface, ionization of Cu hydroxides, liberation of free ionic Cu^{+2}, and transit through the gill epithelium (Howarth and Sprague, 1978). The same mechanism is proposed for Zn (Mount, 1966) and could account for the abrupt toxicity curves which have been previously mentioned for fish exposed to Cu or to Zn. Moreover, at low water hardness (12 ppm), the ILC50 of Cu does not change despite a five-fold change in alkalinity (10 to 50 ppm) (Fig. 4). However, at 98 ppm hardness, the same alkalinity changes causes the ILC50 to increase about two-fold (i.e. Cu becomes two times less toxic). These results show the complexity of interactions between water quality parameters and the toxicity of metals in fish, as well as the usefulness of response surfaces.

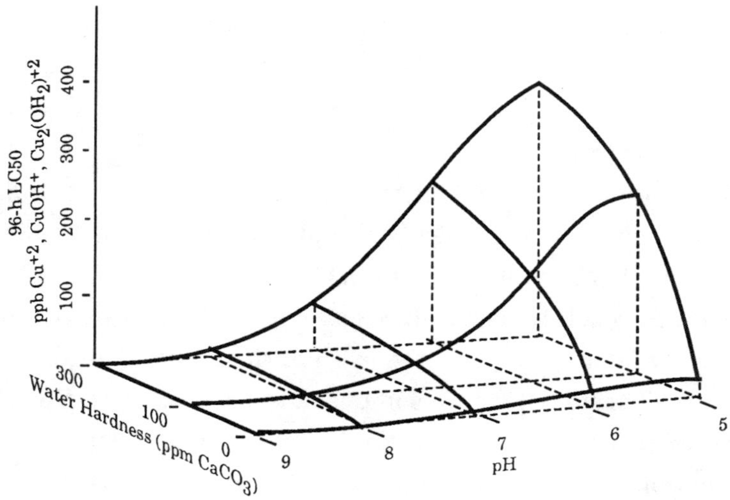

Figure 3. Relationship among pH, water hardness (ppm as $CaCO_3$), and 96-h LC50 for toxic Cu species (Cu^{+2}, $CuOH^{+}$, and $Cu^2(OH)_2{}^{+2}$). (Modified and redrawn after Howarth and Sprague, 1978.)

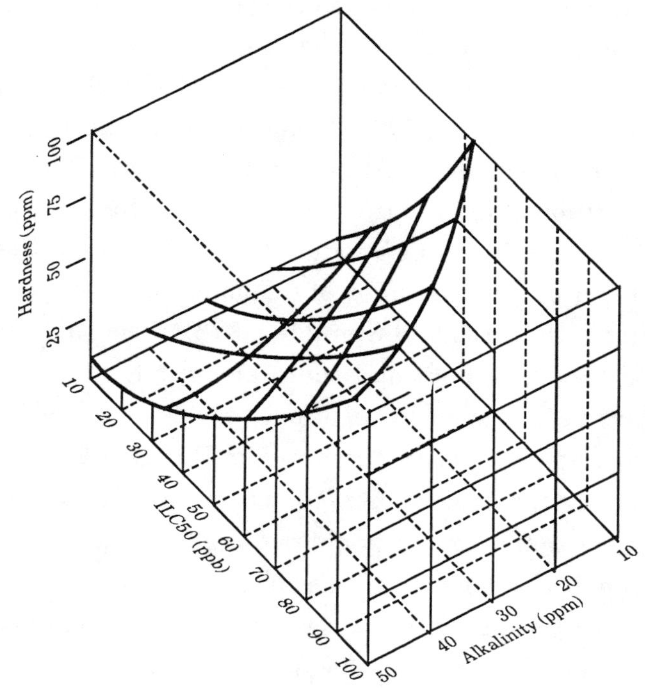

Figure 4. Incipient LC50 for Cu calculated for rainbow trout in water of various total hardness (ppm, as $CaCO_3$) and alkalinity (ppm, as $CaCO_3$). (Modified and redrawn after Miller and Mackay, 1980.)

Summary

In conclusion, this chapter is intended to show the ubiquity and diversity characteristic of exposures of fish to toxic metals and metalloids. Anthropogenic activity has led to either redistribution of natural stores of toxic elements or manufacture of poisonous derivatives prior to the dissemination of those poisons.

This chapter also presents the unique ways in which elements interact with biological ligands. For example, the interactions of Se, As, and Pb tend to be less complicated than those of Zn, Cd, Cu, and Hg. In each of the seven following chapters, these specific elements will be considered as single contaminates. The last chapter will present some of the current information available regarding interactions among elements. The focus of this volume will be placed on quantitative and qualitative assessment of the teleost response to the toxic elements. Toxicity, accumulation, retention, redistribution, histopathological alterations, behavior, and other parameters will be evaluated.

And finally, this chapter presents some specific terminology used to describe teleost response at the organismic level. The brief presentation of terms should facilitate understanding for those less familiar with the terminology of this field.

References

Alabaster, J.S., and R. Lloyd, Fish toxicity testing procedures, Water Quality Criteria for Freshwater Fish, Butterworths, London, (1980) 253.

Badsha, K.S., and C.R. Goldspink, Preliminary observations on the heavy metal content of four species of freshwater fish in NW England, J. Fish. Biol., 21 (1982) 251.

Bohn, A., and B.W. Fallis, Metal concentrations (As, Cd, Cu, Pb and Zn) in shorthorn sculpins, *Myoxocephalus scorpius* (Linnaeus), and Arctic char, *Salvelinus alpinus* (Linnaeus), from the vicinity of Strathcona Sound, Northwest Territories,, Water Res., 13 (1978) 659.

Brown, V.M., The calculation of the acute toxicity of mixtures of poisons to rainbow trout, Water Res., **2** (1968) 723.

Bryan, S.E., and H.A. Hidalgo, Nuclear cadmium-115: Uptake and disappearance correlated with cadmium-binding protein synthesis, Biochem. Biophys. Res. Commun., 68 (1976) 858.

Burton, D.T., E.L. Morga, and J. Cairns Jr., Mortality curves of bluegills (*Lepomis macrochirus* Rafinesque) simultaneously exposed to temperature and zinc stress, Trans. Am. Fish. Soc., 101 (1972) 435.

Bussey, R.E., D.E. Kidd, and L.D. Potter, The concentrations of ten heavy metals in some selected Lake Powell game fishes, University of New Mexico, Albuquerque, Lake Powell Research Project Bulletin, (1976) 1.

Cairns, J., Jr., and A. Scheier, The effects of temperature and hardness of water upon the toxicity of zinc to the common bluegill (*Lepomis macrochirus* Raf.), Not. Nat. Acad. Sci., Phil., 299 (1957) 1.

Cairns, J., Jr., T.K. Bahns, D.T. Burton, K.L. Kickson, R.E. Sparks, and W.T. Waller, The effects of pH, solubility and temperature upon the acute toxicity of zinc to the bluegill sunfish (*Lepomis macrochirus* Raf.), Trans. Kan. Acad. Sci., 74 (1971) 81.

Cappon, C.J., Cadmium and lead in Lake Ontario salmonids, Bull. Environ. Contam. Toxicol., 38 (1987) 695.

Carpenter, K.E., A study of the fauna of rivers polluted by lead mining in the Aberystwyth district of Cardiganshire, Ann. Appl. Biol., 11 (1924) 1.

Carpenter, K.E., On the biological factors involved in the destruction of river-fisheries by pollution due to lead-mining, Ann. Appl. Biol., 12 (1925) 1.

Carpenter, K.E., The lead mine as an active agent in river pollution, Ann. Appl. Biol., 13 (1926) 395.

Carpenter, K.E., The lethal action of soluble metallic salts on fish, Br. J. Exptl. Biol., 4 (1927) 378.

Carpenter, K.E., Further researches on the action of metallic salts on fishes, J. Exp. Zool., 56 (1930) 407.

Chow, T.J., C.C. Patterson, and D. Settle, Occurrence of lead in tuna, Nature, 251 (1974) 159.

Coombs, T.L., T.C. Fletcher, and A. White, Interaction of metals ions with mucus from the plaice (*Pleuronectes platessa* L.), Proc. Biochem. Soc., 128 (1972) 128.

Eichhorn, G.L., Metal ions as stabilizers or destabilizers of the deoxyribonucleic acid structure, Nature, 194 (1962) 474.

El Nabawi, A., B. Heinzow, and H. Kruse, As, Cd, Cu, Pb, Hg, and Zn in fish from the Alexandria region, Egypt, Bull. Environ. Contam. Toxicol., 39 (1987) 889.

Failla, M.L., R.J. Cousins, and M.J. Mascenik, Cadmium accumulation and metabolism by rat parenchymal cells in primary monolayer culture, Biochim. Biophys. Acta, 583 (1979) 63.

Frazier, J.M., and B.S. Kingsley, Kinetics of cadmium transport in the isolated perfused rat liver, Toxicol. Appl. Pharmacol., 38 (1976) 583.

Frazier, J.M., and J. Puglese, Dose dependence of cadmium kinetics in the rat liver following intravenous injection, Toxicol. Appl. Pharmacol., 43 (1978) 461.

Gale, N.L., B.G. Wixson, and M.W. McMenus, Lead concentrations in edible fish filets collected from Missouri's Old Lead Belt, 16th Annual Conference on Trace Substances in Environmental Health, (1982) 12.

Greichus, Y.A., A. Greichus, B.D. Ammann, and J. Hopcraft, Insecticides, polychlorinated biphenyls, and metals in African lake ecosystems. III. Lake Nakuru, Kenya., Bull. Environ. Contam. Toxicol., 19 (1978) 454.

Harvey, S.C., Heavy metals, in Goodman, L.S., and A. Gilman (Eds.), The Pharmacological Basis of Therapeutics, MacMillan, London, (1975) 924.

Horowitz, A., and B.J. Presley, Trace metal concentrations and partitioning in zooplankton, neuston, and benthos from the south Texas Outer Continental Shelf, Arch. Environ. Cont., 5 (1977) 241.

Howarth, R.S., and J.B. Sprague, Copper lethality to rainbow trout in waters of various hardness and pH, Wat. Res., 12 (1978) 455.

Jacobs, E.E., and M. Jacob, Uncoupling of oxidative phosphorylation by cadmium ion, J. Biol. Chem., 223 (1956), 147.

Johnson, M.G., Trace element loadings to sediments of fourteen Ontario lakes and correlations with concentrations in fish, Can. J. Fish. Aquat. Sci., 44 (1987) 3.

Jones, J.R.E., The relation between the electrolytic solution pressures of the metals and their toxicity to the stickleback (*Gasterosteus aculeatus* L.), J. Exp. Biol., 16 (1939) 425.

Jones, J.R.E., A study of the zinc-polluted river Ystwyth in North Cardiganshire, Wales, Ann. Appl. Biol., 27 (1940) 368.

Kleinert, S.J., P.E. Degurse, and J. Ruhland, Concentration of metals in fish, in Surveys of Toxic Metals in Wisconsin, Department of Natural Resources, Tech. Bull. 74, Madison, Wisconsin, (1974) 8.

Lloyd, R., The toxicity of zinc sulfate to rainbow trout, Ann. Appl. Biol., 48 (1960) 84.

Mathis, B.J., and N.R. Kevern, Distribution of mercury, cadmium, lead, and thallium in a eutrophic lake, Hydrobiologia, 46 (1975) 207.

McKim, J.M., and D.A. Benoit, Effects of long-term exposures to copper on survival, growth, and reproduction of brook trout (*Salvelinus fontinalis*), J. Fish. Res. Bd. Can., 28 (1971) 655.

Miller, T.G., and W.C. Mackay, The effects of hardness, alkalinity, and pH of test water on the toxicity of copper to rainbow trout (*Salmo gairdneri*), Water Res., 14 (1980) 129.

Mount, D.I., The effect of total hardness and pH on acute toxicity of zinc to fish, Air Water Pollut. Int. J., 10 (1966) 49.

Mount, D.I., Chronic toxicity of copper to fathead minnows (*Pimephales promelas*, Rafinesque), Water Res., 2 (1968) 215.

Mount, D.I., and C.E. Stephen, Chronic toxicity of copper to the fathead minnow (*Pimephales promelas*) in soft water, J. Fish. Res. Bd. Can., 26 (1969) 2449.

Ney, J.J., and J.H. Van Hassel, Sources of variability in accumulation of heavy metals by fishes in a roadside stream, Arch. Environ. Contam. Toxicol., 12 (1983) 701.

O'Hara, J., Alterations in oxygen consumption by bluegills exposed to sublethal treatment with copper, Water Res., 5 (1971) 321.

Olafson, R.W., Differential pulse polarographic determination of murine metallothionein induction kinetics, J. Biol. Chem., 256 (1981) 1263.

Pagenkopf, G.K., and D.R. Neuman, Lead concentrations in native trout, Bull. Environ. Contam. Toxicol., 12 (1974) 70.

Petres, J., and A. Berger, The effect of inorganic arsenic on DNA-synthesis of human lymphocytes in vitro [in German, English summary], Arch. Derm. Forsch., 242 (1972) 343.

Prasada-Rao, P.V.V., M.K.C. Sridhar, and A.B.O. Desalu, Effect of acute oral cadmium on mitochondrial enzymes in rat tissues, Arch. Environ. Contam. Toxicol., 12 (1983) 293.

Rolfe, G.L., and J.C. Jennett, Environmental lead distribution in relation to automobile and mine and smelter sources, in Forstner, V., and G.T. Wittman (Eds.), Metal Pollution in Aquatic Environments, Springer-Verlag, Berlin, (1979) 231.

Sanderson, R.T., Chemical Periodicity, Reinhold Publ. Co., New York, New York, (1960) 1.

Schmitt, C.J., and S.E. Finger, The effects of sample preparation on measured concentrations of eight elements in edible tissues of fish from streams contaminated by lead mining, Arch. Environ. Contam. Toxicol., 16 (1987) 185.

Settle, D.M., and C.C. Patterson, Lead in albacore: Guide to lead pollution in Americans, Science, 207 (1980) 1167.

Sorensen, E.M.B., Toxicity and accumulation of arsenic in green sunfish, *Lepomis cyanellus*, exposed to arsenate in water, Bull. Environ. Contam. Toxicol., 15 (1976) 756.

Sorensen, E.M.B., Protective effects of calcium in the amelioration of cadmium-induced cytotoxicity in cultured murine parenchymal hepatocytes, in Goldberg, A.M. (Eds.), In Vitro Toxicology: Alternative Methods in Toxicology, Mary Ann Liebert, Inc., New York, New York, (1985) 100.

Sprague, J.B., Lethal concentrations of copper and zinc for young Atlantic salmon, J. Fish. Res. Bd. Can., 21 (1964) 17.

Sprague, J.B., and B.A. Ramsay, Lethal levels of mixed copper-zinc solutions for juvenile salmon, J. Fish. Res. Bd. Can., 22 (1965) 425.

Sprague, J.B., Measurement of pollutant toxicity to fish: I. Bioassay methods for acute toxicity, Water Res., 3 (1969) 793.

Sprague, J.B., Measurement of pollutant toxicity to fish: II. Utilizing and applying bioassay results, Water Res., 4 (1970) 3.

Sprague, J.B., Measurement of pollutant toxicity to fish: III. Sublethal effects and "safe" concentrations, Water Res., 5 (1971) 245.

Sprague, J.B., The ABC's of pollutant bioassay using fish, in Biological Methods for the Assessment of Water Quality, Am. Soc. Test. Mater., ASTM STP 528, Philadelphia (1973) 6.

Stacey, N.H., L.R. Cantilena Jr., and C.D. Klaassen, Cadmium toxicity and lipid peroxidation in isolated rat hepatocytes, Toxicol. Appl. Pharmacol., 53 (1980) 470.

Stoll, R.E., J.F. White, T.S. Miya, and W.F. Bousquet, Effects of cadmium on nucleic acid and protein synthesis in rat liver, Toxicol. Appl. Pharmacol., 37 (1976) 61.

Taylor, D.D., and T.J. Bright, The distribution of heavy metals in reef-dwelling groupers in the Gulf of Mexico and Bahama Islands, Texas A & M University, Sea Grant College, TAMU-SG-73-208 (1973) 1.

Uthe, J.F., and E.G. Bligh, Preliminary survey of heavy metal contamination of Canadian freshwater fish, J. Fish. Res. Bd. Can., 28 (1971) 786.

Vallee, B.L., and D.D. Ulmer, Biochemical effects of mercury, cadmium, and lead, Annu. Rev. Biochem., 41 (1972) 91.

Van Hassel, J.H., J.J. Ney, and D.L. Garling, Seasonal variations in the heavy metal concentrations of sediments influenced by highways of different traffic volumes, Bull. Environ. Contam. Toxicol., 23 (1979) 592.

Van Hassel, J.H., J.J. Ney, and D.L. Garling Jr., Heavy metals in a stream ecosystem at sites near highways, Trans. Amer. Fish. Soc., 109 (1980) 636.

Villarreal-Trevino, C.M., M.E. Obregon-Morales, J.F. Lozano-Morales, and A. Villegas-Navarro, Bioaccumulation of lead, copper, iron, and zinc by fish in a transect of the Santa Catarina River in Cadereyta Jimenez, Nuevo Leon, Mexico,, Bull. Environ. Contam. Toxicol., 37 (1986) 395.

Warner, R.E., Bio-assays for microchemical environmental contaminants, Bull. Wld. Hlth. Org., 36 (1967) 181.

Wilson, R.C.H., Prediction of copper toxicity in receiving waters, J. Fish. Res. Bd. Can., 29 (1972) 1500.

Zamojski, J., S. Smoczynski, K.A. Skibniewska, and R. Amarowicz, Organochlorine insecticides and heavy metals in fish from Mutek Lake, N.E. Poland, Bull. Environ. Contam. Toxicol., 37 (1986) 587.

Zook, E.G., J.J. Powell, B.M. Hackley, J.A. Emerson, J.R. Brooker, and G.M. Knobl Jr., National Marine Fisheries Service preliminary survey of selected seafoods for mercury, lead, cadmium, chromium, and arsenic content, J. Agric. Food Chem., 24 (1976) 47.

CHAPTER II
SELENIUM

Introduction

Exposure concentrations of Se determine whether Se is toxic or essential for good health. Since Se is an essential element, beneficial effects of low levels of Se are well-known (Franke and Potter, 1935; Muth, *et al.*, 1958; Schubert, *et al.*, 1961; Scott, *et al.*, 1967; Cantor, *et al.*, 1975a; Cantor, *et al.*, 1975b; Poston, *et al.*, 1976; Calvin, 1978; Chow and Chen, 1980; Hilton, *et al.*, 1980; Stadtman, 1980; Gunby, 1981). In rainbow trout, dietary levels of 0.07 ppm and/or aqueous levels of 0.0004 ppm are essential (Hilton, *et al.*, 1980). The nutritive role of Se is linked to the presence of four gram atoms of Se per mole of glutathione peroxidase (Flohe, *et al.*, 1973). The presence of trace levels of Se protects lipids and nucleic acids from peroxidation, as well as subsequent hepatocellular lysis (Christophersen, 1968; Little and O'Brien, 1968; Christophersen, 1969; Anundi, *et al.*, 1979). Therefore, Se is thought to act as an intracellular antioxidant to prevent free radical damage to erythrocyte and other cellular membranes (Chow and Chen, 1980; Stadtman, 1980). Manifestations of Se deprivation in fish include increased mortality, reduced growth, hepatic injury, muscle lesions, and depressed glutathione peroxidase activity (Poston, *et al.*, 1976; Hilton, *et al.*, 1980).

However, excess levels of Se are more poisonous than either As or Hg (Copeland, 1971) and represent a threat to teleost survival (Sakurayama, 1960; Glenn, *et al.*, 1964; Adams, 1976; Cardwell, *et al.*, 1976; Neal, 1977; Halter, *et al.*, 1980; Hilton, *et al.*, 1980). The ratio of nutritive to toxic levels (safety factor) for Se is 25; while those of Hg are 500–2000 and of As are 50–500. A 20-wk exposure of rainbow trout to a diet containing >3 ppm Se results in higher mortality, reduced growth rate, and poor feed-consumption-to-weight-gain ratio. Toxic levels of Se are about forty-three times the essential dietary level or one-thousand times the required aqueous level. That is, dietary Se at about 3 ppm or aqueous Se at about 0.40 ppm are toxic to teleosts (Sakurayama, 1960; Adams, 1976; Cardwell, *et al.*, 1976; Halter, *et al.*, 1980). However, much lower aqueous levels of Se (i.e. 3–12 ppb Se) are characteristic of Se-rich habitats from which moribund fish are collected (Sorensen and Bauer, 1984b; Gillespie and Baumann, 1986; Sorensen, 1986) and of Se laboratory solutions from which altered fish are taken (Hunn, *et al.*, 1987). Therefore, Se is considered both an essential element and a villainous killer among trace elements, depending upon the exposure concentration (Wilber, 1980; Gunby, 1981; Goyer, 1986).

This review evaluates various effects of Se on fish—including toxicity, accumulation, retention, biological effects, and others. Effects are compared with the Se levels where possible because Se accumulation is a fairly accurate predictor of prior Se exposure. But, as expected, elemental uptake and effects are more variable in feral fish (Sorensen, *et al.*, 1980; Sorensen, *et al.*, 1981b; Rameriz-Mitchell and Sorensen, 1985; Sorensen, *et al.*, 1985) than in laboratory fish (Sorensen, 1976a; Sorensen, 1976b; Sorensen, 1976c; Sorensen, *et al.*, 1979a; Sorensen, *et al.*, 1979b; Sorensen and Smith, 1981a; Sorensen, *et al.*, 1982c). Environmental variation can result from differences in the oxygen tension of the aqueous habitat, food availability, trophic level of the species under consideration, chemical form, age of fish, dietary composition, population composition and competition, habitat preference, the presence of other metals/metalloids, thermal gradients, and others (Sorensen, 1974; Sorensen and Burkett, 1978). Moreover, as an essential element and constituent of every cell in the vertebrate body, Se is one of the most fascinating and challenging of all elements involved in metal poisoning of fish.

Environmental Exposures

Coal and petroleum combustion are the main sources of Se pollution in recent times. About 70% of Se is redistributed to air antropogenically *via* combustion of these two fossil fuels; 20% is derived from the glass industry; 8% is emitted from metallurgical processes; and 2% is released from mining, milling, xerography, electronics, and miscellaneous sources (Vokal-Borek, 1979). Selenium levels in coal range from 0.1–1.5 ppm for the United States and <0.5–3.7 ppm for Sweden; whereas, petroleum contains about 0.82–0.99 ppm Se.

Coal combustion generates wastes such as bottom ash—which does not reach complete combustion in the boilers. Cyclone-fed boilers consume ~100 tons of crushed or pulverized coal per hour and produce ash which incorporates 0.3% of coal-derived Se (Andren, *et al.*, 1975). Flue gases are cleaned by electrostatic precipitators which discharge ~0.5% of the intake gases through 122-m stacks. Because electrostatically-precipitated fly ash contains 310–1100 times more Se than bottom ash (slag) collected from boilers (Gutenmann, *et al.*, 1976), fly ash is considered the most hazardous waste generated by power plants (Natusch, *et al.*, 1974; Andren, *et al.*, 1975; Gutenmann, *et al.*, 1976). Unfortunately, electric generating companies use fly ash to surface roads, to make charcoal briquettes, and for soap production.

Scrubber sludge is a power plant waste obtained from the scrubbing devices which are used to remove sulfur dioxide from exhaust air emissions. These wastes can be transported by wet sluicing to ash basins, where the particulates are collected by sedimentation. The often-extended period of contact between ash sluice water and particulate wastes can result in considerable solubilization of adsorbed or absorbed metal/metalloid contaminants such as Se.

Discharge of sluice water from ash sedimentation basins into the main basin of Belews Lake and Martin Lake is believed to have resulted in declines in fish populations (Cumbie and Van Horn, 1978a; Sorensen, *et al.*, 1982a; Sorensen, *et al.*, 1982b; Sorensen, *et al.*, 1983a; Sorensen, *et al.*, 1983b; Sorensen and Bauer, 1983c; Sorensen, *et al.*, 1984a; Sorensen and Bauer, 1984b; Sorensen and Bauer, 1984c), resulting from both direct mortality of individuals (Sorensen and Bauer, 1983c; Sorensen, *et al.*, 1984a; Sorensen and Bauer, 1984b; Sorensen and Bauer, 1984c) and from reproductive failure of those individuals which did survive (Cumbie and Van Horn, 1978a;

Sorensen, *et al.*, 1984a). These reservoirs continue to have abnormally high levels of aqueous Se. Moreover, analyses for pesticides, metals, metalloids and physiochemical parameters document negative results (Cumbie and Van Horn, 1978a; Sorensen, *et al.*, 1982a; Sorensen, *et al.*, 1982b; Sorensen, *et al.*, 1983a; Sorensen, *et al.*, 1983b; Sorensen and Bauer, 1983c; Sorensen, *et al.*, 1984a; Sorensen and Bauer, 1984b; Sorensen and Bauer, 1984c), except for the detection of sufficiently high levels of Se in the water to warrant concern (Sorensen and Bauer, 1984b).

Based on thermodynamics, selenates appear to be the potentially predominant chemical form of Se in an alkaline ash basin because of stability and high solubility at alkaline pH (The National Research Council, 1976; Cumbie, 1980). However, following discharge of effluent into the reservoir, redistribution among the chemical forms would be expected; so that, a mixture of inorganic chemical forms of Se would exist in the aquatic ecosystem (i.e., selenates, selenites, selenic acid, *et cetera*). Organic chemical forms also would be present as a result of the microbial transformations of Se (Fleming and Alexander, 1972; Spangler, *et al.*, 1973; Doran and Alexander, 1977; Sarathchandra and Watkinson, 1981).

Expected forms of Se in emissions from coal combustion are Se^0, SeO_2, SeO_3^{-2}, and SeO_4^{-2}. Of the four chemical forms, only Se^0 is water-insoluble and must be oxidized to SeO_3^{-2} before dissolution can occur (Andren, *et al.*, 1975). In fly ash landfills, oxidation of Se^0 to higher valence and more water-soluble forms probably is responsible for increased leachability and availability to plants and animals (Gutenmann, *et al.*, 1976). Nonseleniferous plants (e.g. white sweet clover) growing on beds of fly ash can accumulate Se up to values exceeding 200 ppm (dry weight) (Gutenmann, *et al.*, 1976). Livers of fish in fly ash-contaminated ponds in the area concentrate 4.5–9.0 ppm Se (dry weight), levels considerably higher than normal.

Natusch and others (Natusch, *et al.*, 1974) find that Se, As, Cd, and Pb, but not Cu, are most concentrated in the smallest respirable particles emitted from coal-fired power plants. Particles 1 μm in diameter deposit in the alveoli, where absorption efficiency is 50–80%. Larger particles deposit in the nasal, pharyngeal, and bronchial regions of air-breathing vertebrates. Fly ash samples from eight U.S. power plants contain particles from >74 μm to 1 μm in diameter. These particles contain 140 and 1600 ppm Pb, respectively. Similarly the same size particles accumulate 12 and 59 ppm Se,

180 and 1700 ppm As, 500 and 13,000 ppm Zn, and other elements, respectively. By comparison, major matrix elements in coal fly ash (i.e. Al, Fe, and Ca) show no evidence of concentration dependence on particle size. Volatilization and adsorption (or condensation) probably account for preferential concentration of toxic elements onto the large available surface area per unit mass provided by the small particles. Generally, particulates show higher superficial than internal concentration of elements (Natusch, *et al.*, 1974).

Ratios of Se/S have be used to evaluate Se mobility and sources of atmospheric pollutants. Apparently, in high-temperature oxidation, Se burns in air to form an instable compound (SeO_2) which is readily reduced; whereas, S is oxidized to the relatively stable SO_2 (Weiss, *et al.*, 1971). Analysis of glacial samples dated 800 B.C. to 1965 A.D. from Antarctica and Greenland shows S enrichment relative to Se (Weiss, *et al.*, 1971). Pre-1960 samples have Se/S ratios varying from 2.9–5.9 x 10^{-4}; whereas, post-1960 samples range in value from 1.1–1.6 x 10^{-4}. Therefore, Se is mobilized to a lesser extent than is S because of the differences in chemical and physical properties of the tetravalent oxides.

Ratios of Se/S also have been calculated to establish major sources of atmospheric pollutants in the Boston area. Ratios of Se/S vary from 48 x 10^{-4} for soil, to 4.8 x 10^{-4} for coal, to 0.2 x 10^{-4} for soot (Hashimoto, *et al.*, 1970). Apparently, Boston has considerable atmospheric pollution due to petroleum combustion, as indicated by the coincidence of Se/S ratios for those unique to petroleum and the predominate atmospheric pollutants in the Boston area (Hashimoto, *et al.*, 1970).

Microorganisms in sediments and/or sewage are capable of elemental methylation (Fleming and Alexander, 1972) and demethylation (Spangler, *et al.*, 1973). Fleming and Alexander (Fleming and Alexander, 1972) find that dimethylselenide [$Se(CH_3)_2$] is generated slowly from raw sewage (2.7 μg from a 50 ml sample containing 1,000 μg selenite/ml over 7 d), containing a *Penicillium* fungal strain. Sulfate and methionine can enhance growth of the fungal isolate and $Se(CH_3)_2$ production in media containing selenite, but not selenate. Methionine inhibits $Se(CH_3)_2$ but stimulates fungal growth in selenate-containing solutions. Moreover, *Penicillium* apparently does not alkylate Se as a mechanism of Se detoxification.

Although $Se(CH_3)_2$ demethylation is not measured, Spangler and colleagues (Spangler, *et al.*, 1973) find demethylators for Hg in sediments from the delta area of the St. Clair River in Michigan. These demethylators are capable of degrading CH_3Hg^+ produced in sediments. In 50 d, the CH_3Hg^+ level increases from zero to 0.31 ppm (wet weight), then declines to about 0.05 ppm in another 90 d. In contrast, volatile Hg^0 increases to a plateau in 80 d and remains unchanged over the next 60 d. Therefore, CH_3Hg^+ is degraded by organisms in the sediment. A mixed culture from sediment produces the same effect. Four cultures of *Pseudomonas* are isolated from sediment and show variable capacity to degrade CH_3Hg^+. Therefore, demethylators present in sediments can degrade CH_3Hg^+ to volatile Hg^0, probably single-handedly destroying even trace amounts of CH_3Hg^+ in a wide variety of naturally occurring sediments and in water.

In conclusion, high concentrations of Se are deposited on the surface of the smallest particulate released from power plants. These 1μ-diameter particles are capable of lodging in alveoli and solubilizing toxic elements. Mobilization of Se in the environment is reduced compared with S. Since 1960, a proportionate decrease in the ratio of Se/S indicates that (compared with S) Se is more likely to be deposited at the site of release since 1960. Once deposited, marked technical difficulties surround speciation of forms released during high-temperature combustion of coal. Selenium is methylated and probably demethylated in the environment and cycled through a number of components of the food web, complicating determination of chemical forms available to fish. Even greater problems exist in identifying the chemical form or forms of Se which are toxic to various components of the ecosystem and the fate of each through the food web.

General Toxicology

Generally, Se is considered one of the most toxic of elements to fish. For example, in golden shiners (*Notemigonus crysoleucas*), the 96-h LC50 for Se is 11.2 ppm, compared with those for As, Cr, and Cu (i.e. 12.5, 55.0, and 84.6 ppm, respectively) at pH 7.5 and a water hardness of 72.2 ppm (Hartwell, *et al.*, 1989). Selenium-induced teleost mortality varies with species, temperature, the presence of parasites, specimen age (or life-stage), source, exposure concentration, exposure time, chemical form, and other factors (Sakurayama, 1960; Adams, 1976; Cardwell, *et al.*, 1976; Halter, *et al.*,

1980). Adams (Adams, 1976) reports 48-d LC50 values of 0.4, 0.5 and 1.1 ppm Se, as selenite, at 15°C for bluegills (*Lepomis macrochirus*), rainbow trout (*Salmo gairdneri*), and fathead minnows (*Pimephales promelas*), respectively. In this study, species-specific differences alter these LC50 values by about three-fold. Moreover, 48-d LC50 values for selenate and selenite differ by two-fold (2.0 and 1.1 ppm, respectively) in fathead minnows (Adams, 1976).

Watenpaugh and Beitinger (Watenpaugh and Beitinger, 1985a) determine critical thermal maxima (CTM) for fathead minnows (*Pimephales promelas*) to establish the effect of Se exposure on thermal tolerance of this species. Fish are exposed 24 h to 15, 40, or 59 ppm Se (as selenate) in water (21°C). The 24-h LC50 is 82 ppm Se, as selenate, under conditions of this study. From 21°C, the temperature is elevated at 0.3°C/min until loss of equilibrium (defined as the CTM). About 27–45 min are required to reach CTM under these conditions. The CTM decreases from 35°C in control fish to 29°C for fish exposed 24 h to 59 ppm Se; therefore, CTM is suppressed 6°C by a 24-h exposure to Se at 72% of the 24-h LC50 value for fathead minnows.

To determine if Se protects bass against columnaris disease, MacFarlane and others (MacFarlane, *et al.*, 1986) expose striped bass (*Morone saxatilis*) to aqueous solutions containing low concentrations of five elements singly or as a mixture. Isolates of the bacteria causing columnaris disease (*Flexibacter columnaris*) come into contact with fish for 2 min at 22°C and 265 ppm hardness. Singly, 10 ppb Cu protects against columnaris disease. Cadmium (30 ppb) provides marginal protection but is slightly toxic after 2-d exposures. Arsenic (10 ppb) increases susceptibility to the disease. Lead (10 ppb) and Se (20 ppb) do not alter effects of the columnaris bacteria on striped bass when elements are administered singly.

These results indicate that, of elements present in the mixture, only Cu provides complete protection against columnaris disease in striped bass at 10 ppb and 24-h exposures. Although the mechanism of action whereby Cu protects against columnaris bacteria is not understood, Cu could prevent attachment of bacteria cells, stimulate mucous production, and/or interfere with production of extracellular proteases. Therefore, unlike Cu, Se does not protect against columnaris disease in striped bass at 20 ppb Se.

Hunn and co-workers (Hunn, *et al.*, 1987) find that sac fry are sensitive to Se at levels among the lowest reported in laboratory studies. Sac fry of rainbow trout (*Salmo gairdneri*) are exposed to 8–100 ppb Se (as selenite) for up to 90 d to estimate the no-effect level of water-borne Se at 12°C, pH 7.2, and 40 ppm hardness. At 100 ppb, weights are reduced after 90-d exposures, and levels of 47 or 100 ppb Se reduce 90-d survival. Whole body bioconcentration factors are 200 or less and increase with decreasing exposure levels. Calcium levels in bone are decreased in fish exposed to 12 ppb or more Se; however, proline, hydroxyproline, and collagen levels do not change. Bone toughness (energy-absorbing capacity) is increased in trout exposed to 47 ppb Se—a phenomenon related to increased bone density and bone strength. Only minor bone deformities (i.e. irregularly shaped or highly compacted bone) are recorded.

The most important result of this study is that 12 ppb Se, as selenite, may be a more accurate, no-effect level based on detection of Ca reductions in vertebrae of rainbow trout exposed to aqueous selenite (Hunn, *et al.*, 1987). Correlations between Ca levels in bone and functional parameters have not been clarified. Perhaps the reduction in Ca skeletal levels is within acceptable limits for species survival, since changes in strain, elasticity, rupture, and other parameters are not noted ≤ at 12 ppb Se. At present, elucidation of long-term effects of reduced skeletal Ca in fish would help determine the impact of reduced vertebral Ca on individual (and population) survival in areas of elevated Se.

Dietary sources of Se alter toxicity of the metalloid. Hamilton and colleagues (Hamilton, *et al.*, 1987) provide chinook salmon (*Oncorhynchus tschawytscha*) with two Se-containing diets. The San Luis Drain (SLD) *Gambusia* diet contains 20% fish meal from mosquitofish collected from San Luis Drain. The second diet contains 20% uncontaminated mosquitofish and selenomethionine. Both diets contain 26 ppm Se, wet weight. Swim-up stage fish (about 0.3 g) are exposed 90 d at 12°C, pH 7.9, and 210 ppm hardness in water reconstituted without Se. Both diets cause a similar dose-response effect: percent survival decreases as length of exposure increases. Both diets cause ~80% mortality in 90 d at 13 ppm dietary Se, compared with 40% mortality at 0.7 ppm Se in the diet. The SLD *Gambusia* diet produces more pronounced effects than the selenomethionine diet. At 3.2 ppm Se in this diet, growth is decreased in 90 d; whereas, >13 ppm Se decreases growth for fish receiving the selenomethionine diet.

In a brackish water study (28 ‰, salinity), changes in 2-g chinook salmon are less pronounced. Decreased growth occurs after 90 and 120 d of dietary exposure. A greater response again is noted for the SLD *Gambusia* diet than for the selenomethionine diet. Moreover, in a 10-d seawater-challenge test, reduced survival is observed in fish fed either of the 26 ppm diets. Therefore, as little as 3.2 and 6.5 ppm dietary Se can cause adverse effects in fresh water and brackish water, respectively. The presence of Se could reduce the numbers of parental chinook migrants returning to spawn in the Central Valley of California. Therefore, a lengthy list of factors might alter toxicity of Se, including species, temperature, disease, dietary source of Se, and other factors.

Reproductive Toxicity

Fathead minnow embryos hatch prematurely at >15 ppm Se, as selenite (Adams, 1976). However, hatchability of embryos of zebrafish (Niimi and LaHam, 1975), carp (Huckabee and Griffith, 1974), bluegills (Gillespie and Baumann, 1986), and fathead minnows (Adams, 1976) is not affected by up to 10.0 ppm Se. Amazingly, Se potentiates Hg-induced decreases in hatchability (Huckabee and Griffith, 1974). A level of 3 ppm Hg^{+2} decreases hatchability by 12%, and the presence of Se at increased aqueous Hg levels causes a more pronounced effect. Although high affinity binding of Hg and Se to sulfhydryl groups on the outer embryo membrane probably occurs, death could result from Se and Hg uptake and subsequent tissue injury, rather than simple osmotic imbalance at the embryonic membranes.

Niimi and LaHam (Niimi and LaHam, 1975) show that ascites and death occur in zebrafish (*Brachydanio rerio*). Fish are exposed to 0.5–10.0 ppm Se (as selenite) at 26°C, pH 7.0, and 45–50 ppm total hardness. Newly-hatched larvae and four embryonic stages are tested. Stages 3–5 (2–8 cell ovum), Stage 10 (late blastula), Stage 15 (blastoderm enveloping yolk), and Stage 21 (initiation of heart beat) are exposed. Hatchability is not affected by Se: 40–50% of eggs hatch 120 h after fertilization and 20–30% more hatch within the next 24 h. However, following hatching, mortality increases sharply at concentrations of 3 ppm or more Se. Selenium-induced mortality is correlated with exposure time of zebrafish larvae. A pronounced abdominal enlargement (probably ascites) is the only anomaly consistently observed in Se-exposed larvae. At death, about 40% of larvae show this condition, which is first noted by convex spinal flexure. Death usually follows in 2–3 d. Moreover, the frequency of abdominal enlargement does not appear to be influenced by the Se exposure level or the stage of development in which embryos are exposed. Because hatching requires 4–5 d, use of hatching rates of zebrafish embryos is not considered a reliable toxicological indicator. Interestingly, ascites is noted in bluegills (Gillespie and Baumann, 1986), although it apparently is not diagnosed as such.

Saiki (Saiki, 1986) reports an increase in length of fish born to ovoviviparious fish in areas contaminated by Se-rich irrigation water. The San Luis Drain carries groundwater drainage from 8,000 acres of irrigated land in California into the San Joaquin River. Most Se is present as selenate (Se^{+6}) at about 0.2–0.4 ppm, and boron and molybdenum also are present. Kesterson mosquitofish (*Gambusia affinis*) give birth to about twenty live fry, a 41% decrease compared to fry born to reference females at Volta Wildlife Area. Therefore, percentages of live fry at birth (fry survival) are higher in Volta females. Moreover, fry from Kesterson females are usually longer than those from Volta. Live fry from Kesterson females are 8 mm long compared with 6 mm from Volta females. Possibly, a growth dilution effect in larger Kesterson fry decreases concentration of Se in tissues, such that (in larger fry) protective detoxification mechanisms are not overwhelmed by Se challenge. No abnormalities are noted in mosquitofish fry.

Using selenite, rather than selenate, Hunn and colleagues (Hunn, *et al.*, 1987) observe decreases in length and weight of sac fry of rainbow trout (*Salmo gairdneri*). Fish are exposed to 8–100 ppb Se (as selenite) for up to 90 d to estimate the no-effect level of water-borne Se at 12°C, pH 7.2, and 40 ppm hardness. At 100 ppb, lengths and weights are reduced after 90-d exposures; moreover, levels of 47 or 100 ppb Se reduce 90-d survival. As indicated, the most important result of this study is that 12 ppb Se, as selenite in the water, may be a more accurate, no-effect level based on reduction in Ca skeletal levels in rainbow trout (Hunn, *et al.*, 1987). However, determination of long-term effects of reduced skeletal Ca in fish would help establish the importance of reduced skeletal Ca on survival of fish in Se-contaminated areas.

Accumulation

As part of the National Contaminant Biomonitoring Program, Lowe and colleagues (Lowe, *et al.*, 1985) provide reference levels for seven elements in whole fish in 1978–1981. In the United States, Se normally is present at 0.47 ppm (wet weight) in fish, but whole body levels of about 1 ppm are common (Pakkala, *et al.*, 1972; Nielsen and Nielsen, 1978). However, numerous studies show that Se is found in excess in California, North Carolina, Texas, and other locations (Table 1). In contaminated environmental sites, whole body levels exceed that of the national average for 1980–1981, ranging from a high of 197 ppm (dry weight) (Saiki and Lowe, 1987a) to a low of 4.28 ppm (dry weight) (Saiki, 1987b). Hepatic levels range from 6.7 to 34.0 ppm (wet weight) (Gutenmann, *et al.*, 1976; Freeman, *et al.*, 1978; Sager, 1984; Sorensen, *et al.*, 1984a; Sorensen and Bauer, 1984b). Muscle levels range from 0.69 to 13.00 ppm (wet weight) (Gutenmann, *et al.*, 1976; Freeman, *et al.*, 1978; Sager, 1984; Sorensen, *et al.*, 1984a). Normally, skeletal muscle contains 0.2–0.6 ppm Se. Ovaries vary from 5.9 to 12.1 ppm (dry weight) (Sager, 1984; Sorensen and Bauer, 1984c).

One of the most interesting features of this compilation of accumulation values is the magnitude of the 197 ppm value—a grand mean of 149, 332, 224, 247, 164, 223, 130, and 104 ppm (dry weight) (Saiki and Lowe, 1987a). If a multiplication factor of 0.33 is used for conversion of dry weight to wet weight, the grand mean of 197 ppm (dry weight) becomes a value of 65 ppm (wet weight). The value of 65 ppm (wet weight) is about twice the maximum level of Se in the liver from data compiled in Table 1. The whole body of mosquitofish in the Saiki and Lowe study (Saiki and Lowe, 1987a) is therefore concentrating twice as much Se as the highest hepatic values presently published. Even the parallel Saiki *et al.* study (Saiki, 1987b) presents whole body values at 4.3 ppm (dry weight) for mosquitofish and ≤6.8 ppm for two additional species of fish (Table 1). The possible reason for this deviation is intriguing. Intake of Se-rich particulates during foraging activities (and subsequent gastrointestinal collection) might result in Se-rich particulate with the whole body. Because some marine shales contain so much Se that crystallization of the metalloid occurs, conceivably ground water fluctuations (especially lowering of the water table) could lead to marked increases in aqueous Se reaching the surface for irrigation in

Table 1. Accumulation of Se (in ppm) in fish from Se-contaminated areas. Mean total Se levels are given with numbers of samples analyzed in parenthesis following mean values. Letters in parentheses designate wet weight (w), dry weight (d), or an undesignated (-) basis for analysis of Se levels.

Species	Whole Body	Liver	Skeletal Muscle	Ovaries	Reference
Mosquitofish *(Gambusia affinis)*	4.3 (23) (d)				Saiki, 1987b
	197.0 (24) (d)				Saiki & Lowe, 1987
Bluegill *(Lepomis macrochirus)*	4.7 (6) (d)				Saiki, 1987b
		34.0 (8) (w)	13.0 (22) (w)	12.1 (11) (w)	Sager & Cofield, 1984
Redear Sunfish *(Lepomis microlophus)*		27.2 (4) (w)		5.9 (1) (w)	Sorensen & Bauer, 1984
Green Sunfish *(Lepomis cyanellus)*		21.4 (20) (w)	12.9 (4) (w)		Sorensen *et al.*, 1984
		10.8 (16) (w)			Sorensen *et al.*, 1983
White Catfish *(Ictalurus catus)*		10.8 (10) (w)	5.4 (27) (w)	8.9 (4) (w)	Sager & Cofield, 1984
Channel Catfish *(Ictalurus punctatus)*		11.9 (36) (w)	8.3 (38) (w)	9.9 (4) (w)	Sager & Cofield, 1984
Bullheads *(Ictalurus nebulosus)*		6.7 (2) (d)			Gutenmann *et al.*, 1976
Largemouth Bass *(Micropterus salmoides)*		10.2 (2) (w)	6.7 (2) (w)	10.3 (2) (w)	Sager & Cofield, 1984
Striped Bass *(Morone saxatilis)*	6.8 (6) (d)				Saiki, 1987b
Swordfish *(Xiphas gladius)*			0.69 (47) (-)		Freeman *et al.*, 1978

California. Undoubtedly, some particulate would accompany subsurface drainage.

In their study, Saiki and Lowe (Saiki and Lowe, 1987a) collect water, sediment, aquatic invertebrates, and mosquitofish (*Gambusia affinis*) in Kesterson (contaminated with Se) and Volta (a reference area 10 km south of Kesterson). Generally, Se levels increase from water to sediments to plants to fish. Kesterson and San Luis Drain mosquitofish have whole body Se levels ranging from 140 to 370 ppm (dry weight) and generally are higher in August than May. Volta mosquitofish have Se levels averaging ~1.35 ppm. Additional sampling periods throughout the year might show seasonal or long-term trends at Kesterson.

The absence of linear double logarithmic relationships between Se exposure concentration and concentration factors (CF) in mosquitofish (*Gambusia affinis*) from Kesterson (Se-contaminated area) and Volta (reference area) provides further evidence that mosquitofish have either incorporated or received superficial contamination from some unusually elevated Se source (Fig. 1). Lack of an inverse relationship between CF and elemental exposure concentrations contrasts with the work of others (Eisler, 1971; Eisler, 1974; Adams, 1976; Pascoe and Mattey, 1977). Ultimately, in that portion of the curve expected to be independent of homeostatic regulation (≥3–12 ppb Se), mosquitofish could accumulate Se to a finite limit, contingent upon the ability of fish to concentrate the toxic metalloid. Since the inverse relationship is not obtained (except in the 0.009–0.300 ppm portion of the curve) and since aqueous Se levels as high as 12 ppm have be measured in the water from the Grasslands (Saiki, 1987b), particulate Se in water (taken in during fish drinking) and Se-contaminated sediments in the intestines might be responsible for the absence of a definite trend in these data. Since whole body analyses are conducted without removal of the contents of the gastrointestinal tract, extremely concentrated contents might equalize concentration factors throughout the exposure range. An alternative possibility is that this relationship depicts the consequence of a combination of both homeostatic and nonhomeostatic regulation of the concentration factor of Se in mosquitofish exposed to variable levels of Se in the water. Certainly, dietary, aqueous, and sediment intake make environmental exposures more complicated than laboratory exposures.

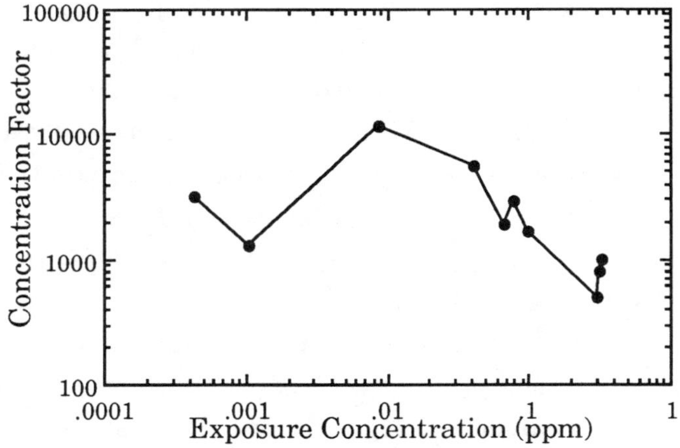

Figure 1. Double logarithmic relationship between Se exposure concentration and concentration factors (CF) in mosquitofish (*Gambusia affinis*) from Kesterson (Se-contaminated area) and Volta (reference area). Lack of an inverse relationship between CF and elemental exposure concentrations contrasts with the work of others (Eisler (1971); Eisler (1974); Adams (1976); Pascoe and Mattey (1977).). (Original figure based on data presented in Saiki and Lowe (1987a).).

In a study originating from North Carolina, Sager and Cofield (Sager, 1984) compare total Se concentration in liver, gonad, and muscle of four species (bluegill, largemouth bass, channel catfish, and white catfish) taken from Hyco Reservoir (Se-contaminated) and Roxboro City Lake (reference lake) in 1980. Generally, livers accumulate more Se than any other tissue or organ, followed in order of decreasing concentration by ovaries, skeletal muscle, and testes. Whereas, tissues from reference fish are <0.2–1.6 ppm Se (wet weight), tissues from Hyco fish range from 1.6–70.0 ppm Se. Liver from Hyco fish has a range of 2.5–70.0 ppm Se. The ranges each of Se levels in ovaries, testes, and skeletal muscle are 7.3–25.0 ppm, 3.0–7.2 ppm, and 1.6–23.0 ppm, respectively. Whereas, white and channel catfish show increasing average Se levels from muscle to ovary to liver, bluegill females show increasing average levels from ovary to muscle to liver. Bass females show similar Se levels for gonads and liver, but lower levels for muscle. Therefore, distribution patterns within tissues, as well as the absolute values of Se accumulated, vary considerably with the species considered.

Freeman and others (Freeman, *et al.*, 1978) determine levels of Se and Hg in skeletal muscle from about fifty swordfish (*Xiphias gladius*) collected from the northwestern Atlantic Ocean. About 80% of fish had muscle levels of Hg above the World Health Organization limit of 0.5 ppm. In those swordfish having muscle levels up to 2.0 ppm Hg, Se levels exceed those of Hg. Therefore, a molar excess of Se to Hg is measured rather than the predicted molar relationship between the two elements. Since nearly 100% of Hg in swordfish muscle is likely to be present as CH_3Hg^+ and since the presence of Se has been shown to lower the toxicity of CH_3Hg^+, swordfish muscle might not be as toxic to humans as it would be without Se. However, the levels of up to 4.4 ppm Hg in swordfish muscle in this study remains a cause for concern. Furthermore, consumption of such elevated levels of mercurials in the CH_3Hg^+ form could be hazardous to human health.

Dietary Se exposures result in higher accumulation of the metalloid than is the case in aqueous exposures (Sandholm, *et al.*, 1973). *Puntius arulius* is selected for testing using aqueous and/or food chain transfer of Se. Phytoplankton is incubated 24 h, zooplankton is added and incubated an additional 24 h before fish (1.5 g each) are added and held 72 h. Thirty times more Se is accumulated by phytoplankton when selenomethionine (rather than selenite) is the source of Se. Also, regardless of dietary source, teleost accumulation of Se from water is low—about 29–44% that of the most complex food chain (Fig. 2). Therefore, greater Se enrichment for this species of fish occurs through the food chain than through aqueous sources alone, as shown in other studies (Glenn, *et al.*, 1964; Adams, 1976; Hilton, *et al.*, 1980). Low epithelial absorption is thought to cause this difference.

Coughlan and Velte (Coughlan and Velte, 1989) use striped bass (*Morone saxatilis*) in a feeding study involving Se-contaminated living prey (20°C, pH 6.9, 30 ppm hardness). Bass are fed Se-enriched red shiners (*Notropis lutrensis*) from Belews Lake and compared with striped bass fed uncontaminated golden shiners (*Notemigonus crysoleucas*) because of a lack of reference red shiners. Bass are fed to satiation three times per week over the 80-d experiment. Whole body Se levels of red shiners are maintained at 9.6 ppm Se (wet weight) compared with 0.3 ppm Se in whole bodies of golden shiners. Golden shiners have similar protein, ash, and moisture content as red shiners; however, fat is 52% higher in red shiners than reference golden shiners. Striped bass which consume red shiners accumulate about 3.8 ppm

Se in skeletal muscle compared with 1.1 ppm in that of control bass (Coughlan and Velte, 1989). Unfortunately tests of renal and hepatic function are not presented, nor are Se levels provided for the extremely damaged hepatic and renal tissues.

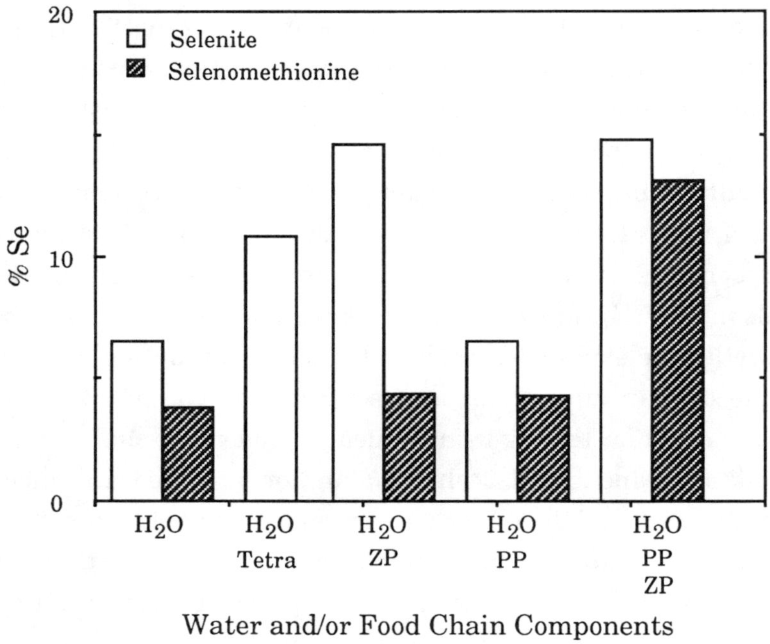

Figure 2. Selenium transfer from water to fish (1.5 g *Puntius arulius*) (H_20), from water to Tetra-Min (a commercial fish food) to fish (H_20, Tetra), from water to zooplankton to fish (H_20, ZP), from water to phytoplankton to fish (H_20, PP), or from water to phytoplankton to zooplankton to fish (H_20, PP, ZP), using Se-75 as a radioisotopic label. Selenium is presented as a percentage of the total available. (This original, relative frequency histogram is derived from numerical data from Sandholm, Oksanen, and Pesonen 1973).)

Generally, the liver is the major organ accumulating Se following either acute (McConnell, 1941; Heinrich, 1955) or chronic exposures (Smith, *et al.*, 1937; Maag, *et al.*, 1960; Sakurayama, 1960; Glenn, *et al.*, 1964; Rosenfeld and Heath, 1964) in a number of vertebrates, including fish. Adams reports levels of 70 and 7 ppm Se in the liver and kidney, respectively, of rainbow trout after 48-d exposure to 0.22 ppm Se (Adams, 1976). Under equilibrium conditions, fathead minnows and rainbow trout show whole body bioconcentration factors (ratios of Se levels in the body to aqueous Se levels) of 4443.0, 29.2 and 10.5 for exposure concentrations of 0.083 ppt, 50.6 ppb, and 0.3 ppm in the water, respectively. Generally, hepatic Se levels in individual fish are sufficiently elevated and consistent for Se analysis of single rather than pooled organs.

The reason for markedly elevated Se levels in fish could be internal recycling (i.e. enterohepatic redistribution). Enterohepatic redistribution can lead to markedly elevated hepatic levels to exacerbate toxicity, as is the case for As (Sorensen, 1976a; Sorensen, 1976b; Sorensen, 1976c; Sorensen, *et al.*, 1979a; Sorensen, *et al.*, 1979b; Sorensen, *et al.*, 1982c). Therefore, hepatic accumulation of Se in vertebrates does not necessarily precede biliary excretion and reduced whole body levels (McConnell and Martin, 1952).

In many species, the liver of Se-exposed fish accumulates higher Se levels than skeletal muscle (Sorensen, *et al.*, 1982a; Sager, 1984; Sorensen, *et al.*, 1984a), ovary (Sager, 1984; Sorensen and Bauer, 1984c) or kidney (mesonephros) (Sorensen, *et al.*, 1982b; Sorensen and Bauer, 1984c). However, Se levels in liver are correlated with Se concentrations in skeletal muscle and other tissues (Sorensen, *et al.*, 1982a; Sorensen, *et al.*, 1982b; Sorensen, *et al.*, 1984a; Sorensen and Bauer, 1984c). Redear sunfish from Se-contaminated sites on Martin Lake, for example, have an average value of 24.8 ppm Se in the liver and 48.4 ppm Se in the testes (Sorensen and Bauer, 1984c). In the same study, redears from a less contaminated site accumulate an average of 10.0 ppm Se in the liver and 21.3 ppm in the testes (Sorensen and Bauer, 1984c). Green sunfish from Belews Lake show the same trend (Sorensen, *et al.*, 1984a). Therefore, hepatic levels of Se are correlated with those of other organs or tissues from the same fish.

Needless to say, determination of the exact environmental level and exposure conditions for each feral fish is impossible under conditions of field studies. Hepatic levels of elements, however, seem to provide a good indication of teleost exposure, determined from laboratory studies using both Se and As (Sorensen, 1976a; Sorensen, 1976b; Sorensen, 1976c; Sorensen, *et al.*, 1979a; Sorensen, *et al.*, 1979b; Sorensen, *et al.*, 1982c; Boecker and Sorensen, 1988).

Assuming that the level of Se in the water at Martin Lake is 5 ppb, bioaccumulation factors of Se in redear sunfish are 0–1180 for ovaries, 2660–23140 for the kidneys, and 4260–10840 for the testes under equilibrium conditions. The hepatopancreas of Se-exposed feral fish accumulates 27.2 ppm Se, yielding average bioaccumulation factors estimated at 1442–2504

(Sorensen, *et al.*, 1982a; Sorensen, *et al.*, 1984a; Sorensen and Bauer, 1984c). Concentrations of Se in skeletal muscle are 33–70% of that in the liver and are correlated with liver concentrations (Sorensen, *et al.*, 1982a; Sorensen, *et al.*, 1984a). Some organs, such as the ovary, do not accumulate exceptionally high levels of Se (Sorensen and Bauer, 1984c), despite apparently debilitating effects upon the function of that organ (Sorensen, *et al.*, 1982a; Sorensen, *et al.*, 1984a). The same is the case for Hg-induced effects on the vertebrate brain (Goyer, 1986). Even though almost identical bioaccumulation factors for congeneric species in two different reservoirs seem to confirm the 5 ppb Se aqueous level as a reasonable estimate of Se in water, considerable uncertainty exists in estimating absolute values of aqueous Se levels in exact habitats occupied by fish. Therefore, strictly controlled laboratory studies are useful in defining concentration factors.

Selenium speciation and fate of Se in the teleost body are areas where much remains to be done. For example, knowledge of the predominate chemical form of Se accumulating in fish tissues could facilitate an understanding of the potential for toxicity. Possibly, methylation prior to compartmentalization of Se might lead to marked skeletal muscle levels of Se, as with Hg (Chapter VIII).

One of the most valuable areas for continued work is that of information storage and retrieval, and this is especially apparent for Se poisoning of fish. Available data from state and federal agencies are not in a form which is amenable to rapid search for specific information. Numerous fish are sacrificed and millions of tax dollars are spent in the collection, analysis, and writing of interagency reports which are practically impossible to locate. In the area of Se bioaccumulation, one might spend several lifetimes scavenging what meager data is available—even at some of the best document repositories.

Biological Effects

Hematology: Anemia is caused by either excess or deficiency levels of Se in vertebrates. The condition is characterized by reduced hematocrit, hemoglobin, and/or erythrocyte counts (Franke, 1934a; Franke and Potter, 1934b; Draize and Beath, 1935; Franke and Potter, 1935; Smith, *et al.*, 1937; Lillie and Smith, 1940; Moxon and Rhian, 1943), but is not always diagnosed in published studies. For example, in rats fed 15–50 ppm Se in the diet, 25%

developed severe anemia (hemoglobin: 20–45%), 25% developed slight anemia (hemoglobin: 50–63%), and 50% showed no change in hemoglobin level (74–95%) (Smith, *et al.*, 1937). Achromia, anisocytosis, polychromatophylia, and/or poikilocytosis occur (Franke and Potter, 1935; Ellis, *et al.*, 1937; Smith, *et al.*, 1937). Smith and others (Smith, *et al.*, 1937) observe marked reticulocytosis and macrocytosis in early stages of Se poisoning, microcytosis in the more advanced stages, and microcytic hypochromic anemia in advanced stages. Paradoxically, both leukopenia and leukocytosis are reported in cases of excess Se (Franke and Potter, 1935; Ellis, *et al.*, 1937).

Excess aqueous levels of Se interfere with hemoglobin formation in fish (Ellis, *et al.*, 1937; Poston, *et al.*, 1976; Hilton, *et al.*, 1980; Sorensen and Bauer, 1983c; Sorensen, *et al.*, 1984a; Sorensen and Bauer, 1984c), as does excess Zn or Cd (Mishra and Srivastava, 1979; Srivastava and Mishra, 1979). Channel catfish administered 3 ppm Se have reduced numbers of erythrocytes and reduced hemoglobin levels (Ellis, *et al.*, 1937). Disproportionate reduction in hemoglobin levels compared to erythrocyte counts indicates that Se interferes with normal hemoglobin formation rather than erythropoiesis *per se*. However, in one study, dietary administration of 0.1–15.0 ppm Se to young rainbow trout does not result in changes in erythrocyte counts, blood iron concentrations, or hematocrits after a 4-wk exposures (Hilton, *et al.*, 1980). The difference between the results of the catfish study (Ellis, *et al.*, 1937) and those from trout (Hilton, *et al.*, 1980) is the use of a dietary supplement of 0.4 IU/g vitamin E in trout diet.

Both Se (trace amounts) and vitamin E prevent anemia and act as antioxidants, preventing peroxidation of fats and fatty acids within membranes (Neal, 1977). Using Atlantic salmon *(Salmo salar)*, Poston and colleagues (Poston, *et al.*, 1976) evaluate interrelationships between dietary vitamin E and Se. Fry are fed supplemental vitamin E (0.5 IU/g diet) and/or 0.1 ppm Se in a 4-wk study. Vitamin E and Se deficiencies result in mortality and in hematocrit values which are 25% of control levels. Selenium supplementation (0.1 ppm) does not improve either parameter; however, addition of 0.5 IU/g vitamin E to the diet elevates hematocrits to 39% in the absence of Se and to 43% in the presence of Se. Therefore, both vitamin E and Se prevent anemia in fish, as in other vertebrates, although vitamin E may play a greater role. Also, Se deficiency can be just as detrimental as Se excess in causing anemia in vertebrates.

Excess Se may be more hazardous for young vertebrates than for mature vertebrates because juveniles are disadvantaged by a reduced storage capacity for vitamin E (Chow and Chen, 1980). Since excess Se induces anorexia (Franke and Potter, 1934b; Draize and Beath, 1935; Franke and Potter, 1935; Smith, *et al.*, 1937; Moxon and Rhian, 1943), the younger animal can be forced into a severe vitamin E deficiency status more rapidly than the adult. However, more prolonged anorexia would cause the same response in mature vertebrates. These studies indicate that Se (and vitamin E) intake levels must be kept within a restricted range for proper blood cell formation in vertebrates.

Regardless of whether fish are collected during the summer (Sorensen, *et al.*, 1984a) or winter (Sorensen and Bauer, 1983c; Sorensen and Bauer, 1984c), hematocrits (packed erythrocyte volumes) vary as a function of Se concentration in the liver. In winter collections, fish accumulating 27.2 ppm hepatic Se have hematocrit values averaging 28%, compared with reference fish accumulating 1.8 ppm Se and having a 43% hematocrit (Sorensen and Bauer, 1983c; Sorensen and Bauer, 1984c). In summer sampling of a congeneric species, hepatic Se levels of 21.4 and 1.3 ppm correspond to 33.0 and 39.5% hematocrits, respectively (Table 2) (Sorensen, *et al.*, 1984a). Therefore, hematocrit is reduced by 35% in the winter and 16% in the summer in these congeneric species. In the winter, reduced dietary supplies, depletion of hepatic glycogen, reduced metabolic rate, and suppression of growth dilution effects probably contribute to Se-induced reduction in hematocrits.

Therefore, studies conducted on electric generating station reservoirs (Sorensen and Bauer, 1983c; Sorensen, *et al.*, 1984a; Sorensen and Bauer, 1984c) demonstrate that fish collected from fresh water containing about 3–8 ppb Se have significantly reduced hematocrit values. Hematocrit values are inversely correlated with hepatic levels of Se. Ease of measurement and a relatively low variability make hematocrits hematologic parameters of choice in the evaluation of the toxic effects of Se on teleosts.

Hemoglobin concentration (measured as g hemoglobin per 100 ml of whole blood) and erythrocyte counts are reduced in sunfish accumulating the

Table 2. Hematocrits for fish having known levels of Se in the liver (i.e. hepatopancreas) and collected July 1981. Selenium levels are measured as ppm on wet weight basis; values are presented as mean (± one standard error of the mean). Sample numbers are provided in parenthesis.

Common Name *Genus and Species*	Lake	Selenium Level in the Hepatopancreas (ppm)	Hematocrit (%)
Redear sunfish	Martin	10.0±1.3 (9)[b]	27.0±1.0(9)[b]
Lepomis microlophus	Reference	2.8±0.4 (4)	33.2±1.4 (4)
Green sunfish	Martin	9.3±1.2 (6)[a]	32.5±2.2 (6)
Lepomis cyanellus	Belews-1	21.4±1.3 (20)[a]	33.0±1.0 (20)[b]
	Belews-2	7.0±0.7 (15)[a]	34.0±1.0 (10)[c]
	Reference	1.3±0.2 (4)	39.5±2.0 (6)

[a]Differs from respective reference fish values, $p \leq 0.001$.

[b]Differs from respective reference fish values, $p \leq 0.01$.

[c]Differs from respective reference fish values, $p \leq 0.05$.

most Se. General trends for erythrocyte counts, hemoglobin concentrations, and hematocrit values are similar and provide triplicate evidence that those sunfish accumulating the highest concentrations of Se are symptomatically anemic, or are becoming so. Correlation analysis of the hepatic level of Se *versus* the hemoglobin concentration show an inverse relationship, with considerable variability (Sorensen and Bauer, 1983c). Therefore, fish concentrating the highest levels of hepatic Se have reduced numbers of erythrocytes and hemoglobin concentrations. Such changes suggest that the presence of Se either directly or indirectly affects hemoglobin formation, the rate of erythropoiesis, and/or erythrocyte destruction. Disproportionate reduction in hemoglobin levels compared to erythrocyte counts indicates that Se interferes with normal hemoglobin formation (Ellis, *et al.*, 1937).

To classify different types of anemia, corpuscular constants are computed from hematocrit values, hemoglobin concentrations, and erythrocyte counts (Best, 1949). For example, mean corpuscular volume (MCV, expressed as μ^3) is determined by the multiplication of percent hematocrit values by ten and division by erythrocyte counts (as 10^6 cells/cc). This constant provides an indication of the average size of individual erythrocytes; therefore, MCV quantitatively distinguishes fish having normal (normocytic), small (microcytic), or large (macrocytic) erythrocytes. Fish with higher hepatic Se levels, have microcytic erythrocytes which average 134 μ^3

(Sorensen and Bauer, 1983c). Reference fish have red blood cells which average 174 μ^3. Therefore, Se-contaminated fish show a 23% reduction in erythrocyte volume. In comparison, microcytic human erythrocytes are less than 80 μ^3; whereas, normocytic human cells are 80–94 μ^3 (mean: 87 μ^3)—a 9% reduction (Best, 1949). The 23% reduction in erythrocyte size for Se-contaminated fish is about two times greater than the 9% reduction found in humans. Therefore, most erythrocytes from Se-contaminated fish are judged to be microcytic, based on MCV values.

Mean corpuscular hemoglobin concentration (MCHC) is another corpuscular constant—in this case providing the relative hemoglobin saturation of erythrocytes, as g of hemoglobin per 100 ml of packed erythrocytes. The MCHC values are computed by multiplication of hemoglobin concentrations by 100 and division by percent hematocrit values. In the human, for which a massive data base regarding the symptoms of anemia have been compiled, full saturation occurs at about 38% and hypochromic erythrocytes show MCHC levels of less than 30%—a 21% reduction (Best, 1949). Reference redear sunfish have MCHC values about 20%, which was 12% less than that computed for reference fish (Sorensen and Bauer, 1983c). In conclusion, despite variation, data derived from evaluation of corpuscular constants is similar to those observed for humans and indicates that sunfish accumulating the greatest hepatic levels of Se are already anemic or are becoming so.

Percentages of mature, immature, and old (or fragile) erythrocytes are compared in two environmental studies (Sorensen and Bauer, 1983c; Sorensen, *et al.*, 1984a). Whereas, the percentage of mature, circulating erythrocytes remains constant in fish (88%) regardless of hepatic levels of Se, circulating immature erythroblasts decrease 37% in fish with high Se. Reduced rates of erythropoiesis in Se-exposed fish may account for this decline. In addition, a 75% increase in numbers of old or fragile erythrocytes (i.e. nuclear shadows—Grizzle and Rogers, 1976) is indicative of accelerated erythrocyte lysis. Both reduced rates of erythropoiesis and accelerated erythrocyte lysis can reduce oxygen availability to tissues. Moreover, optical examinations show poikilocytic (i.e. abnormally shaped) and hypochromic erythrocytes. Hypochromia (reduced staining affinity in erythrocytes) is caused by reduced hemoglobin levels—also a precursor to hypoxia.

Leukocyte differentials identify the trend for increasing percentages of circulating hemoblasts (immature leukocytes) in redear sunfish collected from Se-contaminated areas in Martin Lake (Sorensen and Bauer, 1983c), but not green sunfish from Belews Lake (Sorensen, *et al.*, 1984a). In contrast, numbers of lymphocytes increase in green sunfish (Sorensen, *et al.*, 1984a), but not in redear sunfish (Sorensen and Bauer, 1983c) for specimens with the highest hepatic Se levels. Percentages of thrombocytes and neutrophils in Se-exposed sunfish remain unchanged. In conclusion, erythrocyte or leukocyte differentials are not consistently correlated with Se accumulation in the liver; therefore, these parameters can be eliminated in evaluation of the Se-induced toxicity responses of fresh-water teleosts. Measurement of hematocrit, hemoglobin, and erythrocyte counts yield much more meaningful information, at least in the species evaluated for this review. Of hematologic parameters evaluated, hematocrits provide the most consistent and useful trends in Se-induced hematological effects.

Histopathology: One of the most remarkable features of Se poisoning in vertebrates is the permanence of tissue injury (Draize and Beath, 1935; Smith, *et al.*, 1937; Miller and Schoening, 1938; Rosenfeld and Beath, 1947). Most published research in this area, however, involves vertebrates other than fish. For example, after several years of grazing on nontoxic areas, young cattle which had previously consumed seleniferous forage, still are stunted and show permanent liver and kidney damage (Dudley, 1936). Two months after cessation of a 60-d, daily exposure to 0.67 ppm Se, histopathological lesions persist and ascites and edema of the extremities are observed in rats (Rosenfeld and Beath, 1947). Smith and others (Smith, *et al.*, 1937) find that discontinuation of oral administration does not reverse Se-induced histopathological injury in a number of organs in cats. In cases of acute Se poisoning of vertebrates, cardiac and renal tissues are permanently injured (Draize and Beath, 1935).

Obviously, the best-studied organs in cases of Se poisoning are those involved in detoxification and elimination (i.e. liver and kidney). These organs generally accumulate the highest concentrations of Se because of these roles (McConnell, 1941; Rosenfeld and Beath, 1946; Adams, 1976). Hicks and co-workers (Hicks, *et al.*, 1984) observe overt renal lesions in rainbow trout (*Salmo gairdneri*) fed Se (as selenite) in the diet at 0.6, 6.6, or 11.4 ppm for 16 wk (15°C and pH 7.7–7.9). Body weight decreases, feed:gain ratios increase, and mortality increases in fish exposed 16 wk to 11.4 ppm Se in the diet ($p<0.05$). The liver and kidney accumulate maximum averages of 51.3 and 11.2 ppm (wet weight), respectively, for the same exposure. The liver accumulates 4.4–12.7 fold more Se than does the kidney. Also hepatic levels of Mg increase, as do renal Ca and Mg levels in trout with overt renal lesions. Renal Ca levels are 4.4 times higher in fish (without overt renal lesions) consuming 11.4 ppm Se in the diet than fish consuming 0.6 ppm Se in the diet. However, no ascites or edema are observed.

Often preceded by proliferative kidney disease, nephrocalcinosis is diagnosed by the presence of large chalky deposits filling dilated ureters (Hicks, *et al.*, 1984). Severe and mild lesions are present 40 and 50% of the time, respectively, for fish given 11.4 ppm Se in the diet for 20 wk. The 1–2 mm diameter off-white deposits are visible to the unaided human eye and affect the second convoluted tubules through to the bladder. Casts are formed, and renal tubule cells are necrotic and have pyknotic nuclei. Advanced cases show chronic inflammatory foci (macrophages, epitheloid cells, and fibroblasts) within the interstitium. Apparently, rainbow trout and brook trout are more susceptible to this condition than are other fish species.

Coughlan and Velte (Coughlan and Velte, 1989) also observe marked histopathological changes in striped bass (*Morone saxatilis*) fed Se-enriched red shiners (*Notropis lutrensis*) from Belews Lake to satiation three times per week over 80 d (20°C, pH 6.9, 30 ppm hardness). Whole body Se levels of red shiners are 9.6 ppm (wet weight) compared with 0.3 ppm Se in whole bodies of golden shiners. About 3.8 ppm Se is concentrated by skeletal muscle of striped bass compared with 1.1 ppm in control bass (Coughlan and Velte, 1989). Although exocrine pancreas remains unchanged, parenchymal hepatocytes show fat vacuoles, swollen nuclei, glycogen devacuolization, and granular basophilic cytoplasm. Proximal tubules show casts, hydropic vacuolation, tubular cell degeneration, and/or acidophilic tubular inclusions. Tests of renal and hepatic function are not conducted, but the histopathological alterations can result in uricemia and altered liver function to such an extent that death results. However, respiratory distress also might contribute to mortality, as suggested by the coughing and gulping of air in Se-exposed bass prior to death. Corneal cloudiness and apparent disregard for obstacles indicates that visual disturbances also may occur in striped bass under conditions of this study.

Parenchymal hepatocytes from Se-exposed fish exhibit changes characteristic of the permanent, histopathological alterations reported for other vertebrates exposed to excess Se. The most striking change in the Se-exposed hepatopancreas is hepatocyte degeneration around the central veins (i.e. central necrosis) with collapse of the central vein, apparently resulting in the dilation of these veins (Sorensen, 1986). This feature is regarded as an important diagnostic feature because it is considered an early sign of cirrhosis (Sorensen, *et al.*, 1983b). Remaining hepatocytes show a condition referred to as "unrest" (i.e. parenchymal cell disruption). Focal necrosis (parenchymal hepatocyte degeneration throughout the liver) also is observed in fish accumulating excessive concentrations of Se in the hepatopancreas (Sorensen, 1986). The liver of Se-exposed fish is characterized by reduced rough endoplasmic reticulum and glycogen granules (Sorensen, *et al.*, 1983b), by nuclear and mitochondrial abnormalities (Sorensen, *et al.*, 1983a; Sorensen, *et al.*, 1983b), and by increased numbers of lysosome-like structures in parenchymal hepatocytes (Sorensen, *et al.*, 1983a; Sorensen, *et al.*, 1983b).

Reticulum stains show hyperplasia of the normally-delicate, supportive cellular framework—another diagnostic of early hepatic cirrhosis (Sorensen, *et al.*, 1983b). Lymphocyte infiltration, increased numbers of Kupffer cells, areas of granular cytoplasm, extensive vacuolation of parenchymal hepatocytes around central veins, increased numbers of perisinusoidal lipid droplets, and greater numbers of pleomorphic nuclei occur in Se-exposed sunfish (Sorensen, *et al.*, 1982a; Sorensen, *et al.*, 1983a; Sorensen, *et al.*, 1983b; Sorensen, *et al.*, 1984a).

In addition, exocrine pancreas cells are disrupted architecturally, with reductions in rough endoplasmic reticulum, increased cisternal space, and possibly increased smooth endoplasmic reticulum (Sorensen, *et al.*, 1983b). Hyperplasia, but not hypertrophy, of exocrine pancreas cells is observed in optical (Sorensen, *et al.*, 1983b) and planimetric analyses (Sorensen and Bauer, 1984b). These studies convincingly document a close relationship between Se accumulation in fish and hyperplasia of exocrine pancreas cells. Planimetric analysis of the ratio of areas of exocrine pancreatic cells to that of parenchymal hepatocytes; however, reveals marked specimen variation, with no significant correlation between this ratio and Se accumulation in the hepatopancreas (Sorensen and Bauer, 1984b).

In other studies, livers of vertebrates are severely damaged by excess Se. For example, catfish administered a single intraperitoneal dose of 3 ppm Se are diagnosed as having central necrosis, as well as hepatocyte disruption, edema, and hyaline degeneration of hepatic arterial walls (Ellis, *et al.*, 1937). Chronically-exposed pigs show central necrosis, focal necrosis, fatty degeneration, lymphocyte infiltration, deposition of bile pigment, and proliferation of interstitial tissue in livers containing 3–10 ppm Se (Miller and Schoening, 1938). Horses and mules which accumulate hepatic levels of 2–8 ppm Se following acute exposures, exhibit diffuse fatty degeneration of the liver (Miller and Williams, 1940). Livers of sheep with symptoms of Se toxicosis manifest central necrosis, fatty degeneration, obliteration of the normal lobular architecture, and early fibrosis in chronic cases (Rosenfeld and Beath, 1946). Moreover, early fibrosis with fibroblast and bile-duct proliferation is observed in subacute cases; and massive necrosis, sinusoid dilation, hemorrhages, and congestion are observed in acute cases (Rosenfeld and Beath, 1946). Draize and Beath (Draize and Beath, 1935) observe cloudy swelling of hepatocytes, necrosis and loss of hepatocytes especially around the central veins in livestock diagnosed with acute Se poisoning. Chronic cases are characterized by occasional focal necrosis, cirrhosis, leukocyte infiltration, and even greater cellular deterioration around central veins. Rats receiving 7.5 ppm dietary Se for 6-mo exhibit fatty degeneration of the liver in earlier stages, followed by focal necrosis, widening of blood spaces, stasis, and finally periportal cirrhosis in the more advance stages (Smith, *et al.*, 1937). Therefore, Se causes severe liver damage which is generally manifested as marked hepatocyte necrosis.

Histopathological examinations of the livers of Se-exposed fish and other vertebrates repeatedly document severe, if not permanent, changes. These abnormalities appear to represent a general, nonspecific necrotic response, rather than one which is unique to Se toxicosis. However, the presence of this severe histopathological damage with elevated hepatic Se levels (in the absence of elevations of other xenobiotics) provides presumptive evidence for Se-induced toxicity in fresh-water teleosts.

No less impressive is the severity of renal changes following Se exposure. However, a brief explanation of renal morphology of fish is appropriate at this point because of anatomic intricacies. The mesonephros of unexposed fish resembles the vertebrate counterpart, with minor exceptions. Glomeruli normally occur in groups with mesangial cells having a minimum of matrix and with endothelial cells which are easily distinguished (Sorensen, 1986). Reference sunfish have glomeruli similar to those of the human, except for a decreased pedicle diameter and a resultant somewhat-spiked appearance of the podocyte foot processes. Normal fish kidneys are laminated; that is, floccular material is located between the fenestrated endothelial lining and the basement membrane. Occasional effacement, or fusion of pedicles at the slit pores, occurs normally (Sorensen, *et al.*, 1982b). Otherwise, mesangial cells, podocytes, and endothelial cells of the fish glomerulus are similar to those of the human. In addition, it is noted that the tubular cells of reference fish are morphologically normal, and do not differ to a substantial degree from those of other vertebrates (Sorensen, 1986).

In Se-exposed fish, tubular changes are evident at both the optical and ultrastructural levels (Sorensen, *et al.*, 1982b; Sorensen, *et al.*, 1984a). Fish which accumulate the highest levels of Se in the liver consistently show tubular cell injury, such as severe cytoplasmic vacuolation (Sorensen, *et al.*, 1982b), casts (Sorensen, *et al.*, 1984a), necrosis (Sorensen, *et al.*, 1982b), reduced cell height (Sorensen, *et al.*, 1984a), and desquamation (Sorensen, *et al.*, 1984a). Glomerular abnormalities in fish include periglomerular fibrosis (Sorensen, *et al.*, 1984a) and increased numbers of mesangial cells which excessive mesangial cell matrix (Sorensen, *et al.*, 1982a; Sorensen, *et al.*, 1982b; Sorensen, *et al.*, 1983b; Sorensen, *et al.*, 1984a), which is diagnostic of proliferative glomerulonephritis. Mesangial cell proliferation after exposure to a variety of chemicals is believed to result in permeability changes within the renal corpuscle (Sorensen, *et al.*, 1983b). Uremia, aminoaciduria, and loss of vital minerals and other nutrients can result from such mesangial cell changes. As is the case for Se-induced hepatic histopathological changes, renal changes appear to the nonspecific and to involve more than one cell type (e.g. the glomerular and tubular cells).

Selenium-induced renal effects are similar in other vertebrates. Glomerulonephritis, intertubular hemorrhage, degeneration of collecting tubule cells, and hyaline degeneration of convoluted tubule cells occur in livestock (Draize and Beath, 1935). Severe renal damage, including glomerulonephritis, is noted by Moxon and Rhian (Moxon and Rhian, 1943). Mild tubular cell degeneration is documented in rats administered 7.5 ppm

Se in the diet for 6 mo (Smith, *et al.*, 1937). Cortical hemorrhages and friability, renal capillary engorgement with hemorrhages, tubular cell degeneration, fatty degeneration and cytoplasmic vacuolation of tubular cells, and epithelial exfoliation are observed (Miller and Williams, 1940). Pig kidneys which concentrate 3–25 ppm Se following acute exposures are edematous and have atrophied glomeruli, marked epithelial cell degeneration and desquamation, pigment deposits, and fatty changes with vacuolation of tubular cells (Miller and Schoening, 1938). Catfish administered single intraperitoneal doses of 3 mg/kg have pale tubular cells, interstitial accumulation, and capsular fluid accumulation (Ellis, *et al.*, 1937). The presence of striking glomerular and tubular injuries, therefore, is characteristic of vertebrates exposed to excess Se.

Fish accumulating high levels of hepatic Se are diagnosed with pericarditis and myocarditis (Sorensen, *et al.*, 1982a; Sorensen, *et al.*, 1984a), which is characterized by the presence of increased numbers of inflammatory cells within the pericardial space and between myocardial cells, respectively. Reference fish, in comparison, have few inflammatory cells within the pericardial sac or the mycardium. Inflammatory infiltration, of the type observed in fish accumulating excessive hepatic Se, could be a result of a number of direct or indirect mechanisms, including direct action of Se on the mesonephros with induction of glomerulonephritis and subsequent uremia, which can indirectly result in pericarditis and myocarditis (Sorensen, *et al.*, 1982a; Sorensen, *et al.*, 1984a).

Normally, gill lamellae (i.e. the actual site of gaseous exchange) from reference fish tend to have a spiked appearance following formaldehyde fixation (Grizzle and Rogers, 1976; Sorensen, *et al.*, 1982a; Sorensen, *et al.*, 1984a). Gill lamellae from reference fish are composed of a thin epithelial cell layer covering pillar cells which line blood sinusoids (Grizzle and Rogers, 1976). In contrast, lamellae of fish with elevated Se are noticeably swollen and vacuolated. The physiological significance of this change is unknown but may represent either an edematous response or some adaptation to stabilize ion or water influx into gill vasculature (Sorensen, *et al.*, 1982a). Detailed ultrastructural evaluations of the lamellae of Se-exposed fish have not yet been fully described; however, based on preliminary work, many changes may resemble those in fish following Zn (Skidmore and Tovell, 1972; Matthiessen and Brafield, 1973) or Cu (Bilinski and Jonas, 1973) exposures.

Female fish accumulating the highest levels of hepatic Se have abnormal numbers of degenerating or necrotic egg follicles (Sorensen, *et al.*, 1982a; Sorensen, *et al.*, 1984a; Sorensen and Thomas, 1988a; Sorensen, *et al.*, 1988b). Follicles are normal in the ovaries of reference female sunfish, but many of those from females with elevated levels of tissue Se are atretic or show early changes which can lead to atresia. The necrotic process is marked in some follicles and characterized by an indistinct zona radiata, yolk vacuole coalescence and release from the zona radiata, and increased numbers of fat vacuoles in some instances. These histopathological abnormalities in the ovaries of fish with elevated concentrations of hepatic Se provide an explanation for the reproductive failure of teleosts in Belews Lake (Cumbie and Van Horn, 1978a).

Other Alterations: Behaviorally, fish do not respond to the presence of excess Se in water, a result which might be related to lack of detection ability or stupefaction. Using a steep-gradient, laminar-flow chamber, Hartwell and colleagues (Hartwell, *et al.*, 1989) compare acute lethal concentrations and avoidance thresholds in golden shiners (*Notemigonus crysoleucas*) for five elements (pH 7.5 and water hardness 72.2 ppm). The 96-h LC50 for Cd, Se, As, Cr, and Cu are 2.8, 11.2, 12.5, 55.0, and 84.6 ppm. Shiners avoid Cu, As, and Cr at 0.026, 0.028, and 0.073 ppm, but do not avoid Cd or Se at levels up to 68 and 3489 ppb, respectively. Therefore, behavioral responses are not necessarily indicative of toxicological responses. Moreover, very toxic yet non-repulsive elements (e.g. Se and Cd) may be more damaging than less toxic but repulsive elements (e.g. As and Cu). The absence of repulsion or attraction to Se in shiners at 31% of the 96-h LC50 for Se (as selenite) may indicate that olfactory systems do not respond to the elevated levels of Se or that fish are stupefied and cannot respond to excess Se.

Other behaviorists find essentially the same results. For example, Watenpaugh and Beitinger (Watenpaugh, 1985b) find that fathead minnows (*Pimephales promelas*) do not avoid up to 11.2 ppm Se (as selenate) in flow-through avoidance systems (16°C, pH 7.7, 51 ppm water hardness). Test solutions and control solutions flow into either end of a plexiglass gradient chamber, and a steep gradient is produced at the central drain of the chamber. Fathead minnows are not attracted to or repelled by lethal aqueous levels of 0.3–11.2 ppm Se, as selenate. Therefore, neither fathead minnows or golden shiners respond to the presence of toxic levels of aqueous Se (as selenate or selenite) under conditions of these two studies (Watenpaugh, 1985b; Hartwell, *et al.*, 1989).

Normally, however, Se could be expected to coexist with a number of other very poisonous elements, increasing the possibility the fish will be repelled by the mixture. Perhaps species survival requires olfactory adaptations for recognition of one of four or five elements (for example), rather than every single element of the mixture. Compared with less endowed species, fish with this ability would have a greater chance of avoiding noxious chemicals and reproducing successfully.

The K_{TL}-factor or condition-factor (a measure of body weight per unit body length) is another measure of biological effects used to determine the general health of fish. Condition factors are calculated as follows:

$$K_{TL} = \frac{\text{weight (in g) x } 10^5}{\text{total length (in mm}^3\text{)}}$$

Generally, healthier fish have higher condition-factors (Sorensen, 1974). Winter collections of Se-contaminated redear sunfish have lower condition-factors (mean: 1.38) than reference redear sunfish collected at the same time (mean: 1.59) (Sorensen and Bauer, 1984c). Corresponding values for hepatic Se are 8.4–27.2 ppm (wet weight) in Se-exposed redear sunfish (Sorensen and Bauer, 1984c). Therefore, for redear sunfish, the reduction in K_{TL} with increased Se hepatic levels provides an important indicator. Redear sunfish accumulating high levels of Se appear to be in worse health than redear sunfish controls. This difference affects not only condition-factors, but also relative organ weights and microscopic observations. Redear sunfish do not become edemic when Se accumulates in tissues (Sorensen, *et al.*, 1983b; Sorensen and Bauer, 1984b). However, based on K_{TL} values, edemic fish may appear to be healthier than reference fish (Sorensen, *et al.*, 1982a; Sorensen, *et al.*, 1984a; Sorensen and Bauer, 1984b; Sorensen and Bauer, 1984c). This is the case for green sunfish which accumulate high levels of Se (Sorensen, *et al.*, 1982a; Sorensen, *et al.*, 1984a).

Selenium-exposed redear sunfish from Se-contaminated areas in Martin Lake have liver weight to body weight ratios which are significantly lower than those of reference fish (Sorensen and Bauer, 1984c). Relative liver weights decrease significantly as hepatic Se levels increase. In contrast, green sunfish from Se-contaminated areas in Belews Lake have livers larger than normal (Sorensen, *et al.*, 1984a). For example, green sunfish accumulating an average of 21.4 ppm hepatic Se have livers which are 77% larger than those sunfish which accumulate 7.0 ppm Se in the liver. Histopathological examination of the liver preceding evaluation of relative liver weights or condition-factors is useful in establishing whether an edematous condition exists for fish. However, extreme edema is required for accurate diagnosis of the condition.

Those fish species which become edemic essentially "dilute Se" concentrated by various organs. Selenium-exposed green sunfish have lower average concentrations of Se in the liver than do nonedematous redear sunfish, despite the fact that some of these fish were collected from the same sites on the same date (Sorensen, *et al.*, 1982a; Sorensen, *et al.*, 1984a; Sorensen and Bauer, 1984c). Moreover, as with arsenic accumulation in green sunfish, enterohepatic redistribution of Se may place the liver of this species at more risk than is the case for other species since the green sunfish liver is almost continuously exposed and re-exposed under such circumstances (Sorensen, *et al.*, 1979b). Obviously, to circumvent the problems with variability in responses of congeneric species, reference fish for each species must be used as the basis for comparison. In the case mentioned here, reference fish from each species have unique Se uptake, condition-factors, relative organ weights, and histopathological signs of edema despite collection from the same area on the same day (Sorensen, *et al.*, 1982a).

At present, no known inherent physiological and/or biochemical mechanism accounts for the difference in response of the two congeneric species to Se. Green sunfish may be susceptible to increased capillary permeability and subsequent generalized edema; whereas, redear sunfish may not to be predisposed to this condition. (Sorensen, *et al.*, 1982a; Sorensen, *et al.*, 1984a; Sorensen and Bauer, 1984b; Sorensen and Bauer, 1984c). In Se-contaminated green sunfish, fluids of peripheral circulation may escape into surrounding tissues to produce edema, as in many other studies (Franke, 1934a; Franke and Potter, 1934b; Franke and Potter, 1935).

Other possibilities exist. Consumption of different food items (with variable Se levels and Se speciation) could bring about the difference between the two congeneric species. Redears eat a considerable quantity of snails; whereas, green sunfish tend to consume large quantities of small fish. Hence, redears are more akin to the bottom-feeders and green sunfish more like the upper level predators of the Jernelöv and Lann study (Jernelöv and Lann, 1971) than the halosaurs and macrourids of the Barber *et al.* study (Barber, *et al.*, 1972). Elemental levels in the latter fish probably differ because of metabolic differences rather than differences in dietary items.

Other published reports show similar results, although data interpretation is sometimes complicated. For example, chronic, low-level Se exposure of rainbow trout and goldfish, results in anorexia and poor feed-consumption-to-weight-gain conversion ratios (Ellis, *et al.*, 1937; Hilton, *et al.*, 1980). Rats, cats, hamsters, pigs, and various livestock avoid Se-contaminated food or water (Franke, 1934a; Franke and Potter, 1934b; Franke and Potter, 1935); however, growth weight reductions are not completely accounted for by reduced food consumption (Franke and Potter, 1934b). Selenium exposure can lead to reductions in organ weight (Miller and Schoening, 1938; Rosenfeld and Beath, 1946; Halverson, *et al.*, 1966), but Se-induced edema increases organ weight to complicate the interpretation of relative organ weight data (Franke, 1934a; Franke and Potter, 1934b; Franke and Potter, 1935). Use of both weight and dry weights for calculations of Se tissue levels facilitates detection of the edematous condition. All of these studies show that biological organisms respond to Se challenge in intricate and unique ways.

Summary

Sunfish exposed to aqueous Se accumulate Se in the liver (7–27 ppm, wet weight) and skeletal muscle, although levels in skeletal muscle are about half those in the liver (Sorensen, *et al.*, 1982a). At aqueous exposure levels of 3 to 8 ppb Se in the environment, bioaccumulation factors are 1190–2504 for the two tissues, assuming equilibrium conditions. Selenium analysis of other tissues provides less consistent information.

Those fish accumulating the highest hepatic Se have the lowest hematocrits, lowest hemoglobin levels, lowest erythrocyte numbers, abnormally small and irregularly-shaped erythrocytes, and reductions in both mean corpuscular volumes and mean corpuscular hemoglobin concentrations.

Both redear and green sunfish are diagnosed as having proliferative glomerulonephritis, hematuria, and/or necrosis of convoluted tubule cells of the kidney; central necrosis and/or increased numbers of phagocytic reticuloendothelial cells in hepatic tissue; and marked hypertrophy of exocrine pancreas tissue with ultrastructural changes including architectural disorganization, reduced rough endoplasmic reticulum, and proliferation of smooth endoplasmic reticulum.

Green sunfish with the highest hepatic Se have pericarditis and myocarditis, which could be secondary to glomerulonephritis-induced uremia. Numerous necrotic, degenerating ovarian follicles are present in female green sunfish. These may be pathognomonic for reproductive failure of the population of this species endemic to the main basin of Belews Lake (Cumbie and Van Horn, 1978a). No prominent histopathological changes are observed using optical microscopy for the spleen, testes, stomach, or skeletal muscle.

Generally, Se-exposed green sunfish have higher condition-factors and relative liver weights than controls because Se-exposed green sunfish become edematous. Redear sunfish, on the contrary, do not become edematous when exposed to excess Se and have slightly higher average hepatic concentrations of Se than observed for green sunfish. In conclusion, excess Se (even as low as 3–8 ppb aqueous levels) can cause numerous life-threatening changes in feral fresh-water teleosts.

References

Adams, W.J., The toxicity and residue dynamics of selenium in fish and aquatic invertebrates, Ph. D. Thesis, Michigan State University, (1976) 1.

Andren, A.W., D.H. Klein, and Y. Talmi, Selenium in coal-fired steam plant emissions, Environ. Sci. Techn., 9 (1975) 856.

Anundi, I., H. Hogberg, and A.H. Stead, Glutathione depletion in isolated hepatocytes: Its relation to lipid peroxidation and cell damage, Acta Pharmacol. Toxicol., 45 (1979) 45.

Barber, R.T., A. Vijayakumar, and F.A. Cross, Mercury concentrations in recent and ninety-year-old benthopelagic fish, Science, 178 (1972) 636.

Best, R., A hematologic slide rule for calculating the corpuscular constants, J. Lab. Clin. Med., 34 (1949) 434.

Bilinski, E., and R.E.E. Jonas, Effects of cadmium and copper on the oxidation of lactate by rainbow trout (*Salmo gairdneri*) gills, J. Fish. Res. Bd. Can., 30 (1973) 1553.

Boecker, C.S., and E.M.B. Sorensen, Histopathological alterations in teleosts following selenate exposure, Toxicologist, 8 (1988) 109.

Calvin, H.I., Selective incorporation of selenium-75 into a polypeptide of the rat sperm tail, J. Exp. Zool., 204 (1978) 445.

Cantor, A.H., M.L. Langevin, T. Noguchi, and M.L. Scott, Efficacy of selenium in selenium compounds and feedstuffs for prevention of pancreatic fibrosis in chicks, J. Nutr., 105 (1975a) 106.

Cantor, A.H., M.L. Scott, and T. Noguchi, Biological availability of selenium in feedstuffs and selenium compounds for prevention of exudative diathesis in chicks, J. Nutr., 105 (1975b) 96.

Cardwell, R.D., D.G. Forman, T.R. Payne, and D.J. Wilbur, Acute toxicity of selenium dioxide to freshwater fishes, Arch. Environ. Contam. Toxicol., 4 (1976) 129.

Chow, C.K., and C.J. Chen, Dietary selenium and age-related susceptibility of rat erythrocytes to oxidative damage, J. Nutr., 110 (1980) 2460.

Christophersen, B.O., Formation of monohydroxypolyenic fatty acids from lipid peroxides by a glutathione peroxidase, Biochim. Biophys. Acta, 164 (1968) 35.

Christophersen, B.O., Reduction of linolenic acid hydroperoxide by a glutathione peroxidase, Biochim. Biophys. Acta, 176 (1969) 463.

Copeland, R., Selenium: The unknown pollutant, Limnos, 3 (1971) 7.

Coughlan, D.J., and J.S. Velte, Dietary toxicity of selenium-contaminated red shiners to striped bass, Trans. Am. Fish. Soc., 118 (1989) 400.

Cumbie, P.M., and S.L. Van Horn, Selenium accumulation associated with fish mortality and reproductive failure, Proc. Ann. Conf. S.E. Assoc. Fish Wildl. Agencies, (1978a) 612.

Cumbie, P.M., Effects of pH adjustment and SO_3 fly ash conditioning on the Belews Creek Steam Station Ash Basin, Duke Power Company, Technical Report Series No. 80-01, Charlotte, North Carolina, (1980) 1.

Doran, J.W., and M. Alexander, Microbial transformations of selenium, Appl. Environ. Microbiol., 33 (1977) 31.

Draize, J.H., and O.A. Beath, Observations on the pathology of blind staggers and alkali disease, J. Am. Vet. Med. Assoc., 86 (1935) 753.

Dudley, H.C., Toxicology of selenium. I. A study of the distribution of selenium in acute and chronic cases of selenium poisoning, Am. J. Hygiene, 23 (1936) 169.

Eisler, R., Cadmium poisoning in *Fundulus heteroclitus* (Pisces: Cyprinodontidae) and other marine organisms, J. Fish. Res. Bd. Can., 28 (1971) 1225.

Eisler, R., Radiocadmium exchange with seawater by *Fundulus heteroclitus* (L.) (Pisces: Cyprinodontidae), J. Fish. Biol., 6 (1974) 601.

Ellis, M.M., H.L. Motley, M.D. Ellis, and R.O. Jones, Selenium poisoning in fishes, Proc. Soc. Exp. Biol. Med., 36 (1937) 519.

Fleming, R.W., and M. Alexander, Dimethylselenide and dimethyltelluride formation by a strain of *Penicillium*, Appl. Microbiol., 24 (1972) 424.

Flohe, L., W.A. Gunzler, and H.H. Shock, Glutathione peroxidase: A selenoenzyme, FEBS Lett., 32 (1973) 132.

Franke, K.W., A new toxicant occurring naturally in certain samples of plant foodstuffs. I. Results obtained in preliminary feeding trials, J. Nutr., 8 (1934a) 597.

Franke, K.W., and V.R. Potter, A new toxicant occurring naturally in certain samples of plant foodstuffs. III. Hemoglobin levels observed in white rats which were fed toxic wheat, J. Nutr., 8 (1934b) 615.

Franke, K.W., and V.R. Potter, A new toxicant occurring naturally in certain samples of plant foodstuffs. IX. Toxic effects of orally ingested selenium, J. Nutr., 10 (1935) 213.

Freeman, H.C., G. Shum, and J.F. Uthe, The selenium content in swordfish (*Xiphias gladius*) in relation to total mercury content, J. Environ. Sci. Hlth., A13 (1978) 235.

Gillespie, R.B., and P.C. Baumann, Effects of high tissue concentrations of selenium on reproduction by bluegills, Trans. Am. Fish. Soc., 115 (1986) 208.

Glenn, M.W., J.L. Martin, and L.M. Cummings, Sodium selenate toxicosis: The distribution of selenium within the body after prolonged feeding of toxic quantities of sodium selenate to sheep, Amer. J. Vet. Res., 25 (1964) 1495.

Goyer, R.A., Toxic effects of metals, in Klaassen, C.D., M.O. Amdur, and J. Doull (Eds.), Casarett and Doull's Toxicology, Macmillian Publ. Co., New York, New York, (1986) 582.

Grizzle, J.M., and W.A. Rogers, Anatomy and histology of the channel catfish, Auburn Univ. Agric. Exp. Sta. Publ., Auburn, (1976) 1.

Gunby, P., Selenium may act as cancer inhibitor, J. Amer. Med. Assoc., 246 (1981) 1510.

Gutenmann, W.H., C.A. Bache, W.D. Youngs, and D.J. Lisk, Selenium in fly ash, Science, 191 (1976) 966.

Halter, M.T., W.J. Adams, and H.E. Johnson, Selenium toxicity to *Daphnia magna, Hyallela azeteca*, and the fathead minnow in hard water, Bull. Environ. Contam. Toxicol., 24 (1980) 102.

Halverson, A.W., I.S. Palmer, and P.L. Guss, Toxicity of selenium to post-weanling rats, Tox. Appl. Pharmacol., 9 (1966) 477.

Hamilton, S.J., K.J. Buhl, and N.L. Faerber, Toxicity of selenium in the diet to chinook salmon, Symposium IV, Selenium and Agricultural Drainage, (1987) 22.

Hartwell, S.I., J.H. Jin, D.S. Cherry, and J. J. Cairns, Toxicity versus avoidance response of golden shiner *Notemigonus crysoleucas*, to five metals, J. Fish. Biol., 35 (1989) 447.

Hashimoto, Y., J.Y. Hwang, and S. Yanagisawa, Possible source of atmospheric pollution of selenium, Environ. Sci. Techn., 4 (1970) 157.

Heinrich, M., Jr., Studies on selenium metabolism: The distribution of selenium in the tissues of the mouse, J. Pharmacol. Exp. Ther., 114 (1955) 28.

Hicks, B.D., J.W. Hilton, and H.W. Ferguson, Influence of dietary selenium on the occurrence of nephrocalcinosis in the rainbow trout, *Salmo gairdneri* Richardson, J. Fish Dis., 7 (1984) 379.

Hilton, J.W., P.V. Hodson, and S.J. Slinger, The requirement and toxicity of selenium in rainbow trout (*Salmo gairdneri*), J. Nutr., 110 (1980) 2527.

Huckabee, J.W., and N.A. Griffith, Toxicity of mercury and selenium to the eggs of carp (*Cyprinus carpio*), Trans. Am. Fish. Soc., 103 (1974) 822.

Hunn, J.B., S.J. Hamilton, and D.R. Buckler, Toxicity of sodium selenite to rainbow trout fry, Water Res., 21 (1987) 233.

Jernelöv, A., and H. Lann, Mercury accumulation in food chains, Oikos, 22 (1971) 403.

Lillie, R.D., and M.I. Smith, Histogenesis of hepatic cirrhosis in chronic food selenoisis, Am. J. Pathol., 16 (1940) 223.

Little, C., and P.J. O'Brien, An intracellular GSH-peroxidase with a lipid peroxide substrate, Biochem. Biophys. Res. Commn., 31 (1968) 145.

Lowe, T.P., T.W. May, W.G. Brumbaugh, and D.A. Kane, National contaminant biomonitoring program: Concentrations of seven elements in freshwater fish, 1978-1981, Arch. Environ. Contam. Toxicol., 14 (1985) 363.

Maag, D.D., J.S. Orsborn, and J.R. Clopton, The effect of sodium selenite on cattle, Amer. J. Vet. Res., 21 (1960) 1049.

MacFarlane, R.D., G.L. Bullock, and J.J.A. McLaughlin, Effects of five metals on susceptibility of striped bass to *Flexibacter columnaris*, Trans. Am. Fish. Soc., 115 (1986) 227.

Matthiessen, P., and A.E. Brafield, The effect of dissolved zinc on the gills of the stickleback *Gasterosteus aculeatus* (L.), J. Fish Biol., 5 (1973) 607.

McConnell, K.P., Distribution and excretion studies in the rat after a single subtoxic subcutaneous injection of sodium selenate containing radioselenium, J. Biol. Chem., 141 (1941) 427.

McConnell, K.P., and R.G. Martin, Biliary excretion of selenium in the dog after administration of sodium selenate containing radioselenium, J. Biol. Chem., 194 (1952) 183.

Miller, W.T., and H.W. Schoening, Toxicity of selenium fed to swine in the form of sodium selenite, J. Agr. Res., 56 (1938) 831.

Miller, W.T., and K.T. Williams, Minimum lethal dose of selenium as sodium selenite for horses, mules, cattle and swine, J. Agr. Res., 60 (1940) 163.

Mishra, S., and A.K. Srivastava, Hematology as index of sublethal toxicity of zinc in a freshwater teleost, Bull. Environ. Contam. Toxicol., 22 (1979) 695.

Moxon, A.L., and M. Rhian, Selenium poisoning, Physiol. Rev., 23 (1943) 305.

Muth, O.H., J.E. Oldfield, L.F. Remmert, and J.R. Schubert, Effects of selenium and vitamin E on white muscle disease, Science, 128 (1958) 1090.

Natusch, D.F.S., J.R. Wallace, and C.A. Evans Jr., Toxic trace elements: Preferential concentration in respirable particles, Science, 183 (1974) 202.

Neal, K.G., Vitamins and Minerals, Elot Publ. Co., Long Beach, (1977) 1.

Nielsen, M.G., and G.G. Nielsen, Sensitivity of trout to chronic and acute exposure to selenium, Agr. Environ., 4 (1978) 85.

Niimi, A.J., and Q.N. LaHam, Selenium toxicity on the early stages of zebrafish (*Brachydanio rerio*), J. Fish. Res. Bd. Can., 32 (1975) 803.

Pakkala, I.S., W.H. Gutenmann, D.J. Lisk, G.E. Burdick, and E.J. Harris, A survey of the selenium content of fish from 49 New York State waters, Pestic. Monit. J., 6 (1972) 107.

Pascoe, D., and D.L. Mattey, Studies on the toxicity of cadmium to the three-spined stickleback *Gasterosteus aculeatus* L., J. Fish Biol., 11 (1977) 207.

Poston, H.A., G.F. Combs Jr., and L. Leibovitz, Vitamin E and selenium interrelations in the diet of Atlantic salmon (*Salmo salar*): gross histological and biochemical deficiency signs, J. Nutr., 106 (1976) 892.

Rameriz-Mitchell, R., and E.M.B. Sorensen, The three-dimensional morphology of arsenic inclusions: HVEM versus STEM versus CTEM, J. Electron Micros. Techn., 2 (1985) 563.

Rosenfeld, I., and O.A. Beath, Pathology of selenium poisoning, U. Wyom. Agr. Exp. Stat., Bull. 275, (1946) 1.

Rosenfeld, I., and O.A. Beath, The influence of various substances on chronic selenium poisoning, J. Pharm. Exp. Therap., 91 (1947) 218.

Rosenfeld, I., and O.A. Heath, Geobotany, Biochemistry, Toxicity and Nutrition, Geobotany, Biochemistry, Toxicity and Nutrition, Academic Press, New York, (1964) 141.

Sager, D.R., and C.R. Cofield, Differential accumulation of selenium among axial muscle, reproductive and liver tissues of four warmwater fish species, Water Res. Bull., 20 (1984) 359.

Saiki, M.K., Effects of selenium in agricultural drainage on the aquatic environment: U.S. Fish and Wildlife Service Research in 1985, Symposium III, Selenium in Agricultural Drainage, (1986) 15.

Saiki, M.K., and T.P. Lowe, Selenium in aquatic organisms from subsurface agricultural drainage water, San Joaquin Valley, California, Arch. Envir. Contam. Toxicol., 16 (1987a) 657.

Saiki, M.K., Selenium and other trace elements in fish from the San Joaquin Valley and Suisum Bay, 1985, Symposium IV, Selenium and Agricultural Drainage, (1987b) 35.

Sakurayama, H., Studies on the selenium poisoning, Sikoku Igaku Zasshi, 16 (1960) 122.

Sandholm, M., H.E. Oksanen, and L. Pesonen, Uptake of selenium by aquatic organisms, Limnol. Oceanogr., 18 (1973) 496.

Sarathchandra, S.U., and J.H. Watkinson, Oxidation of elemental selenium to selenite by *Bacillus megaterium*, Science, 211 (1981) 600.

Schubert, J.R., O.H. Muth, J.E. Oldfield, and L.F. Remmert, Experimental results with selenium in white muscle disease of lambs and calves, Fed. Proc., 20 (1961) 689.

Scott, M.L., G. Olson, L. Krook, and W.R. Brown, Selenium-responsive myopathies of myocardium and smooth muscle in the young poult, J. Nutr., 91 (1967) 573.

Skidmore, J.F., and P.W.A. Tovell, Toxic effects of zinc sulfate on the gills of rainbow trout, Water Res., 6 (1972) 217.

Smith, M.I., E.F. Stohlman, and R.D. Lillie, The toxicity and pathology of selenium, J. Pharmacol. Exp. Ther., 60 (1937) 449.

Sorensen, E.M.B., Thermal effects on the biological magnification of arsenic in green sunfish, *Lepomis cyanellus*, Ph. D. Dissertation, The University of Texas at Austin, (1974) 1.

Sorensen, E.M.B., Thermal effects on the accumulation of arsenic in green sunfish, *Lepomis cyanellus*, Arch. Environ. Contam. Toxicol., 4 (1976a) 8.

Sorensen, E.M.B., Toxicity and accumulation of arsenic in green sunfish, *Lepomis cyanellus*, exposed to arsenate in water, Bull. Environ. Contam. Toxicol., 15 (1976b) 756.

Sorensen, E.M.B., Ultrastructural changes in the hepatocytes of green sunfish, *Lepomis cyanellus*, exposed to solutions of sodium arsenate, J. Fish Biol., 8 (1976c) 229.

Sorensen, E.M.B., and R.D. Burkett, Respiratory responses to instant thermal stress under oxygen-enriched conditions in centrarchids, Tx. J. Sci., 30 (1978) 337.

Sorensen, E.M.B., R.E. Henry, and R. Ramirez-Mitchell, Time-dependent localization of arsenic in subcellular liver fractions, J. Environ. Pathol. Toxicol., 2 (1979a) 1161.

Sorensen, E.M.B., R.E. Henry, and R. Ramirez-Mitchell, Arsenic accumulation, tissue distribution and cytotoxicity in teleosts following indirect aqueous exposures, Bull. Environ. Contam. Toxicol., 21 (1979b) 162.

Sorensen, E.M.B., R. Ramirez-Mitchell, C.W. Harlan, and J.S. Bell, Cytological changes in fish liver following chronic, environmental arsenic exposure, Bull. Environ. Contam. Toxicol., 25 (1980) 93.

Sorensen, E.M.B., and N.K.R. Smith, Hemosiderin granules: cytotoxic response to arsenic exposure in channel catfish, Bull. Environ. Contam. Toxicol., 27 (1981a) 645.

Sorensen, E.M.B., L.L. Wenz, B.C. Windsor, and R. Ramirez-Mitchell, Stereological analysis of arsenic-induced cytological structures, J. Tenn. Acad. Sci., 56 (1981b) 131.

Sorensen, E.M.B., T.L. Bauer, J.S. Bell, and C.W. Harlan, Selenium accumulation and cytotoxicity in teleosts following chronic, environmental exposure, Bull. Environ. Contam. Toxicol., 29 (1982a) 688.

Sorensen, E.M.B., C.W. Harlan, and J.S. Bell, Renal changes in selenium-exposed fish, Amer. J. Forensic Med. Pathol., 3 (1982b) 123.

Sorensen, E.M.B., N.K.R. Smith, and R. Ramirez-Mitchell, Electron probe X-ray microanalysis of arsenic inclusions in fish, Arch. Environ. Contam. Toxicol., 11 (1982c) 469.

Sorensen, E.M.B., C.W. Harlan, J.S. Bell, T.L. Bauer, and A.H. Pradzynski, Hepatocyte changes following selenium accumulation in a freshwater teleost, Amer. J. Forensic Med. Pathol., 4 (1983a) 25.

Sorensen, E.M.B., J.S. Bell, and C.W. Harlan, Histopathological changes in selenium-exposed fish, Amer. J. Forensic Med. Pathol., 4 (1983b) 111.

Sorensen, E.M.B., and T.L. Bauer, Hematological dyscrasia in teleosts chronically exposed to selenium-laden effluent, Arch. Environ. Contam. Toxicol., 12 (1983c) 135.

Sorensen, E.M.B., P.M. Cumbie, T.L. Bauer, H.S. Bell, and C.W. Harlan, Histopathological, hematological, condition-factor, and organ-weight changes associated with selenium accumulation in fish from Belews Lake, North Carolina, Arch. Environ. Contam. Toxicol., 13 (1984a) 153.

Sorensen, E.M.B., and T.L. Bauer, Planimetric analysis of redear sunfish (*Lepomis microlophus*) hepatopancreas following selenium exposure, Environ. Toxicol. Chem., 3 (1984b) 159.

Sorensen, E.M.B., and T.L. Bauer, A correlation between selenium accumulation in sunfish and changes in condition factor and organ weight, Environ. Pollut., 34 (1984c) 357.

Sorensen, E.M.B., R. Ramirez-Mitchell, A. Pradzynski, and T.L. Bauer, Stereological analyses of hepatocyte changes parallel arsenic accumulation in the livers of green sunfish, J. Environ. Path. Toxicol. Oncol., 6 (1985) 195.

Sorensen, E.M.B., The effects of selenium on freshwater fish, in Hodgson, E. (Eds.), Reviews in Environmental Toxicology, Elsevier Press, New York, New York, (1986) 1.

Sorensen, E.M.B., and P. Thomas, Selenium accumulation, reproductive status, and histopathological changes in environmentally exposed redear sunfish, Arch. Toxicol., 61 (1988a) 324.

Sorensen, E.M.B., T.L. Bauer, and M.G. Krause, Multielemental analysis of the hepatopancreas of selenium-exposed sunfish, Toxicologist, 8 (1988b) 108.

Spangler, W.J., J.L. Spigarelli, J.M. Rose, and H.M. Miller, Methylmercury: Bacterial degradation in lake sediments, Science, 180 (1973) 192.

Srivastava, A.K., and S. Mishra, Blood dyscrasia in a teleost fish, *Colisa fasciatus*, associated with cadmium poisoning, J. Comp. Pathol., 89 (1979) 609.

Stadtman, T.C., Selenium-dependent enzymes, Ann. Rev. Biochem., 49 (1980) 93.

The National Research Council, Selenium, National Academy of Sciences, Washington, (1976) 1.

Vokal-Borek, H., Selenium, University of Stockholm, Institute of Theoretical Physics, USIP Report 79-16, (1979) 1.

Watenpaugh, D.E., and T.L. Beitinger, Se exposure and temperature tolerance of fathead minnows, *Pimephales promelas*, J. Therm. Biol., 10 (1985a) 83.

Watenpaugh, D.E., and T.L. Beitinger, Absence of selenium avoidance by fathead minnows (Pimephales promelas), Water Res., 19 (1985b) 923.

Weiss, H.V., M. Koide, and E.D. Goldberg, Selenium and sulfur in a Greenland ice sheet: Relation to fossil fuel combustion, Science, 172 (1971) 261.

Wilber, C.G., Toxicology of selenium: A review, Clin. Toxicol., 17 (1980) 171.

CHAPTER III
ARSENIC

Introduction

Clinical studies indicate that arsenic is a carcinogen in humans; however, currently no established mammalian model simulates As-induced carcinogenicity in humans. In fact, intraspecies comparisons show remarkable differences in the processing of As by humans and rodents. For example, rats excrete less As than humans; however, the reverse is true of rabbits (Ducoff, *et al.*, 1948). Moreover, rat erythrocytes retain more As than do human erythrocytes. Also, no common laboratory species consistently produces tumors following As exposure; whereas, humans develop a variety of tumors following exposure to this metalloid (National Research Council, 1977).

In the green sunfish (*Lepomis cyanellus*), however, As induces what can be considered early neoplastic alterations in the liver (Sorensen, 1976a; 1976b; 1976c; Sorensen, *et al.*, 1982a; 1982b; 1979a; 1979b; 1980; 1981b; 1985a; 1985b; Mitchell and Sorensen, 1987). Bizarre morphological alterations are produced under both laboratory and environmental conditions. In 1986, the Museum of Natural History of the Smithsonian Institution incorporated published research on As-induced changes in this species (Sorensen, *et al.*, 1985a; 1985b) in a permanent collection entitled Tumors of Lower Vertebrates.

Although As-induced hepatocyte injury in green sunfish will be discussed in greater detail in the histopathology section, a brief description at this time is appropriate. Changes include fatty infiltration, central or focal necrosis, cirrhosis, cytoplasmic vacuolation, necrosis, and nonspecific autolytic changes (Sorensen, *et al.*, 1980; 1981b). Hemosiderin granules, necrotic bodies, and fibrous bodies also appear (Sorensen, *et al.*, 1980; Sorensen and Smith, 1981a; Sorensen, *et al.*, 1981b; 1985a). However, the most striking As-induced cellular alterations are nuclear and cytoplasmic As-inclusions (Sorensen, 1976a; 1979a; 1980; 1981b), which probably represent initial sites of biochemical lesions ultimately leading to carcinogenicity.

Compared with other elements evaluated in this volume, only a few published reports document As-induced changes in fish. However, both quantitative and qualitative studies indicate that As is extremely poisonous to fish and that it is accumulated to elevated levels in fish. Hepatic As levels of 67 ppm (wet weight) are found in laboratory fish exposed 4 d to 60 ppm arsenate (Sorensen, *et al.*, 1979b) and 64 ppm As (dry weight) is measured in feral fish (Sorensen, *et al.*, 1985a). Enterohepatic redistribution might contribute to the elevation of hepatic levels of As (Sorensen, 1976a; 1976b) and cause both toxicity and the appearance of As-inclusions.

Environmental Exposures

Arsenicals reach aquatic ecosystems by a variety of sources including manufacturing companies, mineral or strip mines, smelting operations, and electric generating stations (power plants) (Cherry and Guthrie, 1977; Cumbie, 1978; Ryther, *et al.*, 1979; Cumbie, 1980; Sorensen, *et al.*, 1980; 1981b; 1984a; 1985b). One major agricultural source of As is the manufacture and use of arsenical defoliants and pesticides (Sandhu, 1977; Sorensen, *et al.*, 1980; 1981b; Shukla and Pandey, 1984a; Sorensen, *et al.*, 1985a). Arsenicals also have been used to kill aquatic plants to reduce the difficulty encountered during hook-and-line fishing of areas overgrown with aquatic vegetation. The metalloid can persist for many years following cessation of application in these areas (Surber, 1943; Ullmann, *et al.*, 1961; Crosby and Tucker, 1966; Foley, *et al.*, 1978).

Increased electricity production by coal-fueled, electric power generating stations (power plants) has increased production of As-contaminated particulate in the United States. Approximately 50 million tons of particulate is trapped by electrostatic precipitators annually, based on one estimate (Anonymous, 1973). Of various particulate wastes (e.g. fly ash, scrubber sludge, or bottom ash), that of greatest concern is fly ash which is collected by conventional control equipment (e.g. electrostatic precipitators).

Coal used by power plants is contaminated with a number of highly concentrated metals and metalloids. During high temperature combustion, particulate waste from power plants concentrate these elements on fly ash which is released from the stack (Natusch and Wallace, 1974). That particulate eventually becomes associated with surface waters, which solubilize the elements. In this way, As contaminates surface waters, as does Ag, Hg, Cu, Zn, Se, and other elements (Sorensen, *et al.*, 1982a; 1982b; 1983a; 1983b; Sorensen and Bauer, 1983c; Sorensen, *et al.*, 1984a; Sorensen and Bauer, 1984b; 1984c; Sorensen, 1986). For example, aqueous and sedimentary levels of As in Belews Lake reservoir in North Carolina are 2–7 ppb and 14–33 ppm, respectively (Cumbie, 1978; 1980; Sorensen, *et al.*, 1984a). Elemental contamination generally increases with the duration of power plant operation.

The manufacture of cotton defoliants has been problematic in the southern United States since arsenicals reach aquatic habitats from the improper handling of wastes from manufacture and from dispersal of defoliants (Sorensen, *et al.*, 1980; 1981b; 1985b). Normally, arsenic acid (H_3AsO_4) is applied to cotton plants to desiccate leaves prior to combining (i.e. the mechanical collection of cotton fibers). Once cotton leaves wilt and fall off plants, cotton harvesters (i.e. combines) collect cotton fibers with less leaf-associated debris. Arsenic contaminated defoliant wastes are released from unreinforced waste treatment ponds, such as those at Finfeather Lake in central Texas. Here, As wastes consisted of a mixture of trivalent and pentavalent arsenicals (e.g. calcium arsenate, arsenic acid, and arsenic trioxide). Under these conditions, biomethylation undoubtedly forms organic arsenicals, as well (McBride and Wolfe, 1971; Wood, 1974). Studies at Finfeather Lake report As levels of 4700, 240 and >0.5 ppm (dry weight) in sediments, water, and fish tissues, respectively (Sorensen, *et al.*, 1980). At these levels, maximum permissible levels of As in drinking water (0.05 ppm), the U.S. Public Health Service recommended drinking water standard (0.01 ppm), and the maximum permissible levels of As in food (i.e. 0.5 ppm, dry weight) are exceeded. Sustained release of As killed most life forms in

Finfeather Lake, and litigation settlements have forced drainage of the lake and burial of As-laced sediments in clay-lined pits near the original site of Finfeather Lake.

Improper applications of defoliants, pesticides, and fertilizers also can result in detrimental effects upon teleosts. Sandhu (1977) reports an extensive fish kill in a reservoir following aerial spraying of an adjoining cotton field with As defoliants. In this report, skeletal muscle from catfish (*Ictalurus punctatus*) contains 5 and 12 ppm after 5 h- and 7 wk-exposures, respectively. Disposal of effluents from a fertilizer plant in India is reported to result in increases in the concentration of As in Lake Chilwa to levels as high as 2 ppm As (Shukla and Pandey, 1984a).

Arsenicals were used commonly in the 1940's to control aquatic vegetation which impaired hook-and-line fishing (Surber, 1943). Repeated herbicidal applications could elevate aqueous levels to 10 ppm As (Crosby and Tucker, 1966). Aquatic plants such as water lilies, bulrushes, and rooted aquatic plants die, along with numerous game fish (e.g. smallmouth bass). Three years following annual use of arsenite at one location, calico bass accumulate 0.1 to 1.0 ppm As (wet weight) in skeletal muscle and viscera (Ullmann, *et al.*, 1961). And other studies clearly show that As can persist in lake water for 13 yr or more following termination of herbicide application to pose a hazard to aquatic life (Foley, *et al.*, 1978).

From the above studies, the ubiquity of As is understandable. The remainder of this chapter focuses on As toxicity in fish, tissue and subcellular localization patterns, histopathological changes, and other biological effects. Because of the similarity of results of As-induced effects on teleosts with those of other vertebrates, some comparisons using species other than fish will be made. Such comparisons are important in evaluating the adaptive significance of the cellular processing of As.

Toxicity

Arsenicals are toxic to fresh-water teleosts (Table 1). For example, 50% of bluegill sunfish, trout, catfish, or goldfish succumb after 24–96 h exposures to 15–60 ppm arsenite (trivalent) (Clemens and Sneed, 1959; Cope, 1966; Crosby and Tucker, 1966; Gilderhus, 1966). However, two- to ten-fold higher levels of the pentavalent form (e.g. arsenate) are required to produce

Table 1. Arsenic toxicity in freshwater fish. (Modified from Sorensen, 1987.)

Chemical Form	Species	LC50 Time	Concentration (ppm)	Temp (°C)	Reference
Sodium Arsenite ($NaAsO_3$)	Bluegill *Lepomis microchirus*	96 h	35.0	12°C	Gilderhus, 1966
		48 h	44.0	24°C	Crosby & Tucker, 1966
		48 h	44.0	24°C	Cope, 1966
	Rainbow trout *Salmo gairdneri*	96 h	25.6	12°C	Gilderhus, 1966
		48 h	36.5	13°C	Cope, 1966
		48 h	60.0	13°C	Crosby & Tucker, 1966
	Channel catfish *Ictalurus punctatus*	72 h	15.0	25°C	Clemens & Sneed, 1959
		24 h	47.9	25°C	Clemens & Sneed, 1959
	Goldfish *Carassius auratus*	96 h	34.0	12°C	Gilderhus, 1966
Sodium Arsenate ($NaAsO_2$)	Green sunfish *Lepomis cyanellus*	48 h	150.0	20°C	Sorensen, 1976

the same effect (Sorensen, 1974; 1976a; Klaassen, 1985). Therefore, although results from various species are difficult to compare, pentavalent inorganic arsenicals are considered much less toxic to fish than are trivalent inorganic arsenicals. Moreover, as expected, inverse correlations are obtained for the magnitude of the LC50 values and the designated time of collection of LC50 data (Table 1).

Toxicity of arsenicals is a function of numerous factors, including the size of fish, chemical form, and concentration of the metalloid. The lethal times for 50% mortality (LT50) for small, intermediate, and large fish exposed to 100 ppm are 39, 55, and 73 h, respectively (Sorensen, 1976b). Therefore, smaller sunfish (0.2–0.7 g) succumb more quickly to 100 ppm arsenate than do intermediate (0.8–1.3 g) or larger (1.4–6.4 g) fish. However, size differences do not affect the time of death following exposures to 500 or 1000 ppm arsenate. Although the exacerbated effect of As on smaller fish is undoubtedly the result of increased metabolic rate of the smaller fish, more research remains to be conducted in this area.

Toxicity of arsenicals also is a function of the chemical form and concentration of As. Rainbow trout (*Salmo gairdneri*) fry are given As in the diet as in the inorganic and organic forms, as well as in the trivalent and pentavalent forms (Cockell and Hilton, 1988). Arsenic trioxide (AT) at 200–1600 ppm As, disodium arsenate (DSA) at 120–960 ppm, dimethylarsinic acid (DMA) at 200–1600 ppm As, or arsanilic acid (AA, at 200–1600 ppm) are administered in the diet (15°C and pH 7.5–7.6). Organic forms are least toxic, with less than 4% mortalities over an 8-wk exposure to a maximum of 30 ppm/d in the diet. No-effect levels for DMA and AA are set at about 1497 ppm; whereas, no-effect levels for DSA and AT are almost equivalent at about 1–200 ppm. Based on 10–13% mortality and and negative growth from fry in the higher exposures, Cockell and Hilton (1988) consider the pentavalent DMA more toxic than the trivalent AT. Differences in accumulation of the trivalent and pentavalent arsenicals probably control variation in toxicity. Of course, many other factors also affect As-induced mortality in fish—including temperature, water hardness, pH, exposure time, exposure level and other factors (Sorensen, 1974; 1976a).

Cause of teleost death in acute exposures has not been elucidated; however, several possibilities exist. One postulate is that mucus builds up on the gill surface and causes suffocation due to inadequate oxygen uptake at the secondary lamellae during countercurrent exchange (Sorensen, 1974; 1976a). Diagnosed as "coagulation film anoxia" (Carpenter, 1927; 1930), the condition is characterized by irregular mandibular and opercula movements and a thickened mucous film over the gills. The 250% reduction in CO_2 evolution and the 20% increase in opercular rate (Carpenter, 1927) provide additional proof that suffocation causes death of fish in acute As poisoning. In chronic exposures, the mucous film can reduce aqueous levels of ions below detection limits, and fish are capable of complete recovery after shedding the mucous-metalloid complex. Apparently, the irritant-like effect of As accelerates mucus production in sunfish exposed in the laboratory. A thick mucous coat is secreted over the entire body surface, but especially over the gill surface (Sorensen, 1974; 1976a).

Accumulation and Retention

Pentavalent and trivalent arsenicals each react with biological ligands (Klaassen, 1985) in different ways. Trivalent forms react with sulfhydryl protein groups, resulting in enzyme inactivation, structural damage, and a number of functional alterations. Pentavalent arsenicals, however, do not appear to react with sulfhydryl groups. Arsenates can competitively inhibit phosphate insertion into nucleotide chains of the DNA of cultured human lymphocytes, causing false formation of DNA because of the instability of the arsenate esters (Petres and Berger, 1972). Dark repair mechanisms also are inhibited, leading to the persistence of these errors in the DNA molecule. Binding differences for the trivalent and pentavalent forms (Petres and Berger, 1972; Klaassen, 1985) probably lead to differences in accumulation (Cockell and Hilton, 1988).

Trivalent and pentavalent arsenicals do accumulate differently (Cockell and Hilton, 1988). Rainbow trout (*Salmo gairdneri*) fry are exposed to dietary As as arsenic trioxide (AT) at 200–1600 ppm As or disodium arsenate (DSA) at 120–960 ppm As. After 8 wk, the whole body levels following administration of the trivalent form are about ~21 ppm, compared with ~14 ppm following administration of the pentavalent form (Fig. 1). Also, rainbow trout given low dietary levels (≤5 ppm/d) of either form, accumulate the pentavalent arsenical to a higher level than the trivalent form. The reverse is true for fry in the 5–10 ppm As/d dietary range. Trivalent As is accumulated to ~33% higher levels than the pentavalent form in this range.

In contrast, the organic arsenicals (arsanilic acid and dimethylarsinic acid) are accumulated to lower levels (7 and 12 ppm As) despite the administration of up to 30 ppm As/d in the diet. The results of this study suggest that toxicity is a function of As accumulation (Cockell and Hilton, 1988) and that uptake of inorganic forms is dictated by both valence state and concentration.

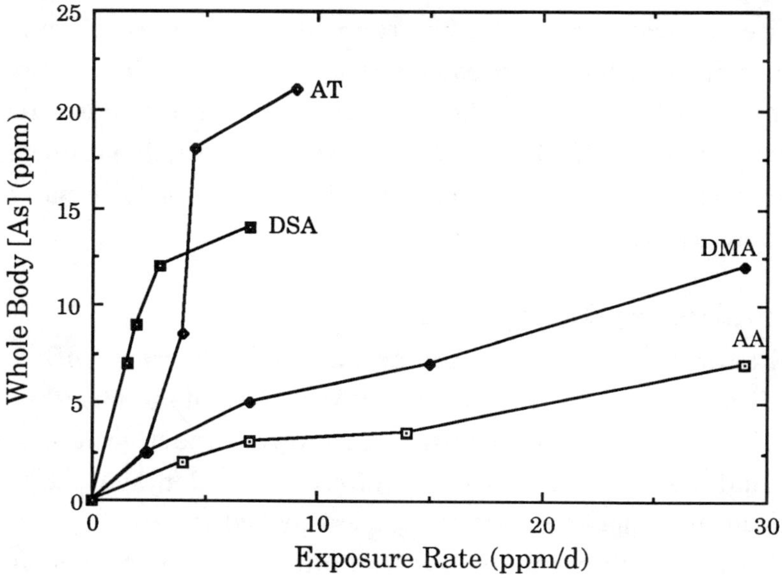

Figure 1. Levels of As in the whole body of juvenile rainbow trout (*Salmo gairdneri*) fed from the fry stage, four times daily to satiety. Maximum accumulation (~21 ppm) is achieved when fish are fed arsenic trioxide (AT, a trivalent inorganic arsenical). Fish fed disodium sodium arsenate (DSA, a pentavalent inorganic arsenical), dimethylarsinic acid (DMA, an organic arsenical), and arsanilic acid (AA, an organic arsenical used as a growth promoter in swine and poultry diets) accumulate less arsenic. (An original figure derived from Cockell and Hilton (1988).)

Whole body uptake of As by fish is a function of exposure concentration and exposure time. At aqueous exposure levels of 100, 500, and 1000 ppm, mean As uptake in the whole body of fish is 33.4, 541.2, and 581.6 ppm, respectively (Sorensen, 1974; 1976a). Concentration factors (CF) are 0.33, 1.08, and 0.58 for fish exposed to 100, 500, and 1000 ppm exposures to arsenate. As is the case with other elements, CF values are inversely proportional to aqueous As levels. However, As accumulation is not correlated with fish total length, dry weight, wet weight, or condition factors. Results of As studies (Sorensen, 1974; 1976a) are similar to those of Cd

(Lovett, *et al.*, 1972), Zn (Mount, 1967), and Cs (Kevern, 1966) uptake: generally elemental uptake and fish size are not correlated.

Concentration factors for As in various fish tissues can be calculated from a number of studies (Sorensen, 1976a; 1976c; Sorensen, *et al.*, 1979a). In the gastrointestinal tract, CF maximizes at 4.2 at 60 ppm arsenate exposures for 2 wk at 20–30°C (Sorensen, 1976a). Corresponding values for skeletal muscle and liver are 1.3 and 24.3, respectively. After a 6-d exposure to 60 ppm, CF for the bile (and gall bladder), liver, spleen, and kidney are 2.6, 0.8, 0.3, and 0.2, respectively (Sorensen, *et al.*, 1979a). Evaluation of hepatic CF for As in fish is warranted. In fact the liver is considered the major target organ in cases of As poisoning because of the detoxification role of that organ (Kodama, *et al.*, 1976; Klaassen, 1985). Moreover, when CF values are 6–28 following 1-wk exposures to 60 ppm aqueous arsenate or 2-wk exposures to 30 ppm aqueous arsenate, the As-inclusion appears in nuclei and cytoplasm of fish (Sorensen, 1976c).

The liver actively excretes As as a component of bile (a fat emulsifier) (Klaassen, 1974), and biliary As increases from 35 to 78 to 159 ppm during 2-, 4-, and 6-d exposures of sunfish to 60 ppm (Sorensen, *et al.*, 1979a). Administration of ^{74}As to brown trout (*Salmo trutta*) shows biliary excretion of both organic and inorganic arsenicals during a 24-h period (Penrose, 1975). Organic, but not inorganic, As appears to be excreted against a concentration gradient. Following a single intravenous exposure, mammals excrete As into the bile by active transport mechanisms (Klaassen, 1974). However, enterohepatic redistribution of As minimizes the overall benefit of the biliary excretion of As in mammals. This factor, no doubt, also results in elevation of As to 67 ppm (wet weight) in fish exposed 4 d to 60 ppm As in the water (Sorensen, *et al.*, 1979b).

In fish, arsenicals can induce Fe storage diseases, which are characterized by lysosomal accumulation of soluble and insoluble deposits of Fe (Bradford, *et al.*, 1969; Crichton, 1971; Trump, *et al.*, 1973a; Sorensen and Smith, 1981a). In catfish, exposed for 6 mo to 15 ppm As, ferritin (a non-heme, Fe-apoferritin complex and chief storage form of Fe in the body) and insoluble hemosiderin granules accumulate in the cytoplasm (Sorensen and Smith, 1981a). Both Fe and As have been positively identified in hemosiderin granules. Long-term storage of large quantities of As as an insoluble deposit would reduce cell injury associated with the accumulation of normally fatal levels of the ionized form of the metalloid.

Arsenic accumulation is considered a function of the detoxification and clearance roles of individual organs (Sorensen, *et al.*, 1979a). The hepatopancreas (i.e. liver and associated exocrine pancreas), spleen, and mesonephros (i.e. kidney) accumulate increasing concentrations of As with time following 6-d exposures to 60 ppm arsenate. Hepatic, renal, and splenic levels double in 6 d. Ratios of hepatic to splenic As vary from 1.8 to 2.5. In North Atlantic finfish, the ratio of As in the liver to that in the spleen ranges from 0.6 to 6.1 (Windom, *et al.*, 1973). Similar results exist for the liver and spleen, but not kidney, of rats administered 1 mg/kg intravenous doses of As (Klaassen, 1974). In this study, the ratio of As in the rat liver to that in the spleen is 0.3 to 0.8 for 7 d and 2 h, respectively, post-injection.

Many factors are responsible for the accumulation of As in the spleen (Sorensen, *et al.*, 1979a). Arsenic-induced erythrocyte lysis decreases the life span of circulating erythrocytes (Klaassen, 1985). Filtration of blood through the spleen, as well as the blood storage function of the spleen, undoubtedly play important roles in the deposition of As in that organ. Trapping of erythrocytes also may increase As accumulation in the spleen. Levels of As in the spleen range from 10–20 ppm (wet weight) following 2- to 6-d exposures (Sorensen, *et al.*, 1979a).

Accumulation of renal As exhibits dramatic functional and morphological changes—including decreased state three respiration, decreased respiratory ratios, swollen mitochondria, and increased numbers of lysosome-like organelles (Brown, *et al.*, 1976). Elevation of As in the proximal tubular cells of the kidney is associated with glomerular dilation, swollen mitochondria, cast production, tubular necrosis, and increased numbers of lysosome-like bodies, as well as proteinuria, mild albuminaturia, and hematuria (Brown, *et al.*, 1976; Klaassen, 1985).

School and gummy sharks from Australian waters concentrate As to 5–30 ppm (wet weight) in skeletal muscle (Glover, 1979). Dover sole concentrate As to 2–4 ppm (wet weight) (Fowler, *et al.*, 1975). Skeletal muscle from ten species of marine fish from four locations on the western coast of Canada accumulate up to 16 ppm (wet weight) (Kennedy, 1976). Levels of As in skeletal muscle are low in some areas of moderate contamination from industrialization and sewage release, ranging from ~0.1

to 0.3 ppm (dry weight) (Ramelow, *et al.*, 1989). Whole body levels of As in fish from across the United States in 1978–1981, range from 0.14 to 0.16 ppm (wet weight) (Lowe, *et al.*, 1985). From these and other studies (Sorensen, 1976a), skeletal muscle of fish may not be a major repository of As—opposite the case for Hg (see Chapter VIII).

Gill and gonadal tissues of *Lepomis cyanellus* do not accumulate As to significantly elevated levels during a 6-d exposure (Sorensen, *et al.*, 1979a) or during a field exposure (Windom, *et al.*, 1973). But despite reduced accumulation of As in the testes and ovaries, these organs of *Colisa fasciatus* (a tropical fresh-water perch species) are damaged histopathologically following 15–30 d exposures to 14 ppm (Shukla and Pandey, 1984a; 1984b). Arsenic clearance from gonadal tissues is rapid. In mammals, a ten-fold decrease occurs in testicular As levels following a single intravenous dose of 1 mg As/kg during the 7-d post-administration period (Klaassen, 1974).

Association of fish with sediments can alter As accumulation. For example, pelagic fish from open waters accumulate less As than bottom feeders (LeBlanc and Jackson, 1973; Kennedy, 1976). Pelagic fish typically consume fish which inhabit ecosystems distant from the often-contaminated sediments; whereas, bottom feeders generally consume sediments along with the invertebrates, debris, and flora located in sediments. Published studies show that pelagic fish can accumulate 0.3–3.0 ppm As, but that bottom-feeders from the same general site uptake higher levels of 1.4–55.0 ppm (LeBlanc and Jackson, 1973; Kennedy, 1976).

Lowe *et al.* (1985) find that fish in the United States generally have whole body As levels of 0.14–0.16 ppm (wet weight). Arsenic levels in whole fish in Lake Texoma, on the Texas-Oklahoma border, show concentration of As to levels of 0.1–34.0 ppm (wet weight) (Hunter, *et al.*, 1981). In their study, Hunter *et al.* (1981) find that planktivorous shad (*Dorosoma cepedianum*) accumulate the highest average level (8.1 ppm As), compared with the omnivorous drum (*Aplodinotus grunniens*) and catfish (*Ictalurus punctatus*) which accumulate 2.4 ppm As. The levels of As in whole fish ranges from 3–43 ppb for three species of fish from Lake Michigan, Lake Erie, and Lake Superior (Lucus, *et al.*, 1970). Studies of As accumulated by bloaters (*Coregonus* species) of the Great Lakes show that whole fish accumulate 0.26–0.84 ppb As (wet weight) (Passino and Kramer, 1980).

When compared with food web organisms, fish generally concentrate strikingly low concentrations of As in the skeletal muscles. For example, Ramelow *et al.* (1989) find that sessile organisms (e.g. oysters) tend to accumulate up to 4.5 ppm As; while, the muscle of fish concentrates 0.2 ppm As. Unfortunately, exact water quality data for Calcasieu River/Lake in Louisiana are not published (Fig. 2). In comparison of the levels of As (Fig. 2a) with those of Hg (Fig. 2b) in food webs, Hg is accumulated to 0.2 ppm (dry weight) in fish muscle (Ramelow, *et al.*, 1989). If levels of Hg in fish muscle approximate those in whole body (see Chapter VIII) and if the dry weight to wet weight conversions are made, the level of Hg is about half the national average reported by Lowe *et al.* (1985) for Hg in whole fish. Moreover, undoubtedly Hg levels in muscle are elevated as high as that of oysters in the Ramelow *et al.* (1989) study because muscle acts as a sink for Hg deposition in fish (see Chapter VIII). Moreover, the level of Cu is 2.0 to 2.6 times higher than the national average, suggesting that Cu is plentiful element in the area of Louisiana evaluated by Ramelow *et al.* (1989). Synergistic, antagonistic, and additive effects can be expected under conditions in which levels of several elements are elevated (see Chapter IX).

Intracellular compartmentalization studies show that nuclei are sites of major As deposits in both mammal and fish hepatocytes (Klaassen, 1974; Sorensen, *et al.*, 1979b). For example, following a 2- to 6-d exposure of fish to 60 ppm, As translocates from a cytoplasmic to a predominantly nuclear location. Two days after the initial aqueous exposure, the soluble cytoplasmic fraction of the liver contains the highest level of As (46%); whereas, the nuclear fraction contains the lowest percentage (12%). After 4- and 6-d exposures, levels of cytoplasmic As are reduced to 28 and 17%, respectively. In comparison, after 4 and 6 d, the nuclear fraction increases to 42 and 58%. Therefore, As translocates from a cytoplasmic to a nuclear location in parenchymal hepatocytes of sunfish. Obviously, formation of an insoluble, inert deposit could reduce cellular injury as hepatic storage of As exceeds 67 ppm (wet weight) in laboratory fish exposed 4 d to 60 ppm As (Sorensen, *et al.*, 1979b) and 64 ppm As (dry weight) in feral fish (Sorensen, *et al.*, 1985b). Such hepatic levels of As are considerably higher than those of many other studies.

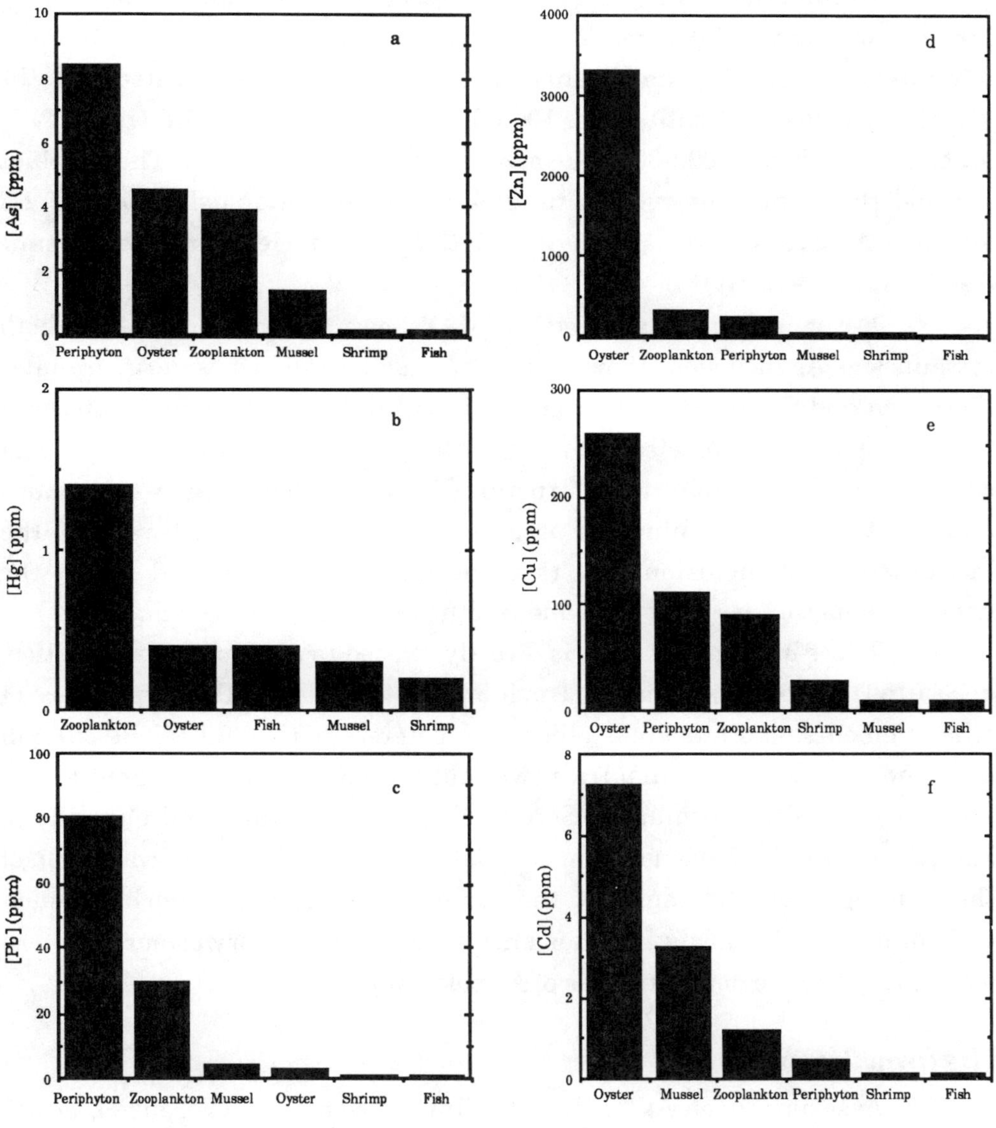

Figure 2. Comparison of the levels of As in skeletal muscle of fish with those of five other elements. The levels of As, Pb, Hg, Zn, Cu, and Cd are presented on a dry weight basis for food web organisms and for the skeletal muscle of fish from the Calcasieu River/Lake in Louisiana. In each case, the order of occurrence of each group depends on the magnitude of elemental enrichment. Sessile organisms (e.g. oysters) tend to accumulate the highest levels of many elements; while, fish muscle accumulates the lowest levels for five of the six elements measured. (An original figure derived from numerical data presented by Ramelow *et al.* (1989).)

In his studies on the rat, Klaassen (1974) administers a 1 mg/kg intravenous dose and determines the subcellular As distribution at 6- or 60-min intervals following the injection. About 50% of As is associated with the soluble cytoplasmic fraction (i.e. 100,000 g supernatant), while only 10% is associated with the 100,000 g nuclear fraction (i.e. the pellet). These values parallel the values observed in the livers of fresh-water fish following 2-d aqueous As exposures (i.e. 12% and 46% in the nuclear and cytoplasmic fractions, respectively).

X-ray microprobe analysis (XRMA) shows the presence of As in both intranuclear As-inclusions (Sorensen, *et al.*, 1982c) and hemosiderin granules (Sorensen and Smith, 1981a). Simultaneous XRMA and imaging of either of the two structures, precisely positions the primary electron beam during collection of X-rays from these structures. This method produces dot maps and point scans confirming the presence of As and S in a 1:6 ratio in the intranuclear As-inclusion and the presence of both Fe and As in the hemosiderin granules (Sorensen and Smith, 1981a; Sorensen, *et al.*, 1982c).

Therefore, uptake of As is directly related to exposure concentration, exposure time, fish size, species, trophic level, feeding habits, the presence of other elements, and numerous other parameters. Biological half-life of As in liver and the gastrointestinal tract (without contents) is about 7 d, but longer than 21 d for skeletal muscle (Sorensen, 1976a). Uptake and elimination kinetics apparently have not been elucidated, nor has a detailed evaluation of the ionoregulatory changes at the lamellar epithelium. Perhaps such evaluations would clarify whether anoxia or ionoregulatory/osmoregulatory disturbances cause death in cases of As poisoning in fish.

Histopathology

Association of As with subcellular organelles (Sorensen, *et al.*, 1979a) could cause considerable morphological disruption, with loss of hepatocyte function. Excess hepatic levels in feral fish probably cause submassive necrosis, focal necrosis, generalized capillary dilation, cirrhosis, marked edema, vacuolation of the cytoplasm of parenchymal hepatocytes, and nonspecific autolytic changes including the appearance and increase in the numbers of necrotic and fibrous bodies (Von Glahn, *et al.*, 1938; Finner and Calvery, 1939; Gilderhus, 1966; Lander, *et al.*, 1975; Sorensen, *et al.*, 1980; 1981b), as well as As-inclusions and hemosiderin granules in laboratory fish

(Sorensen, 1974; 1976c; Sorensen, *et al.*, 1979a; Sorensen and Smith, 1981a; Sorensen, *et al.*, 1982c). Such cellular alterations suggest considerable functional impairment in hepatocytes of As-exposed fish.

In other studies, bile duct proliferation and bile plugs form in some vertebrate livers (Von Glahn, *et al.*, 1938), but not in others (Finner and Calvery, 1939; Byron, *et al.*, 1967). Psoriasis patients treated for 15 to 17 yr with Fowler's solution (a potassium arsenite solution) develop hemangioendotheliomas (i.e. hemangiosarcoma) (Regelson, *et al.*, 1968; Lander, *et al.*, 1975). Two-year exposures of dogs, rabbits, rats, and ferrets, however, do not result in this endothelial cell hypertrophy (Von Glahn, *et al.*, 1938; Byron, *et al.*, 1967). Cirrhosis and fibrosis are observed by some (Franklin, *et al.*, 1950; Buchanan, 1962; Regelson, *et al.*, 1968), but not all (Zachariae, *et al.*, 1974; Huet, *et al.*, 1975; Lander, *et al.*, 1975), investigators. And exposure of humans to the arsenical gas arsine (i.e. AsH_3) results in the appearance of Councilman or acidophilic bodies, which are structures resulting from unicellular degeneration (Neuwirtova, *et al.*, 1961).

Arsenic-Inclusions: As mentioned, the As-inclusion is an unusual morphological alteration of As-exposed fish; progressive increases are observed in the numbers and size of As-inclusions within hepatocytes from fish exposed to higher aqueous concentrations for longer times (Sorensen, 1976c; Sorensen, *et al.*, 1982c; Mitchell and Sorensen, 1987). Intranuclear As-inclusions are observed in about half of hepatocytes examined following 2- to 6-d exposures of fish (Sorensen, *et al.*, 1979a). Using conventional transmission electron microscopy (CTEM) and high voltage electron microscopy (HVEM), the intranuclear As-inclusion is found to have numerous irregular (roughly spherical) granular subunits which lack membranes (Sorensen, *et al.*, 1982c). The As-inclusion is comprised of amorphous puff-like subunits which appear to be linked on a string which twists and coils throughout the volume of the nucleus (Sorensen, *et al.*, 1982c; Mitchell and Sorensen, 1987). The absence of birefringence under polarized light (Sorensen, *et al.*, 1980) indicates that the As-inclusion is not crystalline. Stereographic comparisons of HVEM with CTEM and STEM images corroborate these results (Mitchell and Sorensen, 1987) and indicate that the structure is not associated with the nucleolus or cellular membranes. Arsenic to sulfur ratios are 1:6 in the As-inclusion, based on XRMA data (Sorensen, *et al.*, 1982c). Localization of both As and S within the As-inclusion could mean that As is bound as a metal-protein complex in the nucleus, as is the case of Pb (Goyer, *et al.*, 1970; Moore, *et al.*, 1973; Moore and Goyer, 1974). However, it is not known whether an association between As and protein exists or (if the association exists) whether the protein is histone or nonhistone in nature. One study indicates that the As-inclusion is composed of carbohydrate moieties (Sorensen, 1987).

Intranuclear, electron dense inclusions are observed following the exposure of vertebrates to elements other than As. Morphologically, these intranuclear inclusions are distinct from the As-inclusion. The Pb-inclusion, for example, forms a stacked arrangement of spiked crystals with a spiked periphery and an extremely electron dense core (Goyer, *et al.*, 1970). Another intranuclear inclusion is the dendritic Se-inclusion, which is composed of an intensely-stained, electron dense series of spokes radiating outward toward the nuclear membrane (Groth, *et al.*, 1976; Sorensen, *et al.*, 1983b). And Cd exposure of isolated hepatocytes results in formation of perichromatin granules in the nucleus (Puvion and Lange, 1980; Puvion-Dutilleul and Puvion, 1981). Theoretically, these intranuclear storage deposits serve as sequestration sites for extremely toxic concentrations of metals and metalloids. Storage of these elements in hepatocytes in a nontoxic form could protect more vulnerable membranes, such as the endoplasmic reticulum and mitochondrial membranes.

Since the nucleus is the control center of the cell, it would appear highly unlikely that the nucleus would be involved in storage functions for toxic metals or metalloids. Several possibilities exist to explain the presence of inclusions within the nucleus. Nuclear synthetic functions might be affected, leading to the buildup of precursor molecules and resulting in formation of a visible structure. This is the case following Cd intoxication. Perichromatin granules first appear as electron dense clusters in the nucleus (Puvion and Lange, 1980; Puvion-Dutilleul and Puvion, 1981). The granules are probably formed as a result of the storage of recently synthesized RNA.

Other possibilities exist. Arsenates could substitute for phosphates in nucleotides of the deoxyribonucleic acid (DNA) molecule, thereby forming unstable arsenate esters of DNA (Petres and Berger, 1972). Striking chromosomal changes have been reported in human lymphocytes following As exposures; these include pulverizations, chromatid breaks, and mitotic rate suppression (Shaw, 1970; Petres and Berger, 1972; Nordensen, *et al.*, 1979).

Therefore, the As-inclusion in sunfish could represent a chromosomal alteration. However, no structural continuity is noted for the nucleolus and the As-inclusion. Moreover, the As-inclusion forms within an intact nuclear membrane, so it is unlikely that the inclusion *per se* is of chromosomal derivation since nuclear membranes and chromosomes are not likely to appear together in any one cell stage.

Apparently, no structures even remotely similar to the As-inclusion have been published, except for those of reports listed in this volume. Laboratory exposures of channel catfish (Sorensen and Smith, 1981a) and environmental exposures of sunfish (Sorensen, *et al.*, 1980; 1981b) do not result in the presence of these intranuclear As-inclusions—probably because much lower levels are accumulated. Livers of green sunfish with As-inclusions can accumulate more than 67 ppm (wet weight) (Sorensen, *et al.*, 1979b).

In addition to low As-accumulation in other studies, morphological similarity of small As-inclusions and hepatocyte nucleoli increases the difficulty of differentiation. Conventional transmission electron microscopy was used in first identification of the structure. If inclusions are large, optical microscopy can be used; but as As-inclusion size decreases, the inclusion is more difficult to differentiate. For example, 20% of control hepatocyte nucleoli from green sunfish are indistinguishable from the smallest of the As-inclusions using light microscopy (Sorensen, *et al.*, 1979a). In this case 20% must be deducted from the percentage of As-inclusions observed in As-exposed fish to compensate for the inability to differentiate small As-inclusions from nucleoli. As the hepatic levels of As increase, numbers of As-inclusions within the nucleus and the cytoplasm increase two- to three-fold (Sorensen, 1974; 1976a; 1976b; 1976c). Aberrant mitochondria, lysosomes, and smooth endoplasmic reticulum become more abundant up to a maximum of four times initial values.

<u>Necrotic and Fibrous Bodies</u>: Green sunfish from environmental areas contaminated with wastes from arsenical desiccants lack As-inclusions, but have unusual structures referred to as necrotic and fibrous bodies, as well as other deviations (Sorensen, *et al.*, 1980; 1981b; 1985b). Necrotic bodies average 6 μm in diameter and are membrane-delimited, containing laminar whorls of endoplasmic reticulum, amorphous electron dense material, and numerous vesicles (Sorensen, *et al.*, 1981b). In necrotic bodies, shriveled or

degenerating structures are surrounded by a plasma membrane. Vesicles of various sizes form frothy areas within most necrotic bodies, but these vesiculated areas are distinct from the undulating arrays of bilaminar endoplasmic reticulum (i.e. fingerprints). The necrotic body appears to contain degenerating subcellular organelles and to represent a special product of heterophagy (or cellular engulfing) observed in other studies (Trump, *et al.*, 1973b). Fibrous bodies are smaller, membrane delimited structures, composed of bundles of fibrous material and nucleoid-like, electron dense material (Sorensen, *et al.*, 1981b). The proportion of cellular volume occupied by necrotic bodies and fibrous bodies is a function of the hepatic level of As (Sorensen, *et al.*, 1985b).

Position (compartmentalization) of necrotic and fibrous bodies can be used to show aging of these structures, to compare with the degeneration of parenchymal hepatocytes (Sorensen, *et al.*, 1981b). The established pattern for degeneration and necrosis of terminal hepatocytes begins with the ballooning, then shriveling, and finally the extrusion of the dying cells from a life-long residence site within the hepatic cord. According the Tanikawa's (1968) studies, hepatocytes become increasingly pyknotic prior to extrusion into the sinusoid through the space of Disse. The entire process appears to represent a form of intracellular digestion of dying, As-damaged hepatocytes to minimize nutrient loss from the body, such as that characteristic of heterophagy.

The Sorensen *et al.* (1981b) study shows that, unlike typical degenerating hepatocytes, necrotic bodies become less pyknotic and vesiculated as they move toward the sinusoid. Most necrotic bodies move from a position in the hepatic cord, to the interior of neighboring hepatocytes (by heterophagy), to the space of Disse, and finally into the sinusoid. Necrotic bodies could be fully-aged (i.e. dead) hepatocytes at the time of extrusion into the sinusoid. Therefore, changes in necrotic and fibrous bodies are considered nonspecific degenerative changes occurring under a variety of exposure situations (H. Popper, personal communication). Moreover, other xenobiotics are capable of inducing similar degenerative changes (Norkin, *et al.*, 1960; Neuwirtova, *et al.*, 1961; Biava, 1964; Biava and Mukhlova-Montiel, 1965; Anderson, 1971). And in studies involving other vertebrates, degenerative changes are similar to those reported in fish (U. S. Naval

Medical School, 1902; Kolmer and Lucke, 1921; Schifrin, 1932; Von Glahn, *et al.*, 1938; Gilderhus, 1966; Brown, *et al.*, 1976; Yeh, *et al.*, 1974).

Hemosiderin Granules: An alternate storage form of excess As is the hemosiderin granule (Sorensen and Smith, 1981a). Normally, the presence of large numbers of hemosiderin granules in parenchymal hepatocytes is symptomatic of Fe-overload diseases such as hemosiderosis, hemochromatosis, and hemolytic jaundice. In these disease states, a nonheme form of soluble Fe (i.e. ferritin) is stored in lysosomes and is degraded to insoluble hemosiderin (Bradford, *et al.*, 1969; Crichton, 1971; Trump, *et al.*, 1973a). Hemolytic jaundice can be a product of increased erythrocyte lysis; however, hemochromatosis and hemosiderosis are considered to be disorders of Fe metabolism.

Since arsenicals are sulfhydryl reagents, it is not surprising to find that As and S, as well as Fe, are present together in hemosiderin granules (Sorensen and Smith, 1981a). Association of As with S (Sorensen, *et al.*, 1982c) is of special interest because arsenicals are associated with tissues of high S content—namely hair, nails, and the epidermis (National Research Council, 1977). Arsenic can bind sulfhydryl groups within protein molecules and accumulate to high levels in these tissues compared with others (Peoples, 1964; Lunde, 1970; National Research Council, 1977).

Myelin Figures: Numbers of myelin figures decrease in hepatocytes from sunfish exposed environmentally to As. These membraneous whorls may be formed by the precipitation of suspended lipids, especially phospholipids, during fixation (Sorensen, 1976c). Not observed in living cells (Matthews and Martin, 1971), myelin figures could provide information about unstable, As-containing phospholipids. In fact, chemical similarity between phosphate and arsenate would enable arsenates to follow the same metabolic pathways as phosphates in the production of phospholipids. However, formation of unstable esters would limit the lifetime of the arseno-analogues of phospholipid, which in turn would decrease numbers and/or sizes of myelin figures.

If cationic arseno-analogues of phospholipids (e.g. phosphatidylethanolamine and phosphatidycholine) were formed, the presence of both lipid and non-lipid organic arsenical fractions in fish could be explained (Penrose, 1975; Lunde, 1977). Algae and *Daphnia* produce phospholipid-like metabolites of As which could be hydrolyzed (as could the natural phospholipid) to remove the ß-carboxylic acid in the choline-containing lipid (Frost, 1967). Although definitive answers are not presently available, arsenates may not replace phosphate because of instability under all but anhydrous conditions. Rather, As may form an arsenocholine, taking the place of the nitrogen atom in the choline group, the hydroxyl group of which is bonded to glycerine *via* a phosphate group.

An alternate explanation for As-induced reductions in numbers of myelin figures is that As affects activities of phospholipid biosynthetic enzymes. Uthus *et al.* (1983) publish preliminary data indicating that the level of dietary As can affect the activities of phosphatidylethanolamine methyltransferase, phosphatidyldimethylethanolamine methyltransferase, and choline phosphotransferase. If phospholipid biosynthetic enzymes were altered, the formation of the product (i.e. the phospholipids—visualized as myelin figures) could be reduced. Apparently, arsenicals affect arginine and protein metabolism; however, the mechanism of action of As at the cellular and molecular level has not been elucidated. Much of the difficulty of ascertaining As-induced effects is ascribed to the close association between arsenate and phosphate and the difficulty of chemically separating the two radicals (Frost, 1967). Because traces of As can catalyze the decomposition of disaccharides and polysaccharides (Cohn, 1961), growth stimulatory effects may occur in laboratory experiments involving feeds having normal or deficient levels of As (Uthus, *et al.*, 1983). Frost believes that As in trace levels is a biological catalyst (Frost, 1967).

Gambusia (i.e. top minnows) incorporate 86% As (at 0.01–1.00 ppm) into the lipid fraction, apparently without methylation of the inorganic As prior to incorporation (Woolson, *et al.*, 1976). Of naturally occurring arsenicals in marine species of fish, 10–30% are present as a lipid-soluble form, but little is known about the remaining 70–90% (Lunde, 1977). In addition, at least four chemical forms of As are produced following exposure of fish to arsenate or the alkyl arsenical, methanearsonate (Woolson, *et al.*, 1976). Surprisingly, these compounds are stable despite the use of strong acids and elevated temperatures during extraction.

Lunde (1968; 1977) finds arsenolipids in fish tissues following exposures in the marine environment. These arsenolipids resemble phospholipids more so than neutral lipids. Similarities between the phospholipids and arsenolipids include their ability to produce hydrolysis products including arsenate (instead of phosphate produced by phospholipids), a positive Ninhydrin reaction (indicating the presence of amines), and molecular weights similar to those of phospholipids. Moreover, the arsenolipids of marine oils are present as polar compounds which can be separated from neutral lipids by chromatography (Lunde, 1968). However, lipid-soluble organic arsenicals do not appear to be bound to cell membranes in the same way as phospholipids (Lunde, 1973).

Several independent studies question whether fish are able to convert inorganic arsenicals to organic arsenicals. In one dietary study, inorganic As is converted to organic As (Penrose, 1975). However, poor or negligible conversion of inorganic to organic forms occurs when As is administered by intramuscular injection (Penrose, 1975; Penrose, *et al.*, 1977). Results suggest that the first stage of As conversion in fish may be microbial in origin. Those organisms residing within the gastrointestinal system of fish seem to be responsible for conversion of inorganic arsenicals to organic arsenicals. Also, organisms at each trophic level have a greater or lesser ability to convert arsenicals from the inorganic form to the presumably detoxified organic form (or forms) (Penrose, *et al.*, 1977). Apparently fish and other organisms from higher trophic levels rapidly detoxify and excrete the organic As, utilizing biliary excretory routes for elimination of this chemical form (Penrose, 1975).

Areas of future research could purpose to determine the chemical structure, physiological significance, and biological effects of arsenolipids as these forms are transferred to the higher trophic levels in the food web. At present, identification of the many inorganic and organic forms of As available to and accumulated by teleosts is incomplete. Because As transformation in the environment involves oxidation, reduction, methylation, and volatilization (Woolson, 1977), a number of chemical forms is available and cause a complex series of cellular and tissue responses. In fact, methylated arsenicals produce the so-called "garlic odor" and are formed by bacteria (McBride and Wolfe, 1971), fungi (Challenger, *et al.*, 1954; Cox and Alexander, 1973), and algae (Irgolic, *et al.*, 1977). Several investigators postulate that the garlic odor represents the products of initial steps in the conversion of inorganic to organic arsenicals and that these steps occur in the intestines as a direct result of the action of intestinal microbes (Penrose, *et al.*, 1977).

Summary

Electric generating stations, manufacturing companies, mineral or strip mines, smelting operations, and agricultural/aquacultural concerns release arsenicals into aquatic habitats. Arsenic is one of the most toxic elements to fish. Acute exposures can result in immediate death because of As-induced increases in mucus production, causing suffocation, or direct detrimental effects on the gill epithelium. Chronic exposures can result in the accumulation of the metalloid to toxic levels. The detoxification role of the liver places that organ at considerable risk; in fact, the liver is considered the major target in most cases of As exposure in vertebrates. Renal and splenic tissues also concentrate elevated levels of As because of respective roles in urinary filtration and erythrocyte filtration/blood storage.

Compartmentalization studies involving fish hepatocytes show that As localizes within the cytoplasm within 2 d, but later translocates to the nuclear compartment. X-ray microanalysis indicates that nuclear As deposits are found in a 1:6 ratio with S. Hepatic degenerative changes in fish exposed to As include submassive necrosis, focal necrosis, bile duct proliferation, bile plugs, and acidophilic bodies. Other changes include the formation of necrotic bodies, fibrous bodies, and cytoplasmic and intranuclear As-inclusions within the parenchymal hepatocytes. These structures are more numerous and of a larger size at higher concentrations and following prolonged exposures to As. Hemosiderin granules probably represent an alternate storage site for arsenicals in teleosts.

References

Anderson, W.A.D., Pathology, C. V. Mosby Co., St. Louis, (1971) 282.

Anonymous, Increased heavy metals around Sudbury's smelters, Water Pollut. Control, 111 (1973) 48.

Biava, C., Mallory alcoholic hyalin: A heretofore unique lesion of hepatocellular ergastoplasm, Lab. Invest., 13 (1964) 301.

Biava, C., and M. Mukhlova-Montiel, Electron microscopic observations on Councilman-like acidophilic bodies and other forms of acidophilic changes in human liver cells, Amer. J. Pathol., 46 (1965) 775.

Bradford, W.D., J.G. Elchlepp, A.V. Arstila, B.F. Trump, and T.D. Kinney, Iron metabolism and cell membranes, Amer. J. Pathol., 56 (1969) 201.

Brown, M.M., B.C. Rhyne, R.A. Goyer, and B.A. Fowler, Intracellular effects of chronic arsenic administration on renal proximal tubule cells, J. Toxicol. Environ. Hlth., 1 (1976) 505.

Buchanan, W.D., Toxicity of arsenic compounds, Elsevier Publ. Co., Amsterdam, (1962)

Byron, W.R., G.W. Bierbower, J.B. Brouwer, and W.H. Hansen, Pathologic changes in rats and dogs from two-year feeding of sodium arsenite and sodium arsenate, Toxicol. Appl. Pharmacol., 10 (1967) 132.

Carpenter, K.E., The lethal action of soluble metallic salts on fishes, Br. J. Exp. Biol., 4 (1927) 378.

Carpenter, K.E., Further researches on the action of metallic salts on fishes, J. Exp. Zool., 56 (1930) 407.

Challenger, F., D.B. Lisle, and P.B. Dransfield, Biological methylation. XIV. The formation of trimethylarsine and dimethyl selenide in mold cultures from methyl sources containing carbon^{14}, J. Chem. Soc., (1954) 1760.

Cherry, D.S., and R.K. Guthrie, Toxic metals in surface waters from coal ash, Water Res. Bull., 13 (1977) 1227.

Clemens, H.P., and K.E. Sneed, Lethal doses of several commercial chemicals for fingerling channel catfish, U. S. Fish Wildlife Serv., Spec. Sci. Rept. Fish., No. 316, (1959) 10.

Cockell, K.A., and J.W. Hilton, Preliminary investigations on the comparative chronic toxicity of four dietary arsenicals to juvenile rainbow trout (*Salmo gairdneri* R.), Aq. Toxicol., 12 (1988) 73.

Cohn, M., Phosphorylases (Survey), in Boyer, P.D., H. Lardy, and K. Myrback (Eds.), The Enzymes, 2nd ed., Academic Press, New York, (1961) 179.

Cope, O.B., Contamination of the fresh-water ecosystem by pesticides, J. Appl. Ecol., 3 (1966) 33.

Cox, D.P., and M. Alexander, Production of trimethylarsine gas from various arsenic compounds by three sewage fungi, Bull. Environ. Contam. Toxicol., 9 (1973) 84.

Crichton, R.R., Ferritin: Structure, synthesis and function, N. Engl. J. Med., 284 (1971) 1413.

Crosby, D.G., and R.K. Tucker, Toxicity of aquatic herbicides to *Daphnia magna*, Science, (1966) 289.

Cumbie, P.M., Belews Lake environmental study report: Selenium and arsenic accumulation, Duke Power Company, Technical Report Series No. 78-04, Charlotte, North Carolina, (1978).

Cumbie, P.M., Effects of pH adjustment and SO_3 fly ash conditioning on the Belews Creek Steam Station Ash Basin, Duke Power Company, Technical Report Series No. 80-01, Charlotte, North Carolina, (1980) 1.

Ducoff, H.A., W.B. Neal, R.L. Straube, L.O. Jacobson, and A.M. Brues, Biological studies with arsenic. II. Excretion and tissue localization, Proc. Soc. Exp. Med., 69 (1948) 548.

Finner, L.L., and H.O. Calvery, Pathologic changes in rats and in dogs fed diets containing lead and arsenic compounds, Arch. Pathol., 27 (1939) 433.

Foley, R.E., J.R. Spotila, J.P. Giesy, and C.H. Wall, Arsenic concentrations in water and fish from Chautauqua Lake, New York, Environ. Biol. Fish, 3 (1978) 361.

Fowler, B.A., R.C. Fay, R.L. Walter, R.D. Willis, and W.F. Gutknecht, Levels of toxic metals in marine organisms collected from southern California coastal waters, Environ. Hlth. Pers., 12 (1975) 71.

Franklin, M., W.B. Bean, and R.C. Hardin, Fowler's solution as etiologic agent in cirrhosis, Amer. J. Med. Sci., 219 (1950) 589.

Frost, D.V., Arsenicals in biology-retrospect and prospect, Fed. Proc., 26 (1967) 194.

Gilderhus, P.A., Some effects of sublethal concentrations of sodium arsenite on bluegills and the aquatic environment, Trans. Amer. Fish. Soc., 95 (1966) 289.

Glover, J.W., Concentrations of arsenic, selenium, and ten heavy metals in school shark, *Galeorhinus australis* (Macleay), and gummy shark, *Mustelus antarcticus* Gunther, from south-eastern Australian waters, Aust. J. Mar. Fresh. Res., 30 (1979) 505.

Goyer, R.A., P. May, M.M. Cates, and M.R. Krigman, Lead and protein content of isolated intranuclear inclusion bodies from kidneys of lead-poisoned rats, Lab. Invest., 22 (1970) 245.

Groth, D.H., L. Stettler, and G. Mackay, Interactions of mercury, cadmium, selenium, tellurium, arsenic, and beryllium, in Nordberg, G.F. (Eds.), Effects and Dose-Response Relationships of Toxic Metals, Elsvier Sci. Publ. Co., Amsterdam, (1976) 527.

Huet, P.M., E. Guillaume, J. Cote, A. Legare, P. Lavoie, and A. Viallet, Noncirrhotic presinusoidal portal hypertension associated with chronic arsenic intoxication, Gastroenterology, 68 (1975) 1270.

Hunter, R.G., J.H. Carroll, and J.S. Butler, The relationship of trophic level to arsenic burden in fish of a southern great plains lake, J. Fresh. Ecol., 1 (1981) 121.

Irgolic, K.J., E.A. Woolson, R.A. Stockton, R.D. Newman, N.R. Bottino, R.A. Zingaro, P.C. Kearney, R.A. Pyles, S. Maeda, W.J. McShane, and E.R. Cox, Characterization of arsenic compounds formed by *Daphnia magna* and *Tetraselmis chuii* from inorganic arsenate,, Environ. Hlth. Pers., 19 (1977) 61.

Kennedy, V.S., Arsenic concentrations in some coexisting marine organisms from Newfoundland and Labrador, J. Fish. Res. Bd. Canada, 33 (1976) 1388.

Kevern, N.R., Feeding rate of carp estimated by a radio-isotopic method, Trans. Amer. Fish. Soc., 95 (1966) 363.

Klaassen, C.D., Biliary excretion of arsenic in rats, rabbits, and dogs, Toxicol. Appl. Pharmacol., 29 (1974) 447.

Klaassen, C.D., Heavy metals and heavy-metal antagonists, in Gilman, A.G., L.S. Goodman, R.T. W., and F. Murad (Eds.), The Pharmacological Basis of Therapeutics, 7th ed., MacMillan Publ. Co., New York, (1985) 1605.

Kodama, Y., N. Ishinishi, E. Kunitake, T. Inamusu, and K. Nobutomo, Subclinical signs of the exposure to arsenic in a copper refinery, in Nordberg, G.F. (Ed.), Effects and dose-response relationships of toxic metals, Elsevier Sci. Pub., Amsterdam, (1976) 464.

Kolmer, J.A., and B. Lucke, A study of the histologic changes produced experimentally in rabbits by arsphenamine, Arch. Derm. Syph., 3 (1921) 483.

Lander, J.J., R.J. Stanley, H.W. Sumner, D.C. Boswell, and R.D. Aach, Angiosarcoma of the liver associated with Fowler's solution (potassium arsenite), Gastroenterology, 68 (1975) 1582.

LeBlanc, P.J., and A.L. Jackson, Arsenic in marine fish and invertebrates, Mar. Pollut. Bull., 4 (1973) 88.

Lovett, R.J., W.H. Gutenmann, I.S. Pakkala, W.D. Youngs, and D.J. Lisk, A survey of the total cadmium content of 406 fish from 49 New York State Fresh waters, J. Fish. Res. Bd. Canada, 29 (1972) 1283.

Lowe, T.P., T.W. May, W.G. Brumbaugh, and D.A. Kane, National contaminant biomonitoring program: Concentrations of seven elements in fresh-water fish, 1978-1981, Arch. Environ. Contam. Toxicol., 14 (1985) 363.

Lucus, H.F., Jr., D.N. Edgington, and P.J. Colby, Concentrations of trace elements in Great Lake fishes, J. Fish. Res. Bd. Canada, 27 (1970) 677.

Lunde, G., Analysis of arsenic in marine oils by neutron activation evidence of arseno organic compounds, J. Amer. Oil Chem. Soc., 45 (1968) 331.

Lunde, G., Analysis of trace elements in seaweed, J. Sci. Fd. Agric., 21 (1970) 242.

Lunde, G., Analysis of trace elements, phosphorus and sulphur, in the lipid and non-lipid phase of halibut and tunny, J. Sci. Fd. Agric., 19 (1973) 1029.

Lunde, G., Occurrence and transformation of arsenic in the marine environment, Environ. Hlth. Pers., 19 (1977) 47.

Matthews, J.L., and J.H. Martin, Atlas of Human Ultrastructure, Lea and Febiger, Philadelphia, (1971) 382.

McBride, B.C., and R.S. Wolfe, Biomethylation of dimethylarsine by Methanobacterium, Biochem., 10 (1971) 4313.

Mitchell, R.R., and E.M.B. Sorensen, The three-dimensional morphology of arsenic inclusions: HVEM versus STEM versus CTEM, J. Electron Micros. Techn., 22 (1987) 273.

Moore, J.F., R.A. Goyer, and M. Wilson, Lead-induced inclusion bodies: Solubility, amino acid content, and relationship to residual acidic nuclear proteins, Lab. Invest., 29 (1973) 488.

Moore, J.F., and R.A. Goyer, Lead-induced inclusion bodies: Composition and probable role in lead metabolism, Environ. Hlth. Pers., 7 (1974) 121.

Mount, D.I., A simplified dosing apparatus for fish toxicology studies, Water Res., 1 (1967) 21.

National Research Council, Arsenic, National Academy of Sciences, Washington, D.C., (1977) 24.

Natusch, D.F.S., and J.R. Wallace, Toxic trace elements: Preferential concentration in respirable particles, Science, 183 (1974) 202.

Neuwirtova, R., M. Chytil, A. Valek, S. Daum, and V. Valach, Acute renal failure following an occupational intoxication with arsine (AsH_3) treated by the artificial kidney, Acta Med. Scandinavia, (1961) 535.

Nordensen, I., S. Salmonsson, E. Brun, and G. Beckman, Chromosome aberrations in psoriatic patients treated with arsenic, Hum. Genet., 48 (1979) 1.

Norkin, S.A., R. Weitzel, D. Campagna-Pinto, R.A. MacDonald, and K.G. Mallory, "Alcoholic" hyalin in human cirrhosis; histochemical studies, Amer. J. Pathol., 37 (1960) 49.

Passino, D.R.M., and J.M. Kramer, Toxicity of arsenic and PCBs to fry of deepwater ciscoes (Coregonus), Bull. Environ. Contam. Toxicol., 24 (1980) 527.

Penrose, W.R., Biosynthesis of organic arsenic compounds in brown trout (*Salmo trutta*), J. Fish. Res. Bd. Canada, 32 (1975) 2385.

Penrose, W.R., H.B.S. Conacher, R. Black, J.C. Meranger, W. Miles, H.M. Cunningham, and W.R. Squires, Implication of inorganic/organic interconversion on fluxes of arsenic in marine food webs, Environ. Hlth. Pers., 19 (1977) 53.

Peoples, S.A., Arsenic toxicity in cattle, Ann. N. Y. Acad. Sci., 111 (1964) 644.

Petres, J., and A. Berger, The effect of inorganic arsenic on DNA-synthesis of human lymphocytes in vitro [in German, English summary], Arch. Dermatol. Forsch., 242 (1972) 343.

Puvion, E., and M. Lange, Functional significance of perichromatin granule accumulation induced by cadmium chloride in isolated rat liver cells, Exp. Cell Res., 128 (1980) 47.

Puvion-Dutilleul, F., and E. Puvion, Relationship between chromatin and perichromatin granules in cadmium-treated isolated hepatocytes, J. Ultrastruct. Res., 74 (1981) 341.

Ramelow, G.J., C.L. Webre, C.S. Mueller, J.N. Beck, J.C. Young, and M.P. Langley, Variations of heavy metals and arsenic in fish and other organisms from the Calcasieu River and Lake, Louisiana, Arch. Environ. Contam. Toxicol., 18 (1989) 804.

Regelson, W., V. Kim, J. Ospina, and J.F. Holland, Hemangioendothelial sarcoma of liver from chronic arsenic intoxication by Fowler's solution, Cancer, 21 (1968) 514.

Ryther, J., T.M. Losordo, A.K. Furr, T.F. Parkinson, W.H. Gutenman, I.S. Pakkala, and D.J. Lisk, Concentration of elements in marine organisms cultured in seawater flowing through coal-fly ash, Bull. Environ. Contam. Toxicol., 23 (1979) 207.

Sandhu, S.S., Study on the post-mortem identification of pollutants in the fish killed by pollution: Detection of arsenic, Bull. Environ. Contam. Toxicol., 17 (1977) 373.

Schifrin, A., Der Einfluss qualitativ verschiedener Ernahrungsformen auf die durch Salvarsan hervorgerufenen Lebernekrosen, Virchows Arch. Pathol. Anat., 287 (1932) 175.

Shaw, M.W., Human chromosome damage by chemical agents, Ann. Rev. Med., 21 (1970) 409.

Shukla, J.P., and K. Pandey, Impaired ovarian functions in arsenic-treated fresh-water fish, Colisa fasciatus (BL & SCH), Toxicol. Lett., 20 (1984a) 1.

Shukla, J.P., and K. Pandey, Arsenic-induced cellular and biochemical changes during the testicular cycle of a fresh-water perch, *Colisa fasciatus* (BL & SCH), Cell. Mol. Biol., 30 (1984b) 227.

Sorensen, E.M.B., Thermal effects on the biological magnification of arsenic in green sunfish, *Lepomis cyanellus*, Ph.D. Dissertation, The University of Texas at Austin, (1974) 8.

Sorensen, E.M.B., Thermal effects on the accumulation of arsenic in green sunfish, *Lepomis cyanellus*, Arch. Environ. Contam. Toxicol., 4 (1976a) 8.

Sorensen, E.M.B., Toxicity and accumulation of arsenic in green sunfish, *Lepomis cyanellus*, exposed to arsenate in water, Bull. Environ. Contam. Toxicol., 15 (1976b) 756.

Sorensen, E.M.B., Ultrastructural changes in the hepatocytes of green sunfish, *Lepomis cyanellus*, exposed to solutions of sodium arsenate, J. Fish Biol., 8 (1976c) 229.

Sorensen, E.M.B., R.E. Henry, and R.R. Mitchell, Arsenic accumulation, tissue distribution, and cytotoxicity in teleosts following indirect aqueous exposures, Bull. Environ. Contam. Toxicol., 21 (1979a) 162.

Sorensen, E.M.B., R.E. Henry, and R. Ramirez-Mitchell, Time-dependent localization of arsenic in subcellular liver fractions, J. Environ. Pathol. Toxicol., 2 (1979b) 1161.

Sorensen, E.M.B., R. Ramirez-Mitchell, C.W. Harlan, and J.S. Bell, Cytological changes in fish liver following chronic, environmental arsenic exposure, Bull. Environ. Contam. Toxicol., 25 (1980) 93.

Sorensen, E.M.B., and N.K.R. Smith, Hemosiderin granules: Cytotoxic response to arsenic exposure in channel catfish, Bull. Environ. Contam. Toxicol., 27 (1981a) 645.

Sorensen, E.M.B., L.L. Wenz, B.C. Windsor, and R. Ramirez-Mitchell, Stereological analysis of arsenic-induced cytological structures, J. Tenn. Acad. Sci., 56 (1981b) 131.

Sorensen, E.M.B., T.L. Bauer, J.S. Bell, and C.W. Harlan, Selenium accumulation and cytotoxicity in teleosts following chronic, environmental exposure, Bull. Environ. Contam. Toxicol., 29 (1982a) 688.

Sorensen, E.M.B., C.W. Harlan, and J.S. Bell, Renal changes in selenium-exposed fish, Amer. J. Forensic Med. Pathol., 3 (1982b) 123.

Sorensen, E.M.B., N.K.R. Smith, and R. Ramirez-Mitchell, Electron probe X-ray microanalysis of arsenic inclusions in fish, Arch. Environ. Contam. Toxicol., 11 (1982c) 469.

Sorensen, E.M.B., C.W. Harlan, J.S. Bell, T.L. Bauer, and A.H. Pradzynski, Hepatocyte changes following selenium accumulation in a fresh-water teleost, Amer. J. Forensic Med. Pathol., 4 (1983a) 25.

Sorensen, E.M.B., J.S. Bell, and C.W. Harlan, Histopathological changes in selenium-exposed fish, Amer. J. Forensic Med. Pathol., 4 (1983b) 111.

Sorensen, E.M.B., and T.L. Bauer, Hematological dyscrasia in teleosts chronically exposed to selenium-laden effluent, Arch. Environ. Contam. Toxicol., 12 (1983c) 135.

Sorensen, E.M.B., P.M. Cumbie, T.L. Bauer, H.S. Bell, and C.W. Harlan, Histopathological, hematological, condition-factor, and organ-weight changes associated with selenium accumulation in fish from Belews Lake, North Carolina, Arch. Environ. Contam. Toxicol., 13 (1984a) 153.

Sorensen, E.M.B., and T.L. Bauer, Planimetric analysis of redear sunfish (*Lepomis microlophus*) hepatopancreas following selenium exposure, Environ. Toxicol. Chem., 3 (1984b) 159.

Sorensen, E.M.B., and T.L. Bauer, A correlation between selenium accumulation in sunfish and changes in condition factor and organ weight, Environ. Pollut., 34 (1984c) 357.

Sorensen, E.M.B., R. Ramirez-Mitchell, A. Pradzynski, and T.L. Bauer, Stereological analyses of hepatocyte changes parallel arsenic accumulation in the livers of green sunfish, J. Environ. Path. Toxicol. Oncol., 6 (1985a) 195.

Sorensen, E.M.B., R.R. Mitchell, A. Pradzynski, T.L. Bauer, and L.L. Graham, Correlation between stereological analyses of the hepatocyte and arsenic accumulation in the liver of a fresh-water teleost naturally exposed to arsenic, J. Environ. Pathol. Toxicol. Oncol., 6 (1985b) 195.

Sorensen, E.M.B., The effects of selenium on fresh-water fish, in Hodgson, E. (Ed.), Reviews in Environmental Toxicology, Elsevier Press, New York, New York, (1986) 59.

Sorensen, E.M.B., Ultrastructural cytochemical analysis of intranuclear arsenic inclusions, J. Toxicol. Environ. Hlth., 22 (1987) 273.

Surber, E.W., Pond and small lake management, Trans. N. Amer. Widl. Cont., 8 (1943) 132.

Tanikawa, K., Ultrastructural aspects of the liver and its disorders, Springer-Verlag, Berlin, (1968) 88.

Trump, B.F., J.H. Dees, and J.D. Shelburne, The ultrastructure of the human liver cell and its common patterns of reaction to injury, in Gall, E.A., and F.K. Mostofi (Eds.), The Liver, The Williams & Wilkins Co., Baltimore, (1973a) 80.

Trump, B.F., J.M. Valigorsky, A.V. Arstila, W.J. Mergner, and T.D. Kinney, The relationship of intracellular pathways of iron metabolism to cellular iron overload and the iron storage diseases, Amer. J. Pathol., 72 (1973b) 295.

U. S. Naval Medical School, Color Atlas of Pathology, J. B. Lippincott Co., Philadelphia, (1902) 365.

Ullmann, W.W., R.W. Schaefer, and W.W. Sanderson, Arsenic accumulation by fish in lakes treated with sodium arsenite ,J. WPCF, 33 (1961) 416.

Uthus, E.O., W.E. Cornatzer, and F.H. Nielsen, Consequences of arsenic deprivation in laboratory animals, in Lederer, W.H., and R.J. Fensterheim (Eds.), Arsenic, Van Nostrand Reinhold Co., New York, (1983) 173.

Von Glahn, W.C., F.B. Flinn, and W.F. Keim Jr., Effect of certain arsenates on the liver, Arch. Pathol., 25 (1938) 488.

Windom, H., R. Stickney, R. Smith, D. White, and F. Taylor, Arsenic, cadmium, copper, mercury, and zinc in some species of North Atlantic finfish, J. Fish. Res. Bd. Canada, 30 (1973) 275.

Wood, J.M., Biological cycles for toxic elements in the environment, Science, 183 (1974) 1049.

Woolson, E.A., A.R. Isensee, and P.C. Kearney, Distribution and isolation of radioactivity from ^{74}As-Arsenate and ^{14}C-Methane-arsonic acid in an aquatic model ecosystem, Pest. Biochem. Physiol., 6 (1976) 261.

Woolson, E.A., Fate of arsenicals in different environmental substrates, Environ. Hlth. Pers., 19 (1977) 73.

Yeh, S., H.C. Chen, S.W. How, and C.S. Deng, Fine structure of Bowen's disease in chronic arsenicalism, J. Natl Cancer Inst., 53 (1974) 31.

Zachariae, H., H. Sogaard, and A. Nyfors, Liver Biopsy in psoriatics previously treated with potassium arsenite, Acta Derm. (Stockholm), 54 (1974) 235.

CHAPTER IV
LEAD

Introduction

With its face-centered cubic structure, Pb is the only metallic member of Group IVA in the periodic table. Lead behaves more like the alkaline earth metals (i.e. Ca, Sr, and Ba) with respect to uptake, internal distribution, and excretion (Hirao and Patterson, 1974; Settle and Patterson, 1980). Since Pb is a bone-seeking element, it is processed along with Ca because of chemical resemblance to Ca. However, tissues other than bone are storage sites for Pb in fish. After intramuscular injection of Pb acetate, for example, Pb salts deposit as a layer in scales, fin rays, vertebrae, and opercula, which are areas of active calcification (Van Oosten, 1957).

Lead uptake in vertebrates would equal that of Ca were it not for biopurification processes for Ca. More Ca is retained than Pb because Pb is processed as a trace contaminant of Ca (Settle and Patterson, 1980). Biological mechanisms regulating Ca are inoperable for Pb, and five-fold biopurification of Ca with respect to Pb is multiplicative through successive consumer stages (Hirao and Patterson, 1974; Settle and Patterson, 1980). Lead to calcium ratios tend to decline as trophic position increases, a situation which is opposite the increasing concentration patterns of DDT in higher trophic levels. Use of Pb/Ca ratios is more useful than use of absolute concentrations of Pb because of the analytical difficulty in measurement of Pb levels (Chow, *et al.*, 1974; Settle and Patterson, 1980; Schmitt and Finger, 1987).

Loading of aquatic habitats with Pb can result in complex alterations of teleost trophic levels lasting for centuries. Carpenter first recognized and reported the Pb pollution of rivers in Wales (Carpenter, 1924; 1925; 1926; 1927; 1930). Her work showed the consequences of dumping mine ore wastes directly into river systems: they were rendered barren of fish and other aquatic life. Beginning with Carpenter's studies, this review integrates a portion of the large and growing literature reporting Pb poisoning of fish.

Environmental Exposures

In 1919, Carpenter reports that streams in the Aberystwyth District of Wales were poisoned with Pb (Carpenter, 1924; 1925; 1926; 1927; 1930). Metal-rich veins had been mined since pre-Roman times on land which bordered Cardigan Bay and was traversed by several rapid streams. The nineteenth century development of fine-grinding machinery paved the way for serious pollution of the area. Dozens of small mines soon dotted the Cardiganshire area and, according to mining practice of that time, heaps of Pb-rich tailings piles were deposited directly into stream beds. Grey "dumps" soon became a conspicuous feature of the Pb-mining district of Cardigan Bay. Tailings in all stages of division weather, to bring Pb into solution in the soft, slightly acid river water (pH 6.1–6.9). Therefore, despite closure of most mines during 1883–1886, Pb continued to pollute rivers in the 581 square km area. Abundant fish populations in neighboring, unpolluted streams of Leri, Dyfi, and Teifi rivers were a marked contrast to Rheidol and Ystwyth rivers, which were almost barren of all flora and fauna. Severity of Pb-pollution increased until in 1922, when fish are documented to avoid Pb-polluted Cardiganshire rivers.

Southeastern Missouri's Old Lead Belt also receives Pb contaminates from mining wastes (Gale, *et al.*, 1982; Schmitt, *et al.*, 1984; Harwood and Koirtyohann, 1987; Schmitt and Finger, 1987). This area in Missouri led the world in Pb production at one time. For over 200 yr since the first mines were opened, tailings have contaminated the area with Pb and other elements. Although the last mine of the Old Lead Belt was closed in 1972, levels of Pb in edible tissues of sunfish (*Lepomis megalotis*), redhorse (*Moxostoma duquesnei*), and bass (*Micropterus dolomieui*) are reported at 18, 14, and 9 ppm (wet weight), respectively. Human consumption of Pb-contaminated fish from Missouri rivers can exceed maximum daily "safe" levels for Pb in food recommended by the World Health Organization (i.e. 450 µg) (World Health Organization, 1972); therefore, the public has been discouraged from eating fish taken from a portion of Big River from the city of Leadwood downstream to Washington State Park (Gale, *et al.*, 1982).

Highways also pose a threat to fish because of Pb contamination from automobile exhausts (Pagenkopf and Neuman, 1974; Van Hassel, *et al.*, 1979; Van Hassel, *et al.*, 1980; Ney and Van Hassel, 1983). Lead levels in fish from Back Creek in Virginia are correlated with traffic density for blacknose dace (*Rhinichthys atratulus*), bluehead chub (*Nocomis leptocephalus*), and fantail darters (*Etheostoma flabellare*). In areas where traffic volumes are 15,000 vehicles per d, whole body Pb in the three species average 92% higher (14.5 ppm, dry weight) than in areas were the traffic volume is <50 vehicles per d (7.4 ppm). More Pb is reported present in benthic than pelagic fish. For example, blacknose dace and fantail darters (benthic species) accumulate about 17.5 ppm (whole body, dry weight) compared with pelagic species such as redbreast sunfish (*Lepomis arulius*) and rock bass (*Ambloplites rupestris*), which accumulate about 7.6 ppm Pb. Therefore, even moderate traffic density on highways can endanger fish because of resultant Pb pollution.

The ubiquity of Pb-poisoning of fish spans the globe. Industrial and agricultural discharges are the primary source of Pb-poisoning of fish in Egypt (El Nabawi, *et al.*, 1987), Poland (Zamojski, *et al.*, 1986), England (Badsha and Goldspink, 1982), Texas (Horowitz and Presley, 1977), and Lake Ontario (Cappon, 1987). Industrial and/or sewage effluents are the major source of Pb in fish from Africa (Greichus, *et al.*, 1978), Wisconsin and Michigan (Kleinert, *et al.*, 1974), Ontario (Johnson, 1987), and Mexico (Villarreal-Trevino, *et al.*, 1986). Highway or motorboat traffic are the main sources of Pb in fish from Virginia (Van Hassel, *et al.*, 1980; Ney and Van Hassel, 1983), Arizona (Bussey, *et al.*, 1976) and Illinois (Rolfe and Jennett, 1979). Mine and smelting operations are responsible for Pb contamination of fish in Britain (Carpenter, 1924; Carpenter, 1925; Carpenter, 1926; Carpenter, 1927; Carpenter, 1930; Jones, 1940), Missouri (Gale, *et al.*, 1982; Schmitt and Finger, 1987), and Illinois (Rolfe and Jennett, 1979). Feces from waterfowl and airborne particulate are primary sources of Pb in fish in Michigan (Mathis and Kevern, 1975). Even studies of fish from remote areas as the Arctic (Bohn and Fallis, 1978), tropical islands (Taylor and Bright, 1973), and Canada (Uthe and Bligh, 1971) contain sufficiently elevated Pb levels to be cause for concern (Chow, *et al.*, 1974; Settle and Patterson, 1980).

Toxicity

Early acute toxicity studies produced lethal levels varying from 0.1 to 10.0 ppm Pb, and lack of standard testing procedures lead to some difficulty in data interpretation. For example, Carpenter (1927) reports 0.3 ppm as the lethal Pb concentration for minnows (*Phoxinus phoxinus*), sticklebacks (*Gasterosteus aculeatus*), and brown trout (*Salmo trutta*) in 12–36 h exposures. Jones (1938) reports that 0.1 ppm Pb is the lethal concentration for sticklebacks in 336 h; moreover, 10 ppm is the lethal concentration for goldfish (*Carassius auratus*) after 1–2 h. The lethal Pb concentration after 96 h for fathead minnows (*Pimephales promelas*) is set at 2.4 ppm; 1 ppm is said to be the lethal Pb concentration for rainbow trout (*Salmo gairdneri*) after 100 h (Jones, 1964a). Obviously, conclusions are difficult to draw because data are collected at different times using different species. However 0.1–10 ppm Pb represents toxic levels for several species based on early studies. Variation is due to many factors including life stage, water quality, and the presence of different elements, as well as other factors.

All life stages are sensitive to the toxic effects of Pb; however, embryos are more sensitive to Pb than are later juvenile stages (Davies, *et al.*, 1976). Unlike Cd and some other metals, Pb reduces embryonic resistance with subsequent exposures. Previous exposure of embryos causes a two-fold increase in Pb sensitivity, based on maximum acceptable toxicant concentrations (MATC). In laboratory studies using six species, Carpenter (1930) reports that smaller fish are more sensitive to Pb than are larger fish because small fish have higher metabolic rates. For every 0.1 g increase in fish weight, survival increases an average of 4 min for common shiners (*Notropis cornutus*) in the 3.0–3.6 g range. Metabolic rate, as measured by CO_2 evolution, is correlated with sensitivity of fish to Pb. Therefore, fish of earlier life stages are more sensitive to Pb than are fish of later life stages.

Water quality (especially water hardness) also alters acute Pb toxicity. The presence of 50 ppm Ca, for example, reduces the toxicity of Pb from an

LC50 of 6.6 ppm to one of 110.0 ppm Pb for goldfish (*Carassius auratus*) (Weir and Hine, 1970). Jones (Jones, 1938; Jones, 1964b) finds that addition of 50 ppm Ca increases survival times of stickleback (*Gasterosteus aculeatus*) from 1–4 d to more that 10 d when Ca is added to solutions containing 0.4 or 1.0 ppm Pb. Davies *et al.* (1976) indicates that Pb is less toxic to rainbow trout (*Salmo gairdneri*) in hard water (353 ppm as $CaCO_3$) than in soft water (28 ppm as $CaCO_3$). Bioassays give 96-h LC50 values of 1.17 ppm Pb for soft water and 471.0 ppm Pb for hard water. Holcombe *et al.* (1976) publishes 96-h LC50 values of 4.1 ppm Pb in soft water (44 ppm, as $CaCO_3$) in brook trout (*Salvelinus fontinalis*). The MATC values are set at 120–360 ppb in hard water (353 ppm, as $CaCO_3$) and 4–8 ppb in soft water (28 ppm, as $CaCO_3$) for rainbow trout (Davies, *et al.*, 1976) and 58–119 ppb for brook trout in soft water (44 ppm, as $CaCO_3$) (Holcombe, *et al.*, 1976). Therefore, acute Pb toxicity is inversely proportional to water hardness.

The nemesis of work involving Pb is the precipitation of Pb in hard water (Jones, 1938; Jones, 1964b). At 27 ppm water hardness (as $CaCO_3$), an 8 ppm Pb solution begins to precipitate, and at 53 ppm hardness, only 1.6 ppm Pb remains in solution. Removal of the precipitate by filtration and use of the filtrate (Dilling, *et al.*, 1926) seems less appropriate than use of the solutions with the precipitate. Characterization of the solubility of Pb (II) in soft water and hard water at different pH values, shows that Pb solubility depends on carbonate concentration, which depends upon the pH and partial pressure of CO_2 (Davies, *et al.*, 1976). In hard water (353 ppm), only a fraction of 30 ppb is calculated to be soluble (i.e. a dissolved species). Most Pb is present as a colloid and a precipitate. Carpenter mentions the opalescence of Cardiganshire streams due to Pb carbonate precipitation (Carpenter, 1930), an observation which suggests that multiple forms of Pb exist in environmental exposures. Therefore, in experimental design of laboratory studies, which replicate (as much as possible) the environmental conditions of Pb-exposure, filtration of Pb solutions is not an acceptable alternative.

The presence of elements other than Ca, also complicates toxicity testing. The flood-waters in Cardiganshire, for example, are reported more poisonous than prepared solutions in laboratory studies despite the fact that Pb levels are identical. Fish survive one-third as long in the Pb-contaminated flood-waters as in laboratory solutions. Although she does not measure concentrations of Zn, Carpenter realizes that Pb at this site probably occurs in combination with other metals, particularly Zn (Carpenter, 1925). She conducts experiments showing that combined Pb and Zn solutions produce similar symptoms as does Pb alone. Jones confirms these tests (Jones, 1964b). Also, relationships are found between uptake of Pb and Cd in red snapper and between Pb and Cr in salmon (Zook, *et al.*, 1976). Lead interacts with other elements synergistically or antagonistically (see Chapter IX). Sandstead indicates that known or suspected interactions exist between Pb and Ca, Zn, Cd, Fe, Cu, and S (Sandstead, 1976). Therefore, present understanding of Pb toxicity in fish is derived primarily from acute toxicity tests. Additional reviews documenting early studies of Pb toxicity to fish are provided by Doudoroff and Katz (1953) and Aronson (1971).

Accumulation and Distribution

Like most elements, Pb is absorbed by the epithelium of both the gills and intestine; therefore, fish provide a unique system for study of the regulatory mechanisms involved in Pb uptake and distribution. Varanasi and Gmur find that gills are the major and most efficient site of Ca and/or Pb uptake (Varanasi and Gmur, 1978), while Somero *et al.* (1977) indicates that the intestine is the major site of uptake of Pb and all divalent cations. This difference probably represents a species-specific difference, for Varanasi and Gmur work with a pelagic, anadromous species while Somero and colleagues study a benthophagic, estuarian fish. Inherent osmoregulatory ability, therefore, probably alters species response to Pb, as Pb is absorbed at the two "independent" sites (i.e. gill and intestine).

In evaluating studies using different species, pH level, temperature, Pb concentration, and exposure time; variation must be expected. However, some commonalties occur. As for Cd, kidney, gill, and liver tissues tend to accumulate the highest level of Pb during aqueous or dietary exposures (Holcombe, *et al.*, 1976; Merlini and Pozzi, 1977; Varanasi and Gmur, 1978; Reichert, *et al.*, 1979). Lead concentrations in these tissues range from 2.7–4.1 ppm (Merlini and Pozzi, 1977; Reichert, *et al.*, 1979); however, considerably lower values (i.e. 0.07–0.40 ppm, (Holcombe, *et al.*, 1976)) and higher values (i.e. 12–100 ppm, (Somero, *et al.*, 1977)) are reported. Concentration factors (CF; the ratio of tissue concentrations to aqueous concentrations) range from 2.7 to 21.0 for the gill, liver, and kidney in pumpkinseed (*Lepomis gibbosus*) and Coho salmon (*Oncorhynchus kisutch*)

(Reichert, *et al.*, 1979). In bluegill (*Lepomis macrochirus*) from soft water ponds, the whole body CF value is 33, whereas the CF in a hard water lake is 230 (Wiener and Giesy, 1979). Therefore, variation seems an inherent theme for Pb tissue accumulation in fish.

Tissues possessing coverings of mucous membranes (e.g. gills, fins, skin, and intestine) have the greatest uptake and depuration rates (Somero, *et al.*, 1977). Only the spleen is the exception to the rule that tissues of high Pb uptake have mucous linings. Somero *et al.* (1977) and Van Oosten (1957) speculate that a complex forms between Pb and mucus. Varanasi *et al.* (1975) suggest that the metal-complexed mucus (on the gill and body surfaces) is sloughed and subsequently replaced by metal-free mucus. Lead depuration is a function of mucous turnover; however, although it is effective in binding Pb and protecting fish in acute exposures (Carpenter, 1927; Carpenter, 1930), surface mucus does not prevent the death of fish.

Fish excrete Pb rapidly, therefore, Pb depuration generally reduces organ levels of Pb (Holcombe, *et al.*, 1976; Merlini and Pozzi, 1977; Somero, *et al.*, 1977). However, in a 12-wk depuration study, Holcombe *et al.* (1976) show that liver continues to accumulate Pb for the first 2 wk while the gill and kidney levels decrease. At the end of 12-wk of depuration; however, Pb levels in all tissues have decreased. Reichert *et al.* (1979) show that 37 d of depuration decreases Pb levels in all tissues except the kidney. Following 35 d of depuration, skeletal Pb levels are elevated compared with skeletal Pb levels at the beginning of depuration (Holcombe, *et al.*, 1976).

Lead levels in skeletal muscle generally are lower than those of other tissues due to the low binding rate for Pb to sulfhydryl groups in muscle, as well as low solubility and restricted relocation of Pb salts (Moore and Ramamoorthy, 1984). Also, usually marine species accumulate less Pb than freshwater species. In a worldwide study screening thirty-four commonly consumed seafoods, mean Pb levels in edible tissues (mostly skeletal muscle) are uniform among various species, with only a few species averaging over 0.6 ppm (Zook, *et al.*, 1976). No species exceeded 1 ppm, and Pb levels average 0.49 ppm in fresh fish muscle. Low Pb levels in muscle and marine fish could make canned tuna (and other canned marine fish meats) an excellent low-Pb food source. Such is not the case. Rather, Pb soldering of cans results in a four-hundred-fold increase in Pb in these tissues (Settle and Patterson, 1980). Anthropogenic addition of Pb to canned tuna fish, anchovies, and other meats is an excellent reason to eliminate such foods from the diet. However, at present, excess Hg levels in processed edible marine fish could be far more hazardous to mankind than elevated Pb levels (see Chapter VIII).

Several investigators address the question of whether age (or size) is related to Pb accumulation in the whole body or individual body tissues. Variation in Pb uptake (possibly due to genetic differences in feral fish) seems to override the metabolic differences which normally would be expected to control accumulation levels. Lead levels in the whole body are not correlated with fish size in yellow perch (*Perca flavescens*), black crappie (*Pomoxis nigromaculatus*), bluegills (*Lepomis macrochirus*), and black bullheads (*Ictalurus melas*) from the Fox River in Illinois (Vinikour, *et al.*, 1980). However, Bohn and Fallis (1978) find that smallest shorthorn sculpins (*Myoxocephalus scorpius*) accumulate the highest hepatic Pb concentrations. Moreover, skeletal (bone) Pb levels in white suckers (*Catostomus commersoni*) from highway-contaminated streams are age related for fish between ages 2 and 5—increasing consistently to about 30 ppm (dry weight) (Ney and Van Hassel, 1983). In view of the fact that Pb is a bone-seeking element, measurement of age-related differences in skeletal Pb uptake might yield useful information. The value of skeletal Pb measurements in chronic Pb pollution could outweigh the difficulty of collecting a clean sample of fish bone.

Varanasi and Gmur (1978) speculate that increased aqueous or dietary Ca suppresses synthesis of Ca-binding protein in the gut and results in reduced uptake of both Ca and Pb. However, rather than the gut, only the liver and posterior kidney of Pb-exposed coho salmon (*Oncorhynchus kisutch*) are evaluated for specific Pb-binding proteins in the Reichert *et al.* study. Specific Pb-binding proteins are associated with the 55,000 dalton fraction (rather than cytosolic fraction) in renal and hepatic tissues. This is an area requiring additional research.

Lead uptake depends upon exposure time, aqueous concentration, pH, temperature, salinity, diet, and other parameters. Gill, liver, kidney, and erythrocytes accumulate Pb from aqueous sources in proportion to exposure time and concentration (Holcombe, *et al.*, 1976). For example, during a 38-wk exposure of brook trout (*Salvelinus fontinalis*) to 0.474 ppm Pb, hepatic Pb levels increase from 2 ppm to 100 ppm (dry weight). Moreover, following 38 wk, hepatic Pb concentration increases from about 2 ppm in controls to about

20 ppm, 70 ppm, and 100 ppm for trout exposed to 0.034, 0.119, or 0.474 ppm, respectively (Holcombe, *et al.*, 1976).

Aqueous pH is also important. When pH of the exposure solution is increased from pH 6.0 to 7.5, pumpkinseeds (*Lepomis gibbosus*) accumulate about three times more Pb in tissues (Merlini and Pozzi, 1977).

Temperature affects Pb accumulation. Gobies held at 20–25°C for 42 d accumulate two to twenty times more Pb than do gobies held at 10°C (Somero, *et al.*, 1977).

Salinity alters Pb accumulation. Gobies held in dilute seawater (25% of 34 ‰ salinity) consistently show the greatest accumulation of Pb in tissues compared with gobies held in 50%, 75%, or 100% (of 34 ‰ salinity) (Somero, *et al.*, 1977).

Diet can change Pb uptake. Varanasi and Gmur (1978) report that Pb uptake from water-borne sources is inversely related to Ca levels in coho salmon. However, dietary Ca is more effective in influencing Pb uptake than is water-borne Ca. In addition to exposure time, aqueous concentration, pH, temperature, salinity, and diet; other parameters are important in determining accumulation of Pb—not the least of which is the inherent ionoregulatory/osmoregulatory capacity and metabolism of the species under consideration.

Douben and Koeman (1989) measure relative bioavailability of Pb and Cd in stone loach (*Noemacheilus barbatulus*) for up to 8 d in sediment contaminated with ~27 ppm Cd and ~4720 ppm Pb, as well as to water at ~0.6 ppb Cd and ~8.6 ppb Pb at 16°C. Whole fish levels of Cd and Pb are always higher in fish exposed to contaminated sediments than in controls, and dietary uptake appears less important for Pb than for Cd. The biological half-life for Cd is greater than that for Pb in the stone loach, and Pb is lost more readily following aqueous exposures than following dietary exposures (Vighi, 1981). Water-borne Cd does not contribute substantially to the metal burden of the stone loach. Sedimentary Cd and Pb can accumulate in stone loach, but bioavailability of the two elements remains to be elucidated. Since the stone loach buries itself in the mud, sedimentary metals are more important for this species than for a pelagic species, for example.

Discrepancies in Pb accumulation and internal distribution patterns in different studies may be due to analysis error. Serious doubts are published regarding the validity of most Pb measurements (Chow, *et al.*, 1974; Settle and Patterson, 1980). In 1973, seven superior oceanographic laboratories could not collect and analyze water by atomic absorption or anodic stripping voltametry techniques and report reliable analytical values. Indeed, most published measurements of Pb may be erroneously high by factors of 10 to 100 (Chow, *et al.*, 1974). Accurate detection of Pb requires special "clean laboratory" techniques and stable isotope dilution procedures using thermal ionization source high resolution mass spectrometers or similar instrumentation. Quartz-distilled acids are used to reduce Pb contamination since Pb contaminates most acids used for the wet ashing procedures. The required sophistication of instrumentation has led to a dearth of published Pb accumulation levels.

Another complication is the fact that orchard leaf and bovine liver National Bureau of Standard (NBS) reference materials are said to be contaminated with Pb (Chow, *et al.*, 1974; Settle and Patterson, 1980). Commercial drying and pulverizing processing procedures increase Pb levels four-hundred-fold that of natural materials. Other low-Pb tissues (from uncontaminated areas) are suggested for use—these include tuna, bass, or longear sunfish skeletal muscle (Settle and Patterson, 1980; Schmitt and Finger, 1987). Therefore, published results involving Pb accumulation and distribution in teleosts may contain inaccurate, absolute values for Pb levels. None the less, within a given experiment (e.g. bioavailability studies) relative comparisons yield valuable information.

Biological Effects

Lead poisoning in fish results in hematological, neuronal, muscular, and other effects akin to those observed in mammals. However, a few exceptions are characteristic of fish. These include black tails, lordoscoliosis, pigment pattern alterations, and coagulation of surface mucus.

Hematological Alterations: Changes in the blood of fish exposed to Pb are published (Dawson, 1935; Jackim, 1973; Johansson-Sjobeck and Larsson, 1979; Schmitt, *et al.*, 1984; Tewari, *et al.*, 1987). Direct erythrocyte injury is considered the first and most important sign in Pb poisoning of catfish (Dawson, 1935). For example, catfish (*Ictalurus nebulosus*) exposed to 27 ppm Pb for up to 183 d have large numbers of small, highly refractile, injured erythrocytes which are fragile and easily distorted. The destruction of large number of erythrocytes produces yellow-brown storage deposits (pigments) in the liver, spleen, and mesonephros (Dawson, 1935). Basophilic stippling of erythrocytes is observed in rainbow trout (*Salmo gairdneri*) exposed 30 d to 0.3 ppm Pb in the water (Johansson-Sjobeck and Larsson, 1979). Dawson (1935) finds that catfish exposed 183 d to 27 ppm Pb have increased numbers of atypical thrombocytes, monocytes, and eosinophils; however, Johansson-Sjobeck and Larsson (1979) find no change in leukocytes of rainbow trout exposed 79 d to 0.300 ppm aqueous Pb, despite Pb-induced anemia and basophilic stippling of erythrocytes. Shorter exposure times and lower Pb concentrations in the Johansson-Sjobeck and Larsson study undoubtedly explain this difference. Therefore, erythrocytes of Pb-exposed fish are destroyed in great numbers, while thrombocytes (and sometimes leukocytes) increase in number.

Evaluation of the heme synthesis of Pb-exposed fish provides results similar to those found in other vertebrates. In heme synthesis, the erythrocyte enzyme delta-aminolevulinic acid dehydratase (ALA-D,E.C.4.2.1.24) catalyzes formation of porphobilinogen (PGB) from aminolevulinic acid (ALA). Fish ALA-D is sensitive to the action of Pb, as is the case in mammals (Hernberg, *et al.*, 1970; Hernberg, *et al.*, 1972). Essential sulfhydryl groups in ALA-D are inactivated by Pb and, as a consequence, less hemoglobin is formed (Johansson-Sjobeck and Larsson, 1979; Tewari, *et al.*, 1987). Activity of ALA-D is inversely correlated with Pb levels in the liver and kidney of the mummichog (*Fundulus heteroclitus*) and winter flounder (*Pseudopleuronectes americanus*) (Jackim, 1973); the erythrocytes, spleen, and renal tissue of the rainbow trout (*Salmo gairdneri*) (Johansson-Sjobeck and Larsson, 1979); and in the erythrocytes of thirty-five species of catostomids (Schmitt, *et al.*, 1984). Activity of ALA-D is negatively correlated with blood Pb concentration ($r=-0.66$) in longear sunfish (*Lepomis megalotis*) from a Missouri stream contaminated with Pb-rich mine tailings (Dwyer, *et al.*, 1988). Moreover, the absence of ALA-D inhibition after exposure of fish to Cu, Cd, Zn, and Hg makes ALA-D measurement a valuable diagnostic indicator of recent Pb poisoning in fish (Jackim, 1973).

As indicated, Pb inactivates essential ALA-D sulfhydryl groups, resulting in reduced hemoglobin formation (Johansson-Sjobeck and Larsson, 1979; Tewari, *et al.*, 1987). Therefore, strong ALA-D inhibition is thought to be the contributing factor inducing anemia in Pb-poisoned fish (Johansson-Sjobeck and Larsson, 1979). Both rainbow trout (exposed 30 d to aqueous 0.3

ppm Pb) and *Barbus conchonius* (exposed 30–60 d to 0.05 ppm aqueous Pb) show anemia (Tewari, *et al.*, 1987).

Physiologically, fish could compensate for Pb-induced disturbance of erythrocyte formation by releasing stored erythrocytes or by stimulation of erythropoiesis. The teleost spleen (a storage organ for mature erythrocytes) apparently releases a large number of erythrocytes upon Pb-challenge. Release is evidenced by a sudden decrease in the spleen somatic index (SSI, the spleen weight to body weight ratio). The SSI is decreased in rainbow trout (*Salmo gairdneri*) exposed 30 d to 0.075 ppm Pb, but not in fish exposed 30 d to 0.3 ppm (Johansson-Sjobeck and Larsson, 1979). A second compensatory mechanism appears to be the observed stimulation of erythropoiesis in various hematopoietic tissues of catfish (*Ictalurus nebulosus*) (Dawson, 1935).

Hodson *et al.* (1977) find that rainbow trout recover normal erythrocyte ALA-D activity following depuration; whereas, Johansson-Sjobeck and Larsson (1979) do not observe this recovery. The discrepancy may be due to the difference in exposure of fish to Pb. Johansson-Sjobeck and Larsson expose trout for 30 d at 0.010, 0.075, and 0.300 ppm, whereas Hodson *et al.* expose fish for only 7 days at 0.120 ppm. Recovery of ALA-D activity undoubtedly is more expeditious if severity of Pb exposure is lessened. Thus, despite slight discrepancies in results from different laboratories, measurement of ALA-D as an indicator of recent Pb poisoning has gained a solid reputation for sensitivity, speed, and accuracy—surpassing that even of Pb analyses of water and/or fish tissues.

Tewari, Gill, and Pant (1987) evaluate hematological parameters, blood glucose, tissue glycogen, and cholesterol analyses in *Barbus conchonius* (a freshwater fish from India) under chronic Pb exposure. After a 60-d exposure to 47 ppb (equivalent to 20% of the 96-h LC50), adults of this species show severe microcytic anemia, 24% lower erythrocyte counts, a 12% decrease in hemoglobin, a 13% reduction in hematocrit, and a 10% reduction in mean corpuscular volume (see Chapter II). The 60-d hyperglycemia (42% increase in blood glucose) and the depletion of glycogen reserves are undoubtedly secondary stress responses mediated through hormones of the pituitary-adrenal axis (Mazeaud, *et al.*, 1977). Energy reserves are mobilized from the liver and muscle to increase blood glucose 54 and 42% after 30- and 60-d exposures, respectively. Also, significant reductions are noted in circulating and tissue levels of cholesterol; however, the significance of this change is unknown. Disruption of plasma membranes and/or alteration of steroidogenesis are possibilities (Tewari, *et al.*, 1987).

Respiratory Changes: Respiratory distress occurs in fish living in rivers receiving Pb mining wastes in England (Carpenter, 1924; 1925; 1926). This distress is characterized by a coughing reflex due to formation of a coagulation mucus over the gills and is reported for Zn-exposed bluegills (*Lepomis macrochirus*) (Sparks, *et al.*, 1972). Fish are thought to be asphyxiated as a result of the mucous coating over the gills. Production of excess mucus also may occur when fish are exposed to Zn, Fe, Cu, Cd, Hg, Mn, Co, Ni, Au, Ag, and Al (National Academy of Sciences, 1972). Death by asphyxiation may be the result of formation of a coagulated mucous film and/or metal-induced damage to the gill epithelium.

Mucin is secreted by mucus cells of the epidermis and contains a glycoprotein which reacts with water to form mucous slime coating the lamellar and body surface. Strong heavy metal complexing sites in these proteins withdraw metals from water. For example, Cu and Zn are bound in a more concentrated form in the mucus of plaice (*Pleuronectes platessa*) than the concentration of the original exposure solutions (Coombs, *et al.*, 1972). Mucous fluidity is increased following Pb binding (Varanasi, *et al.*, 1975), resulting in a more rigid film, which may be impermeable to gases (Van Oosten, 1957; Somero, *et al.*, 1977). The fluidity change also is thought to alter the hydrodynamic pattern of surface water, as well as osmoregulatory and protective functions of mucus. Polar sites of constituent glycoproteins are thought to interact long after metals are removed from the surrounding waters (Varanasi, *et al.*, 1975).

Factors causing production of excess mucous remain to be elucidated; however, clearly this effect does not occur in fish exposed to Pb in hard water (i.e. Ca-rich water). Jones (1938) finds that mucus formation is prevented in the stickleback (*Gasterosteus aculeatus*) by the presence of 50 ppm Ca in solutions of Pb (or Zn). Moreover, mucus collected from eels (*Anguilla anguilla*) remains clear in the presence of Ca and Pb (or Zn) but forms a cloudy solution with mucus strands if Ca is omitted. Therefore, the presence of excess aqueous Ca prevents the Pb-induced mucous precipitation.

Neuronal and Muscular Abnormalities: Lead-induced muscle spasms, paralysis, hyperactivity, and loss of equilibrium in Pb-exposed fish (Davies, *et al.*, 1976; Holcombe, *et al.*, 1976) are similar to the symptoms of encephalopathy in human cases of Pb poisoning (de la Burde and Choate, 1972; Brown, 1975). Muscles of the caudal region of trout are affected by Pb—causing dorsal-ventral spinal flexures (lordosis) and/or bi-lateral spinal flexures (scoliosis) (Davies, *et al.*, 1976; Holcombe, *et al.*, 1976). The condition generally occurs together and is termed lordoscoliosis. Lordoscoliosis could be a logical consequence of elevated Pb levels in the brain. Bream (*Abramis brama*) and roach (*Rutilus rutilus*) accumulate 49 and 37 ppm (dry weight) in the brain, respectively (Badsha and Goldspink, 1982)—unusually elevated levels. However, elevated brain levels of Pb are not noted in other studies (Somero, *et al.*, 1977). None the less, low Pb levels in muscle tissues of bream and roach agree with other published studies and make results of the Badsha and Goldspink (1982) study believable. Measurements of brain Pb levels might provide interesting results for other species; however, these studies should be conducted with the understanding that even trace levels of metallic poisons (e.g. Hg) can cause marked behavioral and neuronal changes (see Chapter VIII).

Behavioral Changes: Aqueous Pb (0.07 ppm) impairs goldfish avoidance behavior, reducing the ability of conditioned goldfish to avoid a 2 s, 14 V electrical shock (Weir and Hine, 1970). Minnows, however, are less sensitive to Pb since they are able to detect and avoid 2, 40, and 400 ppm Pb (Sprague, *et al.*, 1965). Small trout "test" water and avoid solutions containing 10 ppm Pb. Both Jones (1964a) and Weir and Hine (1970) find that Pb is more readily detected and avoided than Hg, despite the greater toxicity of Hg. Therefore, detection and avoidance abilities of teleosts apparently are not correlated with toxicity of Pb.

Several studies show that fish avoid metal-contaminated waters. In New Brunswick, Canada, downstream from active Zn-Cu mines, even sublethal levels of metals are avoided by adult Atlantic salmon (*Salmo salar*) (Sprague, *et al.*, 1965). Studies involving salmon counting fences show that 10–22% of salmon migrating upstream retreat from the metal-polluted water during 4 yr of pollution, compared with 1–3% before pollution. A tagging study determines that 62% of retreating salmon do not reenter a polluted area during the same season. Villarreal-Trevino *et al.*. (1986) find that fish avoid an area in the Santa Catarina River 200 km from Monterrey City in northeastern Mexico. At that site, contamination from domestic and industrial wastes (i.e. metallurgical, petrochemical, refinery, and car battery) result in elevated Pb levels to 23.7 ppm (wet weight) in whole bodies of Mexican tetras (*Astyanax mexicanus*) and 14 ppm in Amazon mollies (*Poecilia formosa*), which are found just outside the 200 km portion of the Santa Catarina River.

Additional Effects: Other effects of Pb poisoning in fish include growth inhibition (Dilling, *et al.*, 1926; Crandall and Goodnight, 1963); degenerated renal tubules (Crandall and Goodnight, 1963); inhibition of glutamic oxalacetic transaminase (GOT) and lactate dehydrogenase (LDH) (Christensen, 1971/72); inhibition of alkaline phosphatase, acid phosphatase, xanthine oxidase, RNAase, and catalase (Jackim, *et al.*, 1970); pigment pattern abnormalities (Ozoh, 1980); and blackened caudal peduncles (Davies, *et al.*, 1976; Holcombe, *et al.*, 1976). The latter effect is unusual. The presence of blackened caudal peduncles precedes scoliosis in trout given 15–27 wk exposures to $\geq$ 0.119 ppm Pb. Under anesthesia, brook trout with recently-developed scoliosis exhibit muscle relaxation and return to normal (Holcombe, *et al.*, 1976). However, fish with scoliosis for several weeks show vertebral calcification and do not return to the undeformed condition under anesthesia. In conclusion, toxic effects of Pb on fish range from mortality to subtle effects on reproduction, growth, and behavior. Additional details regarding Pb-induced effects on fish are reviewed by Katz (1979).

Summary

Despite overwhelming available evidence of the hazard of Pb to fish, some workers believe that there is "no evidence that Pb precipitate in natural waterways is harmful to fish" and that "soluble Pb precipitate is not present in natural waters of the United States in concentrations likely to be toxic to fish" (National Academy of Sciences, 1972). Although Pb has been shown to be less available for uptake from sediments (Wiener and Giesy, 1979; Johnson, 1987), scientists studying Pb-contaminated teleosts (and other components of aquatic ecosystems) warn of Pb toxicity and complicated interactions. Biomethylation of Pb contributes to the existent problem, as is the case for several other priority pollutants (elemental) (Wood, 1974). Thus, the presence of Pb in aqueous ecosystems poses a hazard to fish—the full impact of which probably will never be completely unraveled.

Inhabitants of planet Earth remove and distribute several millions of tons of Pb annually. Logically, because of such extensive anthropogenic redistribution, no site on Earth is so secluded as to remain free of Pb pollution. Past and present handling of industrial, agricultural, and domestic wastes remains irresponsible for the most part. The release of Pb is shrouded in ignorance, with midnight dumping, oceanic releases, and release of poisons into the main basin of power plant reservoirs. Ignorant, careless attitudes could leave those who follow only photographic epitaphs of a thriving planet of many species. While some species could become tolerant to metal poisons (Hamilton and Reash, 1988), many more are decreasing in number (Cumbie and Van Horn, 1978; Goodyear, 1985). Only conscientious handling of metal-contaminated wastes will prolong the existence of teleosts and other living forms. Just as human avarice sometimes pollutes beyond recovery, so can human responsibility salvage an increasingly polluted planet.

References

Aronson, A.L., Biologic effects of lead in fish, J. Wash. Acad. Sci., 61 (1971) 124.

Badsha, K.S., and C.R. Goldspink, Preliminary observations on the heavy metal content of four species of freshwater fish in NW England, J. Fish. Biol., 21 (1982) 251.

Bohn, A., and B.W. Fallis, Metal concentrations (As, Cd, Cu, Pb and Zn) in shorthorn sculpins, *Myoxocephalus scorpius* (Linnaeus), and Arctic char, *Salvelinus alpinus* (Linnaeus), from the vicinity of Strathcona Sound, Northwest Territories, Water Res., 13 (1978) 659.

Brown, D.R., Neonatal lead exposure in the rat: Decreased learning as a function of age and blood lead concentrations, Tox. Appl. Pharm., 32 (1975) 628.

Bussey, R.E., D.E. Kidd, and L.D. Potter, The concentrations of ten heavy metals in some selected Lake Powell game fishes, University of New Mexico, Albuquerque, Lake Powell Research Project Bulletin, (1976).

Cappon, C.J., Cadmium and lead in Lake Ontario salmonids, Bull. Environ. Contam. Toxicol., 38 (1987) 695.

Carpenter, K.E., A study of the fauna of rivers polluted by lead mining in the Aberystwyth district of Cardiganshire, Ann. Appl. Biol., 11 (1924) 1.

Carpenter, K.E., On the biological factors involved in the destruction of river-fisheries by pollution due to lead-mining, Ann. Appl. Biol., 12 (1925) 1.

Carpenter, K.E., The lead mine as an active agent in river pollution, Ann. Appl. Biol., 13 (1926) 395.

Carpenter, K.E., The lethal action of soluble metallic salts on fish, Br. J. Exptl. Biol., 4 (1927) 378.

Carpenter, K.E., Further researches on the action of metallic salts on fishes, J. Exp. Zool., 56 (1930) 407.

Chow, T.J., C.C. Patterson, and D. Settle, Occurrence of lead in tuna, Nature, 251 (1974) 159.

Christensen, G.M., Effects of metal cations and other chemicals upon the in vitro activity of two enzymes in the blood plasma of the white sucker, *Catostomus commersoni* (*Lacepede*), Chem.-Biol. Interactions, 4 (1971/72) 351.

Coombs, T.L., T.C. Fletcher, and A. White, Interaction of metals ions with mucus from the plaice (*Pleuronectes platessa* L.), Proc. Biochem. Soc., 128 (1972) 128.

Crandall, C.A., and C.J. Goodnight, The effects of sublethal concentrations of several toxicants to the common guppy, *Lebistes reticulatus*, Trans. Amer. Microsc. Soc., 82 (1963) 59.

Cumbie, P.M., and S.L. Van Horn, Selenium accumulation associated with fish mortality and reproductive failure, Proc. Ann. Conf. SE Assoc. Fish Wildl. Agencies, (1978) 612.

Davies, P.H., J.P. Goettl Jr., J.R. Sinley, and N.F. Smith, Acute and chronic toxicity of lead to rainbow trout *Salmo gairdneri*, in hard and soft water, Water Res., 10 (1976) 199.

Dawson, A.B., The hemopoietic response in the catfish, *Ameiurus nebulosus*, to chronic lead poisoning, Biol. Bull., 68 (1935) 335.

de la Burde, B., and M.S. Choate Jr., Does asymptomatic lead exposure in children have latent sequelae?, J. Pediatr., 81 (1972) 1088.

Dilling, W.J., C.W. Healey, and W.C. Smith, Experiments on the effects of lead on the growth of plaice (*Pleuronectes platessa*), Ann. Appl. Biol., 13 (1926) 168.

Douben, P.E.T., and J.H. Koeman, Effect of sediment on cadmium and lead in the stone loach (*Noemacheilus barbatulus* L.), Aq. Toxicol., 15 (1989) 253.

Doudoroff, P., and M. Katz, Critical review of literature on the toxicity of industrial wastes and their components to fish--II: The metals, as salts, Sew. Ind. Wastes, 25 (1953) 802.

Dwyer, F.J., C.J. Schmitt, S.E. Finger, and P.M. Mehrle, Biochemical changes in longear sunfish, *Lepomis megalotis*, associated with lead, cadmium and zinc from mine tailings, J. Fish. Biol., 33 (1988) 307.

El Nabawi, A., B. Heinzow, and H. Kruse, As, Cd, Cu, Pb, Hg, and Zn in fish from the Alexandria region, Egypt, Bull. Environ. Contam. Toxicol., 39 (1987) 889.

Gale, N.L., B.G. Wixson, and M.W. McMenus, Lead concentrations in edible fish filets collected from Missouri's Old Lead Belt, 16th Annual Conference on Trace Substances in Environmental Health, (1982) 12.

Goodyear, C.P., Toxic materials, fishing, and environmental variation: Simulated effects on striped bass population trends, Trans. Amer. Fish. Soc., 114 (1985) 107.

Greichus, Y.A., A. Greichus, B.D. Ammann, and J. Hopcraft, Insecticides, polychlorinated biphenyls, and metals in African lake ecosystems. III. Lake Nakuru, Kenya., Bull. Environ. Contam. Toxicol., 19 (1978) 454.

Hamilton, S.J., and R.J. Reash, Bone development in creek chub from a stream chronically polluted with heavy metals, Trans. Amer. Fish. Soc., 117 (1988) 48.

Harwood, J.J., and S.R. Koirtyohann, Effects of cover materials on leaching of constituents from dolomitic lead mine tailings, Water, Air, and Soil Pollution, 34 (1987) 31.

Hernberg, S., J. Nikkanen, G. Mellin, and H. Lilius, ∂-aminolevulinic acid dehydrase as a measure of lead exposure, Arch. Environ. Health, 21 (1970) 140.

Hernberg, S., S. Tola, J. Nikkanen, and S. Valkonen, Erythrocyte ∂-aminolevulinic acid dehydratase in new lead exposure, Arch. Environ. Health, 25 (1972) 109.

Hirao, Y., and C.C. Patterson, Lead aerosol pollution in the High Sierra overrides natural mechanisms which exclude lead from a food chain, Science, 184 (1974) 989.

Hodson, P.V., B.R. Blunt, D.J. Spry, and K. Austen, Evaluation of erythrocyte delta-amino levulinic acid dehydratase activity as a short-term indicator in fish of a harmful exposure to lead, J. Fish. Res. Bd. Can., 34 (1977) 501.

Holcombe, G.W., D.A. Benoit, E.N. Leonard, and J.M. McKim, Long-term effects of lead exposure on three generations of brook trout (*Salvelinus fontinalis*), J. Fish. Res. Bd. Can., 33 (1976) 1731.

Horowitz, A., and B.J. Presley, Trace metal concentrations and partitioning in zooplankton, neuston, and benthos from the south Texas Outer Continental Shelf, Arch. Environ. Cont., 5 (1977) 241.

Jackim, E., J.M. Hamlin, and S. Sonis, Effects of metal poisoning on five liver enzymes in the killifish (*Fundulus heteroclitus*), J. Fish. Res. Bd. Can., 27 (1970) 383.

Jackim, E., Influence of lead and other metals on fish ∂-aminolevulinate dehydrase activity, J. Fish. Board Can., 30 (1973) 560.

Johansson-Sjobeck, M., and A. Larsson, Effects of inorganic lead on delta-aminolevulinic acid dehydratase activity and hematological variables in the rainbow trout, *Salmo gairdneri*, Arch. Environ. Contam. Toxicol., 8 (1979) 419.

Johnson, M.G., Trace element loadings to sediments of fourteen Ontario lakes and correlations with concentrations in fish, Can. J. Fish. Aquat. Sci., 44 (1987) 3.

Jones, J.R.E., The relative toxicity of salts of lead, zinc and copper to the stickleback (*Gasterosteus aculeatus* L.) and the effect of calcium on the toxicity of lead and zinc salts, J. Exp. Biol., 15 (1938) 394.

Jones, J.R.E., A study of the zinc-polluted river Ystwyth in North Cardiganshire, Wales, Ann. Appl. Biol., 27 (1940) 368.

Jones, J.R.E., The metals as salts, Fish and River Pollution, Butterworth & Company, London, (1964a) 66.

Jones, J.R.E., Lead, zinc, and copper: The "coagulation film anoxia" theory, Fish and River Pollution, Butterworth and Co., London, (1964b) 53.

Katz, M., The effects of heavy metals on fish and aquatic organisms, in Forstner, V., and G.T. Wittman (Eds.), Metal Pollution in Aquatic Environments, Springer-Verlag, Berlin, (1979) 25.

Kleinert, S.J., P.E. Degurse, and J. Ruhland, Concentration of metals in fish, in Surveys of Toxic Metals in Wisconsin, Department of Natural Resources, Madison, Wisconsin, Tech. Bull. 74, 8., (1974).

Mathis, B.J., and N.R. Kevern, Distribution of mercury, cadmium, lead, and thallium in a eutrophic lake, Hydrobiologia, 46 (1975) 207.

Mazeaud, M.M., F. Mazeaud, and E.M. Donaldson, Primary and secondary effects of stress in fish: Some new data with a general review, Trans. Amer. Fish Soc., 106 (1977) 201.

Merlini, M., and G. Pozzi, Lead and freshwater fishes: Part I—Lead accumulation and water pH, Envir. Poll., 12 (1977) 167.

Moore, J.W., and S. Ramamoorthy, Lead, Heavy Metals in Natural Waters: Applied Monitoring and Impact Assessment, Springer-Verlag, Berlin, (1984) 100.

National Academy of Sciences, Lead: Airborne Lead in Perspective, National Academy of Sciences, Washington, D.C., (1972) 188.

Ney, J.J., and J.H. Van Hassel, Sources of variability in accumulation of heavy metals by fishes in a roadside stream, Arch. Environ. Contam. Toxicol., 12 (1983) 701.

Ozoh, P.T.E., Effect of lead on pigment pattern formation in zebrafish (*Brachydanio rerio*), Bull. Environ. Contam. Toxicol., 24 (1980) 276.

Pagenkopf, G.K., and D.R. Neuman, Lead concentrations in native trout, Bull. Environ. Contam. Toxicol., 12 (1974) 70.

Reichert, W.L., D.A. Fiderighi, and D.C. Malins, Uptake and metabolism of lead and cadmium in coho salmon (*Oncorhynchus kisutch*), Comp. Biochem. Physiol., 63C (1979) 229.

Rolfe, G.L., and J.C. Jennett, Environmental lead distribution in relation to automobile and mine and smelter sources, in Forstner, V., and G.T. Wittman (Eds.), Metal Pollution in Aquatic Environments, Springer-Verlag, Berlin, (1979) 231.

Sandstead, H.H., Interactions of cadmium and lead with essential minerals, in Nordberg, G.F. (Ed.), Effects and Dose-Response Relationships of Toxic Metals, Elsevier, Amsterdam, (1976) 511.

Schmitt, C.J., F.J. Dwyer, and S.E. Finger, Bioavailability of Pb and Zn from mine tailings as indicated by erythrocyte ∂-aminolevulinic acid dehydratase (ALA-D) activity in suckers (*Pisces: Catostomidae*), Can. J. Fish. Aquat. Sci., 41 (1984) 1030.

Schmitt, C.J., and S.E. Finger, The effects of sample preparation on measured concentrations of eight elements in edible tissues of fish from streams contaminated by lead mining, Arch. Environ. Contam. Toxicol., 16 (1987) 185.

Settle, D.M., and C.C. Patterson, Lead in albacore: Guide to lead pollution in Americans, Science, 207 (1980) 1167.

Somero, G.N., T.J. Chow, P.H. Yancey, and C.B. Snyder, Lead accumulation rates in tissues of the estuarine teleost fish, *Gillichthys mirabilis*: Salinity and temperature effects, Arch. Environ. Contam. Toxicol., 6 (1977) 337.

Sparks, R.E., J. Cairns Jr., R.A. McNabb, and G. Suter II, Monitoring zinc concentrations in water using the respiratory response of bluegills (*Lepomis macrochirus* Rafinesque), Hydrobiologia, 40 (1972) 361.

Sprague, J.B., P.F. Elson, and R.L. Saunders, Sublethal copper-zinc pollution in a salmon river--a field and laboratory study, Int. J. Air Wat. Poll., 9 (1965) 531.

Taylor, D.D., and T.J. Bright, The distribution of heavy metals in reef-dwelling groupers in the Gulf of Mexico and Bahama Islands, Texas A & M University, Sea Grant College, TAMU-SG-73-208, (1973).

Tewari, H., T.S. Gill, and J. Pant, Impact of chronic lead poisoning on the hematological and biochemical profiles of a fish, *Barbus conchonius* (Ham), Bull. Environ. Contam. Toxicol., 38 (1987) 748.

Uthe, J.F., and E.G. Bligh, Preliminary survey of heavy metal contamination of Canadian freshwater fish, J. Fish. Res. Bd. Can., 28 (1971) 786.

Van Hassel, J.H., J.J. Ney, and D.L. Garling, Seasonal variations in the heavy metal concentrations of sediments influenced by highways of different traffic volumes, Bull. Environ. Contam. Toxicol., 23 (1979) 592.

Van Hassel, J.H., J.J. Ney, and D.L. Garling Jr., Heavy metals in a stream ecosystem at sites near highways, Trans. Amer. Fish. Soc., 109 (1980) 636.

Van Oosten, J., The skin and scales, in Brown, M.E. (Ed.), The Physiology of Fishes, Vol. 1, Academic Press, New York, (1957) 207.

Varanasi, U., P.A. Robisch, and D.C. Malins, Structural alterations in fish epidermal mucus produced by water-borne lead and mercury, Nature, 258 (1975) 431.

Varanasi, U., and D.J. Gmur, Influence of water-borne and dietary calcium on uptake and retention of lead by coho salmon (*Oncorhynchus kisutch*), Tox. Appl. Pharm., 46 (1978) 65.

Vighi, M., Lead uptake and release in an experimental trophic chain, Ecotoxicol. Environ. Safety, 5 (1981) 177.

Villarreal-Trevino, C.M., M.E. Obregon-Morales, J.F. Lozano-Morales, and A. Villegas-Navarro, Bioaccumulation of lead, copper, iron, and zinc by fish in a transect of the Santa Catarina River in Cadereyta Jimenez, Nuevo Leon, Mexico,, Bull. Environ. Contam. Toxicol., 37 (1986) 395.

Vinikour, W.S., R.M. Goldstein, and R.V. Anderson, Bioconcentration patterns of zinc, copper, cadmium, and lead in selected fish species from the Fox River, Illinois, Bull. Environ. Contam. Toxicol., 24 (1980) 727.

Weir, P.A., and C.H. Hine, Effects of various metals on behavior of conditioned goldfish, Arch. Environ. Health, 20 (1970) 45.

Wiener, J.G., and J.P. Giesy Jr., Concentrations of Cd, Cu, Mn, Pb, and Zn in fishes in a highly organic soft water pond, J. Fish. Res. Board Can., 36 (1979) 270.

Wood, J.M., Biological cycles for toxic elements in the environment, Science, 183 (1974) 1049.

World Health Organization, Evaluation of mercury, lead, cadmium, and the food additives amaranth, diethylpyrocarbonate, and octyl gallate, World Health Organization, 16th Rep. Joint FAO/WHO Expert Committee on Food Additives, WHO Techn. Rep. Serv. No. 505, (1972).

Zamojski, J., S. Smoczynski, K.A. Skibniewska, and R. Amarowicz, Organochlorine insecticides and heavy metals in fish from Mutek Lake, N.E. Poland, Bull. Environ. Contam. Toxicol., 37 (1986) 587.

Zook, E.G., J.J. Powell, B.M. Hackley, J.A. Emerson, J.R. Brooker, and G.M. Knobl Jr., National Marine Fisheries Service preliminary survey of selected seafoods for mercury, lead, cadmium, chromium, and arsenic content, J. Agric. Food Chem., 24 (1976) 47.

CHAPTER V
ZINC

Introduction

Zinc is found in large quantities in the vertebrate body, second only to Fe. As an element of Group IIB of the periodic table, it forms stable complexes with S, P, and C. Because Zn is an essential element, many functions are recognized (Prosser and Brown, 1966; Williams, 1969). For example, it is an essential part of—or a cofactor for—a number of enzymes, including carbonic anhydrase, carboxypeptidase, superoxide dismutase, lactate dehydrogenase, phosphatase, and glutamate dehydrogenase.

As a component of carbonic anhydrase, Zn plays a role in binding carbon dioxide in teleost cells, combining it with water to form carbonic acid, and releasing carbon dioxide from capillaries at secondary lamellae of the gills. Carbonic anhydrase is also important in maintaining acid-base balance in renal tubule cells. As a cofactor of the protein-splitting enzyme carboxypeptidase, Zn has a key role in protein digestion. As a part of lactate dehydrogenase, Zn plays a role in carbohydrate digestion by interconversion of pyruvate and lactate in the glycolytic pathway for glucose oxidation.

The importance of Zn might be deduced from reductions in the level of Zn in human disease states. For example, the diabetic human pancreas contains about half the normal amount of Zn. It is this Zn which combines with insulin forming a Zn-insulin complex—thought to be the storage form of insulin. Also, in patients with leukemia, about 10% less Zn is found than normal. Although the exact function of Zn in white blood cells in not known, a large quantity of Zn is bound to protein in leukocytes. In human cirrhosis, serum Zn levels are low, urinary excretion is increased, and postmortem studies reveal reduced Zn levels in the liver. Cirrhosis may increase Zn requirements—increasing Zn uptake but nonetheless causing a deficiency state. Therefore, Zn concentration changes in the pancreas, leukocytes, and liver suggest that Zn is important for normal function of these tissues.

In view of the essential nature of Zn and the high concentration of Zn in vertebrate tissues (relative to most other elements), one might speculate that Zn serves only useful functions in fish. This is not the case. Zinc is known to be toxic to fish, causing mortality, growth retardation, tissue alterations, respiratory and cardiac changes, inhibition of spawning, and a multitude of additional detrimental effects which threaten survival of fish species.

Environmental Exposures

Zinc uptake in fish is used as a measure of the extent of contamination of surface waters. However, for essential elements such as Zn and Se, toxicologists debate whether the ecomix concept (Odum, 1960) or homeostatic regulation (Goodyear and Boyd, 1972; Murphy, *et al.*, 1978) is operational. In homeostatic regulation, certain levels and ratios of essential elements are maintained because of physiological requirements for survival. Accordingly, the vertebrate body would need a balanced supply of the metal, despite irregularities in the concentration of Zn or Se in the external medium. However, the ecomix concept argues that nutrient accumulation occurs in the same ratio that these elements occur in the environment. Data exists from field studies to lend credibility to both theories.

Roch and colleagues (1982) and Goodyear and Boyd (1972) obtain results which support the homeostatic regulation concept. In the Roch *et al* (1982) study, rainbow trout (*Salmo gairdneri*) are collected from successively more polluted areas of the Campbell River System on Vancouver Island. Aqueous levels of metals have increased statistically from 1971 to 1981 up to levels of 0.25 ppm Zn. Despite progressively increasing levels of metals (particularly Cu and Zn) in the water, hepatic levels of Zn are not statistically different (123–168 ppb). In contrast, this study shows that hepatic levels of Cu and Cd are increased in more polluted areas. Moreover, Cu and Cd (but not Zn) are bound to metallothionein, despite the much higher aqueous levels of Zn than either Cd or Cu. Interestingly, hepatic metallothionein levels increase from 58 to 269 nmol/g from less polluted to more polluted areas and provide an indication of pollution of each lake in the Campbell River System.

In the Goodyear and Boyd (1972) study, little variation is found in mean whole body concentrations of eleven essential elements (including Zn) in largemouth bass (*Micropterus salmoides*) collected from five locations. In contrast, water concentrations of elements vary markedly. Authors attribute low variability in elemental composition to homeostatic control of body concentrations of these elements. In this case, internal concentrations of the elements would remain constant despite markedly different concentrations of elements in the external environment.

In contrast, homeostatic regulation does not appear to exist in largemouth bass of Palestine Lake (Murphy, *et al.*, 1978). Whole body Zn levels of eleven fresh-water fish species are compared for two parts of Palestine Lake—the West Basin than the East Basin. An electroplating factory discharges metal contaminated wastes into the West Basin, raising aqueous Zn levels from 52 ppm to 293 ppm. Sediments in the West Basin contain an average of 12,800 ppm (dry weight) compared to 320 ppm in the East Basin. Seven of eleven species of fish from the West Basin have higher average whole body Zn concentrations, with maximal values averaging 477 ppm (dry weight) for redears (*Lepomis microlophus*). Redears are in almost continuous contact with contaminated sediments and consume sediment along with benthic organisms. In fact, the gastrointestinal tract of redears is known to contain large quantities of sediment, placing the species more at risk to metal poisoning (Sorensen and Bauer, 1983a; Sorensen, *et al.*, 1983b; ; Sorensen and Bauer, 1984a; 1984b).

Field studies illustrate the complexity of the interactions which can occur in fish which have been exposed to excess levels of Zn. Pippy and Hare (1969) report fish kills involving Atlantic salmon grilse (*Salmo salar*), suckers (*Catostomus commersoni*), and shiners (*Notropis cornutus*) for June 1967 through August 1968 in the Miramiche River in New Brunswick in Canada. A base metal mine area is the source of pollution. Freshnets (floods) apparently are associated with elevated discharge rates of Zn and Cu, causing deaths of at least 153 shiners in June 1967 alone. In contrast, salmon deaths (64) and sucker deaths (308) appear associated with elevated temperature, surges of pollution, and the pseudomonal bacterium *Aeromonas liquefaciens*. This study illustrates interrelationships among temperature, Zn-Cu pollution, and bacterial infections.

From these studies, it is evident that fish under certain conditions can balance internal levels of Zn despite environmental fluctuations (Goodyear and Boyd, 1972; Roch, *et al.*, 1982) or that they can accumulate levels which are altered by the extent of environmental contamination (Murphy, *et al.*, 1978). Others contend that ratios of toxic metals to essential metals bound or unbound to metallothionein are of greatest importance in environmental contamination with Zn and other metals (Roch and McCarter, 1984a; 1984b; Hamilton, *et al.*, 1987).

Toxicity

In vivo Studies:

Temperature: As poikilotherms, fish are influenced more by external temperature changes than are homeotherms. Although poikilotherms have the ability to regulate body temperature to some extent, rapid heat exchange between medium and specimen leaves aquatic cold-blooded organisms more at the mercy of the environment. Therefore, in poikilotherms, body temperature is almost the same as that of the environment because of efficient heat exchange between the two. For example, large fish show little thermal lag even when subjected to rapid temperature changes (1°C/3 min) (Cairns, *et al.*, 1975).

External thermal flux alters body temperature which in turn alters metabolism. Oxygen consumption increases with increasing temperature from 25° to 30°C in both bluegill (*Lepomis macrochirus*) and pumpkinseed sunfish (*L. gibbosus*) (Barrington, 1968). Temperature quotients (Q_{10}) measure metabolic change for every 10°C thermal elevation and are calculated as follows:

$$Q_{10} = \left(\frac{k_1}{k_2}\right)^{\frac{10}{t_1 - t_2}}$$

where k_1 and k_2 represent velocity constants corresponding to temperatures t_1 and t_2 (Prosser and Brown, 1966). A curve depicting the logarithm of reaction velocity *versus* temperature can be used to determine Q_{10} directly from the slope of the line. Normally 10°C thermal elevations produce two-fold higher metabolism (Q_{10}). However, rapid thermal alterations can cause a metabolic overshoot which produces a temporary Q_{10} of greater than two (Cairns, *et al.*, 1975).

Zinc poisoning is more of a hazard at high than low temperatures, as expected, because of the increase in metabolism at the higher temperature. However, use of survival times, rather than LC50 values, provides a more sensitive measure of thermal effects on Zn poisoning in fish. In fact, temperature effects show the insensitivity of 96-h LC50 measurements. Two studies are of special interest (Cairns and Scheier, 1957; Cairns, *et al.*, 1971a).

Cairns and Scheier (1957) expose bluegills (*Lepomis macrochirus*) to Zn in either hard water or soft water at 18°C or 30°C, conditions which normally are not harmful. Two-week acclimation to temperature or water hardness is allowed, with 2-mo acclimation to both hard water and high temperature. The 96-h LC50 values for bluegills at 18°C in soft water are 2.86–3.78 ppm Zn *versus* 1.93–3.63 ppm Zn at 30°C. In comparison, 96-h LC50 values in hard water are 10.13–12.50 ppm Zn at 18°C and 10.15–12.30 ppm Zn at 30°C. Therefore, in this study the 12°C thermal difference does not seem to have an effect. Water hardness impacts bluegill tolerance to Zn more so than does a temperature difference (in this case 18°C *versus* 30°C) using LC50 values as an indicator of Zn toxicity.

Cairns and associates (1971a) expose bluegills (*Lepomis macrochirus*) to soluble and insoluble forms of Zn (Zn sulfate and Zn phosphate, respectively) at two temperatures (21–24°C or 7–9°C) and two pH ranges (5.7–7.0 or 7.3–8.8). For soluble Zn exposures, cold-acclimated bluegills die at a much slower rate and survival time is extended compared with warm-acclimated fish. At 32 ppm soluble Zn, neither pH 6.0 or pH 7.3 nor acclimation temperature has any effect on percent survival (0%) in 96-h toxicity tests for bluegills. At pH 8 or higher, Zn^{++} dissociation is much less and does not cause mortalities (100% survival). Nor do mortalities occur in bluegills exposed to insoluble Zn phosphate for 96 h (13.5–32.0 ppm) at the same temperatures producing 100% mortalities in fish exposed to comparable levels of soluble Zn sulfate.

Using LC50 values alone, it is logical to assume that the two studies (Cairns and Scheier, 1957; Cairns, *et al.*, 1971a) illustrate the absence of thermal effects on Zn toxicity at the higher temperatures. This is not the case, as is observed from Fig. 1. At 32 ppm Zn exposures, survival times are considerably higher if fish are exposed to 7.5°C rather than 21°C. Extreme toxicity of a combination of both elevated temperature and Zn exposure level makes assessment of thermal effects difficult. However, by using survival times, rather than LC50 values, greater sensitivity is assured in monitoring the effect of temperature on increasing Zn toxicity.

The value of survival times rather than LC50 values is established in other studies. Lloyd (1960) evaluated the effect of temperature on Zn toxicity in rainbow trout (*Salmo gairdneri*). Exposure of rainbow trout to water of 12–320 ppm hardness results in a curvilinear relationship for the higher hardness levels (50 and 320 ppm hardness) and linear relationship for the lowest hardness (12 ppm hardness). Apparent threshold concentrations (at which Zn level-median survival curves become asymptotic to the Y axis) are about 1.75 ppm Zn for water of 50 ppm hardness (as $CaCO_3$) and about 3.5 ppm Zn for water of 320 ppm hardness. One of the most important results is that increases in temperature decrease survival time but that threshold concentrations remain the same. A 10° C increase (from 12 to 22°C) in temperature reduces survival of fish by a factor of 2.35 (Q_{10}). Even though thresholds and LC50 values for Zn may or may not be influenced by changes in temperature, survival time is increased from two to ten times by a decrease in temperature (Lloyd, 1960). Apparently, the higher the thermal range tested, the less the effect of temperature on survival time in acute toxicity studies (Cairns and Scheier, 1957; Cairns, *et al.*, 1971a) .

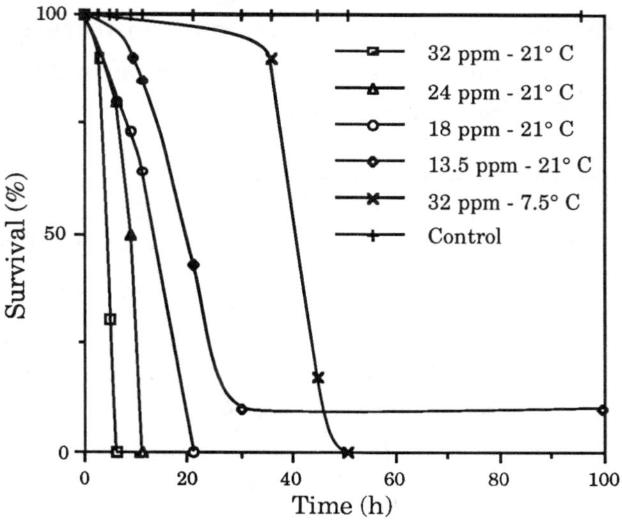

Figure 1. Percent survival for warm- and cold-acclimated bluegill sunfish (*Lepomis macrochirus*) exposed to soluble Zn sulfate at pH 7.3 and 46 ppm average total hardness. From left to right, fish are exposed to 32, 24, 18, and 13.5 ppm Zn at 21°C. The fifth curve represents data from 32 ppm Zn at 7.5°C, and control values are shown as 100% survival at the top. For comparison, no fish succumbed within 96-h if insoluble Zn phosphate is used in comparable toxicity tests. (Modified and redrawn from Cairns *et al.* (1971a).)

Thusfar, discussion has involved exposure of fish in a thermally-stable test solutions. The effects of changing temperature produces striking effects on Zn toxicity. Burton and colleagues (1972a) expose bluegills (*Lepomis macrochirus*) for 96 h to five thermal regimes and five concentrations of Zn (0, 1.0, 5.6, 10.0, and 32.0 ppm). Fish are acclimated and exposed at 20°C or 30°C, or they are subjected to thermal elevation from 20°C to 30°C at rates of 1°C/24 h, 1°C/h, or 1.5°C/10 min. Thermal elevation decreases survival time of bluegills exposed to these levels of Zn. Thermally-acclimated fish exposed to 32 ppm Zn die about three times faster at 30°C than 20°C. Moreover, the LT50 for fish exposed to 32 ppm Zn at 20°C is 12.0 h, but only 2.5 h at a thermal regime of 1.5°C/10 min. Therefore, a 10°C thermal elevation at 1.5°C/10 min reduces survival about five times. A Zn level of 5.6 ppm produces no mortalities at 20°C in 96 h, but is toxic when temperature is increased 10°C at 1.5°C/10 min. Log-probit transformations of 32 ppm Zn curves (but not 10 ppm Zn curves) shows parallelism (Fig. 2), suggesting that only temperature and Zn level interact in this experiment. Generally, the Burton *et al.* (1972a) study indicates that higher temperature and more rapid alterations of temperature are more hazardous to fish exposed to Zn. A thermally-stable environment such as a cold-water spring would provide the most protection of fish from Zn-poisoning.

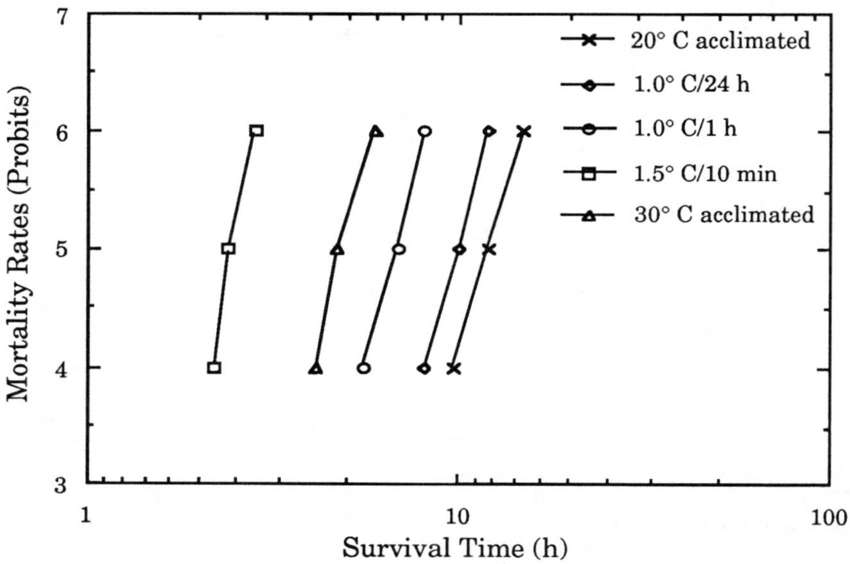

Figure 2. Log-probit transformation of mortality data on bluegills (*Lepomis macrochirus*) exposed to 32 ppm Zn and five thermal regimes. Fish are exposed at 20°C or 30°C, or they are subjected to thermal elevations from 20°C to 30°C at rates of 1°C/24 h, 1°C/h, or 1.5°C/10 min. As indicated, those fish exposed to the most rapid thermal fluctuations succumb first (1.5°C/10 min from 20° C to 30°C), followed by fish exposed to 30°C stable environment, 1°C/1h from 20° C to 30°C, 1°C/24 h from 20° C to 30°C, and 20°C stable environment. Lower temperatures and slower rates of thermal flux prolong survival. (Modified and redrawn from Burton *et al.*(1972a).)

Thermal-Zn interactions in poikilotherms (such as fish) are complex. For fish exposed to Zn, thermal increases reduce survival time (Fig. 1) (Lloyd, 1960; Sprague, 1964; Cairns, *et al.*, 1971a; Burton, *et al.*, 1972a), reduce incipient lethal level (Sprague, 1964), and decrease median tolerance limit (TL_m) (Pickering and Henderson, 1966). However, temperature does not alter 96-h LC50 values for bluegills (Cairns and Scheier, 1957; Cairns, *et al.*, 1971a). Lack of thermal effect here is probably due to lack of sensitivity of the LC50 measurement and what may be a reduced effect of temperature at the higher thermal range. The Burton *et al.* (1972a) study indicates that fish die about twice as fast when thermal increases are suddenly imposed (1.5°C/10 min) than if temperature does not change (Fig. 2). Amplification of this effect is observed at higher Zn exposure levels (i.e. 32 ppm) relative to lower Zn levels (1.0, 5.6, or 10.0 ppm). Therefore, either elevated temperature or elevated Zn alone can produce the death of fish; moreover, the rate of temperature change can markedly alter Zn toxicity. Much remains in completely explaining the full gamut of temperature-Zn interactions.

pH: Zinc induced toxicity in fish is a function of Zn speciation, which is a function of pH (Mount, 1966; Cairns, *et al.*, 1971a; Farmer, *et al.*, 1979; Cusimano, *et al.*, 1986; Everall, *et al.*, 1989). Below pH 5, almost all Zn is present as a free ion; but at pH 5.5, small amounts of $Zn(OH)^+$ appear (Farmer, *et al.*, 1979) (Fig. 3). At pH 7, percentages of Zn^{+2} and $Zn(OH)^+$ are about 85% and 15% of total Zn. Zinc hydroxide [$Zn(OH)_2$] appears at pH>7 and comprises about 5% of total Zn at pH 8 and 42% of total Zn at pH 9.

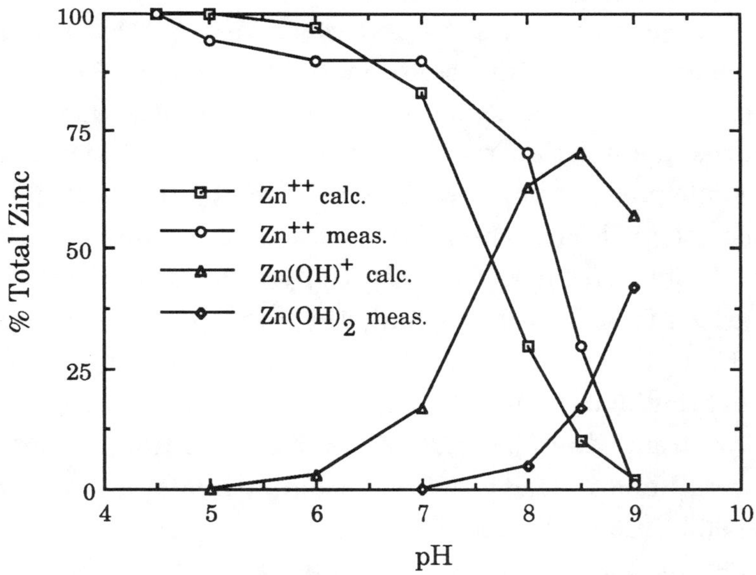

Figure 3. Zinc speciation as a function of pH. The free ion (calculated, calc.; measured, meas.) predominates at pH 4–7; whereas, Zn hydroxides comprise a substantially greater percentage of total Zn at pH 8–9. Free Zn ion is measured in a 10 mg/l solution of zinc sulfate ($ZnSO_4 \cdot 7H_2O$) at hardness 12–24 mg/l using a divalent cation electrode for electrochemical determination. (Redrawn from Farmer *et al.* (1979).)

Visible precipitates of Zn salts have been a nemesis to toxicologists studying Zn toxicity and other effects. For example, Pickering and Henderson (1966) indicate that the addition of Zn to hard water (360 ppm, as $CaCO_3$) at pH 8.2 causes water to become milky with precipitate. Initially, the pH is lowered from 8.2 to 7.4; however, after 24 h, the pH increases to 8.1. Mount (1966) notes that filtration (0.45 μ) removes about 30% of Zn from a solution at pH 8 and having a hardness of 200 ppm at 4–32 ppm Zn aqueous levels. Only 10% is removed at concentrations less than 4 ppm under the same conditions. In contrast, Zn precipitates cannot be filtered from solutions at pH 6. Other workers note speciation differences in test solutions of different pH. Cusimano and associates (1986) indicate that test solutions at pH 4.7, 5.7, and 7.0 are dominated by the free ion (96–97% Zn^{+2}). Therefore, because of the potential for formation of particulate Zn, continuous-flow systems (Mount, 1966; Cusimano, *et al.*, 1986) or recirculating pumps on static systems (Everall, *et al.*, 1989) are required to maintain suspensions.

The questionable results of using static toxicity tests are exemplified by a study conducted by Cairns and co-workers (1971a) using bluegills (*Lepomis macrochirus*). In this study, mortality is similar for fish exposed to soluble Zn at pH 6 and pH 7.3, but not to pH 8. The latter do not succumb in 96 h. Cairns and associates (1971a) reason that dissociation of ionic Zn^{+2} from Zn sulfate probably produces Zn-induced mortalities at pH 6 and pH 7.3. In contrast, at pH 8 or higher, Zn^{++} dissociation is insufficient to cause mortalities. Static testing of Zn (especially in hard water) does not keep precipitated Zn in suspension (Mount, 1966) and produces results opposite those of others which use continuous-flow toxicity testing of Zn.

Use of continuous-flow systems results in increased Zn toxicity at high pH for brown trout (Everall, *et al.*, 1989) or steelhead trout (Cusimano, *et al.*, 1986). Everall and colleagues (1989) use a recirculating pump system to determine 96-h LC50 values for brown trout (*Salmo trutta*) at six pH levels, six Zn concentrations, and in both hard water (204 ppm, as $CaCO_3$, and pH 5) and soft water (10 ppm, as $CaCO_3$, and pH 8). Zinc is most toxic to brown trout in alkaline soft waters and least toxic in acid hard waters (Fig. 4). Increases in toxicity in hard water at pH 7–9 may arise from entrapment of

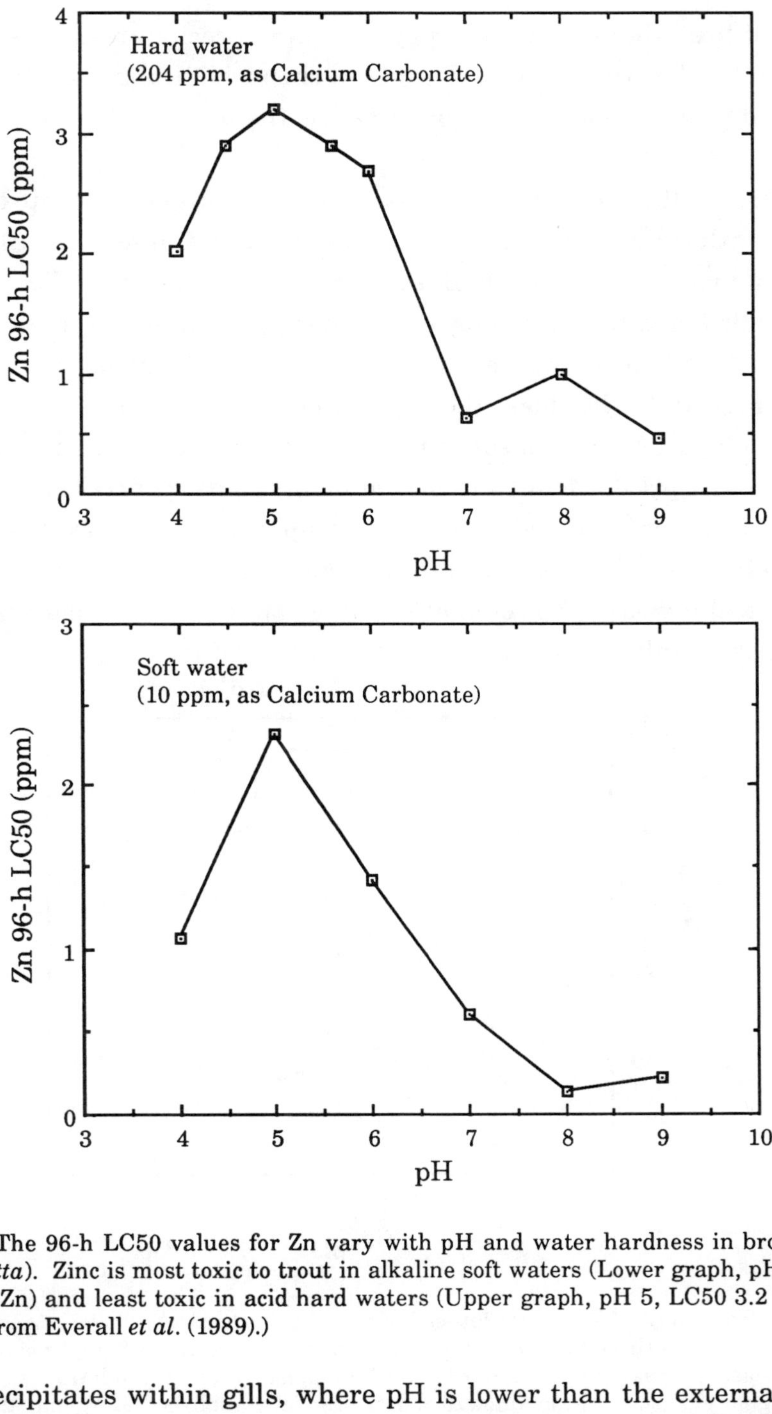

Figure 4. The 96-h LC50 values for Zn vary with pH and water hardness in brown trout (*Salmo trutta*). Zinc is most toxic to trout in alkaline soft waters (Lower graph, pH 8, LC50 <0.14 ppm Zn) and least toxic in acid hard waters (Upper graph, pH 5, LC50 3.2 ppm Zn). (Redrawn from Everall *et al.* (1989).)

metal precipitates within gills, where pH is lower than the external media due to local release of CO_2. Reduced pH at the gill surface could facilitate release of toxic soluble Zn^{+2} from accumulated precipitates and subsequent uptake of Zn^{+2} as the free ion. Therefore, chemical factors such as pH are as important as water hardness in determining the toxicity of Zn to brown trout (Everall, *et al.*, 1989).

Mount (1966) found similar results using continuous-flow toxicity tests of Zn on fathead minnows (*Pimephales promelas*). Curves (Fig. 5) resemble the middle portion of those obtained by Everall *et al.* (1989) (Fig. 4). In this study, Zn is found to be more toxic at higher pH (8) than lower pH (6 or 7) in 4-d tests. Zinc is more toxic at low water hardness (50 ppm) and high pH (8) than at high water hardness (200 ppm) and low pH (6 or 7). These results support Lloyd's (1960) observation that precipitated Zn exerts a toxic influence. At pH 8, Mount's (1966) minnows appear "fuzzy" as a result of epidermal mucous coagulation and water appears "milky" as a result of Zn precipitation. Mechanical Zn collection on the gill lamellae may result in a pH shift and increased Zn solubility to allow Zn to pass gills membranes as a free ion, as postulated as well by Everall *et al.* (1989).

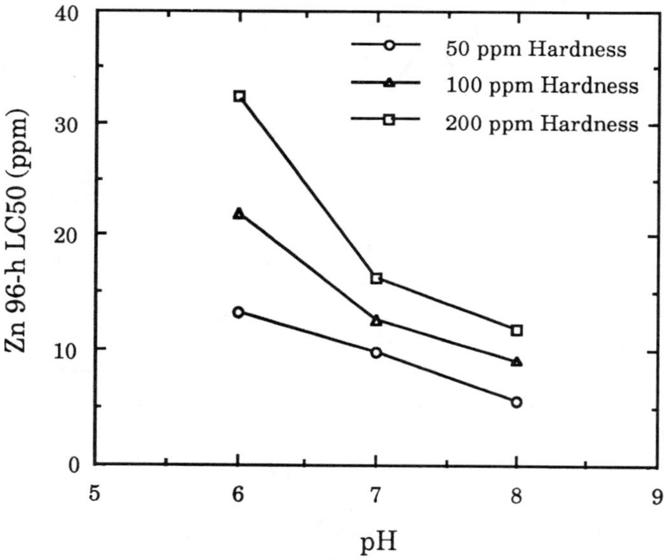

Figure 5. The 96-h LC50 values for fathead minnows (*Pimephales promelas*) exposed to Zn at three pH levels and three hardness levels. Zinc is most toxic at pH 8 and a hardness of 50 ppm and least toxic at pH 6 and a hardness of 200 ppm. At any given hardness, Zn is more toxic at higher pH levels. (From numerical data originally published by Mount (1966).)

Also using continuous-flow testing, Cusimano and associates (1986) report results similar to those of Cairns *et al.* (1971a) and Everall *et al.* (1989). Young (2.36–3.01 g) steelhead trout (*Salmo gairdneri*) are exposed to Zn in soft water at pH 4.7, 5.7, or 7.0 for LC50 determinations at 96- and 168-h. Lower pH levels increase survival. Average LC50 values for Zn are 671, 97, and 66 ppb at pH 4.7, 5.7, and 7.0, respectively. Both 96-h and 168-h test results are similar. Metal speciation differences are eliminated as having significant roles in producing the toxic effects of Zn in steelheads because test solutions at each pH are dominated by the free ion (96–97% Zn^{+2}). Hydrogen ion interference with metal uptake at gills is proposed as the major mechanism for decreased toxicity at acid pH. Net negative charge in positive ion transport channels of gills is decreased following exposure to low pH (McDonald, 1983). High concentrations of H^{+} could neutralize anionic gill sites. Model predictions based on this hypothesis provide a good approximation of the observed pattern of LC50 values for Zn.

Results from published studies such as these attest the importance of pH and the complexity of the interaction between pH and hardness. In studying this interaction, Pagenkoph (1983) reports a Gill Surface Interaction Model (GSIM) as a means of accounting for differences in metal toxicity to fish at variable pH, hardness, and alkalinity. A series of equations are developed to use metal speciation, competitive inhibition of other ions, and gill surface interaction to predict effective toxicant concentration (ETC). In testing GSIM, water chemistry is related to fish toxicity using reported data from Zn, Cu, Cd, and Pb, as well as combinations of metals.

In summary, Zn-induced toxicity in fish is a function of Zn speciation, and Zn speciation is a function of pH. Most free Zn^{+2} ion is present at pH 4–7; whereas, Zn hydroxides (and carbonates) comprise a greater proportion of total Zn at pH 8–9. Use of static or continuous-flow toxicity testing seems to determine the outcome of the pH-induced effects on Zn toxicity in fish. Using static tests, Cairns *et al.* (1971a) and others report increased toxicity at low pH; while, other workers, who use continuous-flow toxicity tests, report increased Zn toxicity at high pH (Mount, 1966; Cusimano, *et al.*, 1986; Everall, *et al.*, 1989). Most likely, circulating Zn precipitates are entrapped within the gill where pH is lower due to release of carbon dioxide. Reduced pH could allow release of toxic, soluble Zn from precipitates and subsequent uptake of Zn as the free ion (Mount, 1966; Cusimano, *et al.*, 1986). Use of a continuous-flow system keeps particulate Zn in solution and simulates Zn exposure of a stream-dwelling fish species.

Life Stage: Zinc toxicity in fish varies with age. Newly hatched fish generally are more sensitive than embryos or intermediate life stages (Skidmore, 1967; Chapman, 1978). For later life stages, increasing Zn tolerance is noted with increasing age in some studies (Chapman, 1978; Goodman, 1951; Chapman, 1978), but not in others (1959; Farmer, *et al.*, 1979; Cairns and Scheier).

Pickering and Vigor (1965) determine LC50 values of Zn in the fathead minnow (*Pimephales promelas*) eggs, embryos, and fry in hard water (174–198 ppm as $CaCO_3$, pH 7.5–7.6, and 20°C). Prior Zn exposure during the embryonic stages confers Zn tolerance. For example, embryos exposed to 1.1 ppm Zn hatch and fry live for at least 3 d. In comparison, newly-hatched fry which have not been exposed previously to Zn die within 24 h after exposure to 1.1 ppm Zn. The 96-h TL_m (median tolerance limit) for 1-d old fry is 0.87 ppm; the corresponding value for embryos is 1.83 ppm. Obviously, newly-hatched fry are more sensitive to Zn than are embryos. At a lower concentration of 0.56 ppm, all 1-d fry survive up to 7 d. Sparks *et al.* (1972) show that 0.235 ppm Zn kills newly-hatched bluegill (*Lepomis macrochirus*) fry within 3 d. Therefore, Zn is considered extremely toxic to fathead minnow embryos and fry—despite use of hard water in testing.

Chapman (1978) exposes four life stages of chinook salmon (*Oncorhynchus tschawytscha*) and steelhead (*Salmo gairdneri*) to Zn in continuous-flow toxicity tests (Fig. 6). Newly-hatched alevins of both species are more tolerant to Zn than intermediate juvenile stages (i.e. swim-up and parr). However, generally, among the later three stages (swim-up, parr, and smolt) increasing Zn tolerance is noted with increasing age, as also observed by Goodman (1951) for 2–8 wk old rainbow trout (*Salmo gairdneri*). In the Chapman (1978) study, the last three stages (swim-up, parr, and smolt) of chinook are more tolerant of Zn than are steelheads (Fig. 6). The 96-h LC50 values for these two species range from 93–815 ppb for Zn at 23 ppm hardness, 12°C, and pH 7.1. Authors suggest that alevins are to be avoided in the selection of a life stage for acute toxicity tests because of marked resistance to Zn (as well as Cu and Cd). Interestingly, both newly-hatched alevins and smolt (about 2-yr-old fish, physiologically ready to migrate to the sea) are equally sensitive to Zn; whereas, the swim-up stage is about seven to eight times more sensitive to Zn.

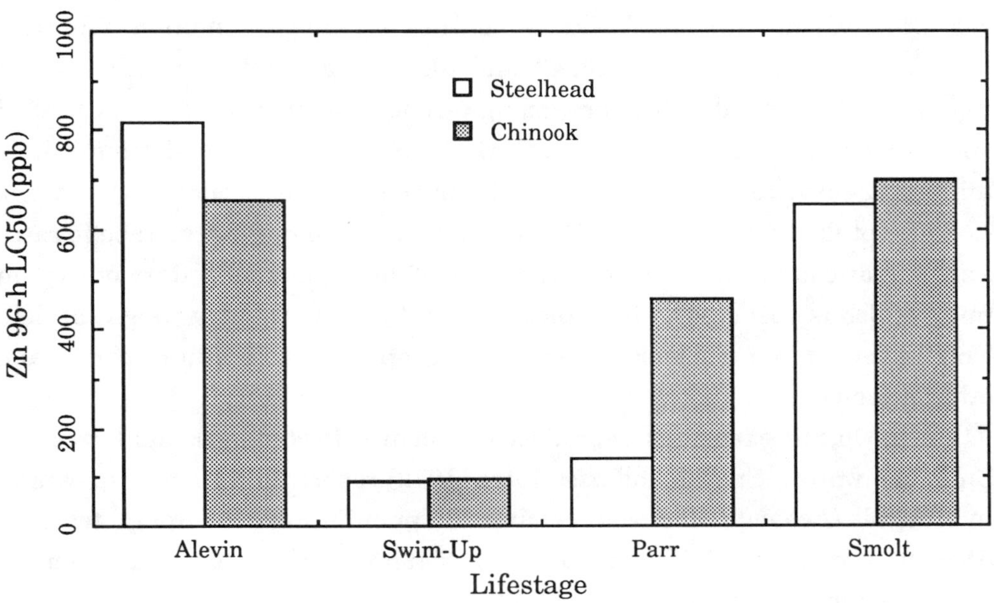

Figure 6. Calculated 96-h LC50 values for steelhead (*Salmo gairdneri*) and chinook salmon (*Oncorhynchus tschawytscha*) exposed to Zn at 23 ppm hardness, 12°C, and pH 7.1. The swim-up stage is most sensitive to Zn; whereas, newly-hatched alevins and smolt (about 2 yr old and ready to migrate to the sea) are approximately equally sensitive. (Constructed from numerical data presented by Chapman (1978).)

This increased sensitivity to Zn in smolt (compared with parr) is not observed in the Farmer *et al.* study (1979). In fact, Atlantic salmon *(Salmo salar)* smolt (especially 16.7 g fish) are at greater risk to Zn poisoning than are parr (Farmer, *et al.*, 1979). The 21-d LC50 is 0.34–0.51 ppm Zn for larger fish (9.2–16.7 g) and 1.45–1.60 ppm Zn for smaller fish (4.8–5.5 g). According to Farmer and colleagues (1979), this unusual result is considered due to the parr-smolt transformation which makes the larger fish (especially those about 16.7 g) at greater risk from Zn poisoning. The 21-d (504-h) LC50 values for Atlantic salmon (340–1640 ppb) (Farmer, *et al.*, 1979) are higher than the 96-h LC50 values for steelhead and chinook salmon (93–815 ppb) (Chapman, 1978).

Results obtained by Skidmore (1967) agree with those of Chapman (1978) in that intermediate sized fish are more susceptible to Zn than are younger and older fish. Survival time of 3–13 d old zebrafish (*Brachydanio rerio*) is markedly reduced compared with younger and older fish; therefore, survival times of Zn are compared with the metabolism of different aged zebrafish. Using Warburg flasks to measure oxygen consumption, zebrafish aged 0, 1, 2, 3, 4, 5, 6, 7, 8, 9, 10, 40, and 100 d are evaluated. Rate of oxygen uptake in 3–13 d old fish is approximately four times greater than that of younger or older fish. Indeed, reaction time (total survival time minus minimum survival time) of zebrafish is inversely proportional to the rate of oxygen uptake (Fig. 7). The 3–13 d old fish have the highest metabolic rates and the lowest reaction times (a measure of survival time). Rate of oxygen uptake also is observed to be directly proportional to the dry weight of fish. This is considered a general phenomenon, applicable to other species and other poisons.

In chronic exposures, reproductive fish are affected by a slight excess of Zn in the water. Sparks and associates (1972) report inhibition of spawning in bluegills (*Lepomis macrochirus*) at 0.235 ppm Zn in flow-through tests (at 51 ppm hardness, 19.7°C, and pH 7.8) for fish brought into breeding condition prior to Zn exposure. Other studies allow fish to come into breeding condition during Zn exposure (Brungs, 1969). Brungs (1969) reports results of both static and continuous-flow tests conducted on fathead minnows (*Pimephales promelas*) in hard water (200 ppm as $CaCO_3$) at 17–23°C and pH 7.7. Chronic Zn exposure is monitored over a 10-mo period. At less than 0.18 ppm, spawning is reduced to about six times less than controls (17%): the average female in 0.18 ppm Zn for 10-mo produces only 263 eggs. Number of eggs produced per female is related to the level of Zn in the water, with 83%, 50%, and 20% reductions in egg production at 0.180 ppm, 0.088 ppm, and 0.042 ppm, respectively. Fish in 1.30, 0.66, and 0.32 ppm Zn do show typical spawning behavior, but they rarely spawn. Spawned eggs are fertile, based on the observance of cell division.

The application factor (i.e ratio of maximum acceptable concentration to the 96-h TL_m value) for Zn-exposed fathead minnows (exposed in hard water) is 0.003 (i.e the mean of controls, 0.030 ppm/the 96-h LD50 value of 9.2 ppm) to 0.020 (0.18 ppm/9.2 ppm) (Brungs, 1969). If tests use an aqueous level of Zn high enough to cause 20–50% reduction in egg production (0.042–0.088 ppm), the Zn application factor would vary from 0.005 to 0.010. For fathead minnows, Zn is different from at least one other metal (i.e. Cu) in that a wide range of Zn concentrations is nonlethal but still limits production (i.e. reproduction, growth, spawning behavior, egg viability, and fry survival). For Cu, only a narrow concentration range exists where neither reproduction nor mortality occurs (Mount, 1968).

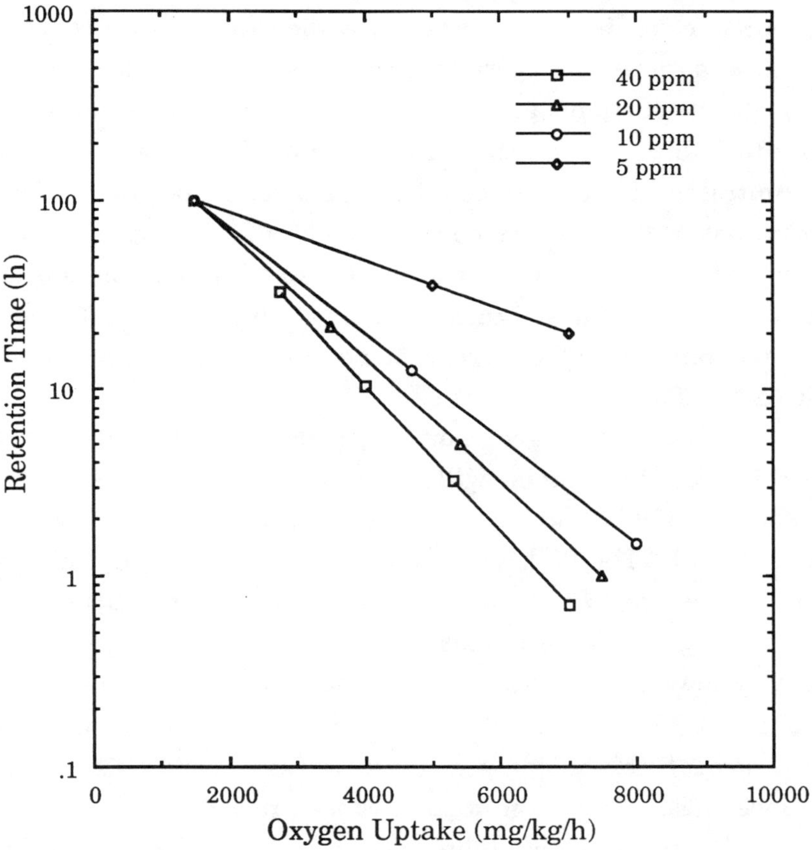

Figure 7. Rate of oxygen uptake *versus* reaction time (total survival time minus minimum survival time) of zebrafish (*Brachydanio rerio*) in four concentrations of Zn. The logarithm of reaction time of zebrafish to Zn poisoning is inversely proportional to the rate of oxygen uptake. The 3–13 d old fish have the highest metabolic rates and the lowest reaction times. Oxygen uptake is measured in terms of the rate of uptake per unit dry weight of fish. (Modified and redrawn from Skidmore (1970).)

In summary, embryos are more resistant than fry to acute exposures of Zn (Pickering and Vigor, 1965). The 96-h LC50 values for fathead minnow 1-d old fry and embryos are 0.87 and 1.83 ppm, respectively. Intermediate life stages (as 3–13 g zebrafish or swim-up and parr stages of steelhead and chinook salmon) are more sensitive to Zn than are younger and older fish of the same species (Skidmore, 1967; Chapman, 1978). This difference is related to the higher metabolic rate of intermediate life stages (Skidmore, 1967). Older fish tend to be progressively more resistant to Zn in some cases (Goodman, 1951; Chapman, 1978), but not in others (Cairns and Scheier, 1959; Farmer, *et al.*, 1979).

Reproductive fish are affected by a wide range of Zn concentrations. Chronic Zn exposures are especially hazardous to the female fathead minnow because 83% reductions in spawning frequency can occur after 10-mo exposure to 0.18 ppm Zn in the water. Egg production is reduced 50% and 20% of control levels at 0.088 and 0.042 ppm Zn, respectively. Aqueous Zn concentrations at 0.180–0.042 ppm Zn do not affect adult survival, growth, hatching, or fry survival. For comparison, growth inhibition is evident only after 30 d at 2.8 ppm Zn in fathead minnows. Use of the 0.180 ppm Zn level as the maximum acceptable Zn concentration results in application factors of 0.003–0.020 for fathead minnows.

Dissolved Oxygen: Low oxygen decreases the tolerance of fish to Zn. Two studies involving bluegills (*Lepomis macrochirus*) show results for short exposures (96-h) and long exposures (20-d) (Cairns and Scheier, 1958; Pickering, 1968). The 96-h LC50 values for Zn at low and high levels of dissolved oxygen are 4.90 and 8.02 ppm Zn, respectively (Cairns and Scheier, 1958). In this particular static toxicity test, periodic low dissolved oxygen is obtained by lowering oxygen to a minimum concentration (2 ppm over a period of about 3 h)—a level which would not injure fish. After maintaining this level for 2 h (by bubbling nitrogen or air depending on whether the oxygen level rose or fell), the oxygen concentration is increased slowly to saturation. Controls are maintained at normal levels throughout this period. As indicated, the 96-h LC50 value for Zn is 8.02 ppm for bluegills held under stable, normal oxygen levels (18°C, soft water); however 4.90 ppm represents the value for fish exposed to periodic low oxygen. Therefore, periodic low oxygen decreased the tolerance of bluegills to Zn compared with constant normal concentrations of 5 to 9 ppm dissolved oxygen. Low oxygen levels and elevated Zn concentrations act together to reduce bluegill survival in this short-term study.

The 20-d LC50 values for Zn at low and high dissolved oxygen are 7.35 and 11.35 ppm Zn, respectively, for bluegills in a study conducted by Pickering and Henderson (1968). Oxygen levels are varied from 1.8–5.6 ppm at Zn concentrations ranging from 1.0–4.0 ppm (370 ppm total hardness, as $CaCO_3$, pH 7.8, and 25°C). Average 20-d LC50 values for low and high test concentrations of dissolved oxygen (1.8 *versus* 5.6 ppm oxygen) differ by a factor of 1.5 (i.e. 7.35 ppm *versus* 11.35 ppm Zn). Comparison of these results with those of Cairns and Scheir (1958) yield a similar factor (i.e. 1.6); therefore, effects of Zn toxicity are similar in short static tests (Cairns and Scheier, 1958) and longer continuous flow-through tests (Pickering, 1968). The results indicate that in some cases, complex flow-through tests may be replaced by simpler, short-term static tests. This is particularly true of studies involving lower concentrations of Zn (as 1–4 ppm Zn in the present study) and/or studies using soft water.

Sex: Increased susceptibility of males to Zn poisoning is recognized in a number of fish species, including platyfish (*Xiphophorus maculatus*, inbred strain 163A) (Rachlin and Perlmutter, 1968a), plaice (Wallace, 1925), mosquitofish (Krumholz, 1948), bullhead, dragonet, coho salmon, and common guppy (Beverton and Holt, 1959). For the inbred platyfish strain, males have a 96-h TL_m of 11.7 ppm Zn compared with 14.1 ppm Zn for the females. In comparison, fathead minnows (*Pimephales promelas*) exposed under the same conditions do not show sex-related differences in 96-h TL_m (7.6 ppm Zn). Increased susceptibility of male platyfish to Zn poisoning may be due to higher Zn accumulation in males than females. An example of this elevated uptake is observed in male dogfish which accumulate up to three times more Zn in gill arches than females (Crespo, *et al.*, 1979). Differences in the action of gonadotrophic hormones might explain the difference in Zn accumulation and the increased susceptibility of males to Zn.

Water Hardness: Soft water environments place fish at greater risk to Zn-poisoning than hard water environments. However, the mechanism of action whereby hard water reduces Zn toxicity in fish is not understood. Mount (1966) clearly shows that increasing total hardness from 50 to 200 ppm (as $CaCO_3$) decreases Zn toxicity by a factor of 1.6 to 2.5, based on 96-h LC50 values for fathead minnows (*Pimephales promelas*) (Fig. 5). The effect exists throughout the 6–8 pH range. The 96-h LC50 value is about 5 ppm Zn at 50 ppm hardness, compared to 12 ppm Zn at 200 ppm hardness (a 2.4-fold decrease in toxicity).

In a comparative study, Solbé (1974) exposes rainbow trout (*Salmo gairdneri*) of 15.0–17.5 cm fork length to Zn at 12–504 ppm hardness (as $CaCO_3$) and 15.5°C and pH 6.6–8.0 to determine LC50 values. Increasing hardness from 12 to 504 ppm causes a three-fold increase in survival time (220 to 660 min, at 7 ppm Zn in water). Solbé (1974) reports decreasing solubility of Zn over this hardness range. At 12 and 50 ppm hardness, all Zn is dissolved; whereas, at 320 and 504 ppm hardness, only 5.9 and 2.6% of Zn is dissolved. The 96-h LC50 for Zn is about 5 ppm for rainbow trout at 504 ppm hardness, pH 7.8, and 15.5°C. Solbé considers hardness (rather than pH) of major importance in Zn toxicity, but other workers recognize the interrelationships which exist between pH and total hardness in causing fish mortality (Mount, 1966).

The mechanism of action explaining the inverse relationship between water hardness and Zn toxicity has been approached using incremental Ca addition (Judy and Davis, 1979) and binding constants for glycine (Zitko and Carson, 1976), to mention just two such studies. In the Judy and Davis (1979) study, fathead minnows *(Pimephales promelas)* are exposed to 0.5–20.0 ppm Zn at 45 ppm hardness, pH 7, and 21° C for 96-h LC50 determination and evaluation of the effect of Ca. The 96-h LC50 is 3.1 ppm Zn (2.2–4.3 ppm confidence limits). Addition of Ca in 25 ppm increments results in progressively decreased Zn toxicity from 65% mortality in soft water (10 ppm Ca) to 0% in water of about twelve times greater hardness (118 ppm Ca). Despite that fact that measured hardness values (obtained by EDTA titration) are double that due to added Ca (possibly due to NO_3^- interference), the presence of Ca is considered to be the determining factor in decreased Zn toxicity as hardness increases.

However, Zitko and Carson's (1976) study shows that Mg competes with Zn, and could explain the effect of increased water hardness in reducing Zn-induced toxicity. In this study, juvenile Atlantic salmon *(Salmo salar)* are exposed to pairs of toxic metals in water of 14 ppm total hardness (as $CaCO_3$). Whereas Ca reduces Zn-induced toxicity only slightly, Mg causes substantial reductions. Binding constants of glycine follow approximately the same pattern as *in vivo* exposures. Log K_1 of Mg for glycine (3.44) is similar to that of Zn (5.52) and Cd (4.80); therefore, Mg can compete with Cd and Zn for binding sites on glycine. However, because Ca has a log K_1 value of 1.31 for glycine, Ca is considered to have little effect on Zn toxicity. Competition between Mg and Zn may explain the effect of increased water hardness in reducing Zn-induced lethality, although binding constants for other components of the gill surface probably differ from that of glycine.

In summary, Zn toxicity spans a wide concentration range depending on species, total water hardness, temperature, pH, and other factors (Table 1). From 0.87 to 40.90 ppm Zn is reported as the 96-h LC50 from just five toxicity studies (Cairns and Scheier, 1957; Cairns and Scheier, 1959; Mount, 1966; Pickering and Henderson, 1966; Rehwoldt, *et al.*, 1971). Extremely low concentrations of Zn (5.6 ppb) are detected and avoided by fish such as rainbow trout (Sprague, 1968). This may provide evidence of the hazard of Zn toxicity to fish; however, fish are unable to detect low concentrations of other elements, despite extreme toxicity (see Chapter II). Much work remains in elucidating factors involved in detection of elemental poisons by fish.

In vitro Studies: Rachlin and Perlmutter (Rachlin and Perlmutter, 1968b; Rachlin and Perlmutter, 1969) report results from use of two cell cultures from fish cell lines—RTG-2 from the rainbow trout and an epithelial cell line derived from fathead minnows. Based on mitotic indices, the fathead minnow cell line is ten times more sensitive to Zn than the rainbow trout cell line. Therefore, fathead minnow cells respond to Zn at concentrations similar to that of living rainbow trout. Closer evaluation of each study seems warranted.

In Rachlin and Perlmutter's (1968b) study, commercially-available cultures derived from fathead minnows (*Pimephales promelas*) are exposed to a series of Zn concentrations (0.00, 0.75, 1.80, 7.50, and 75.00 ppm) for 96 h. The number of mitotic figures in a 1000 cell sample is determined using only metaphase and later stages. The 96-h TL_m for whole fish is 7.6 ppm—a level causing 53% reduction in mitotic index. However, a level 1/10 of the 96-h TL_m (0.75 ppm) for fish exerted no observable effect on cell cultures. Therefore, rather than using the 96-h TL_m levels for whole animals (7.6 ppm), use of 96-h TL_m levels from cell cultures (i.e. 1.8 ppm Zn) may represent a better approximation of a level which represents a "biologically safe concentration" (BSC).

For the rainbow trout cell line (RTG-2), 96-h mitotic indices also are used. Zinc levels up to 10 ppm have no significant effect on mitotic index of

Table 1. Zinc toxicity in fish based on 96-h LC50 data. Total water hardness (in ppm calcium carbonate), temperature, and pH are provided.

Species	LC50 96-h (ppm)	Water Hardness (ppm)	Temperature (°C)	pH	Reference
Banded killifish *(Fundulus diaphanus)*	19.1	53	17	7.8	Rehwoldt *et al*, 1971
Striped Bass *(Morone saxatilis)*	6.7	53	17	7.8	
Pumpkinseed *(Lepomis gibbosus)*	20	53	17	7.8	
White perch *(Morone americana)*	14.3	53	17	7.8	
American eel *(Anguilla rostrala)*	14.6	53	17	7.8	
Carp *(Cyprinus carpio)*	7.8	53	17	7.8	
Fathead minnow *(Pimephales promelas)*	32.25	200	25	6	Mount, 1966
	16.30	200	25	7	
	11.85	200	25	8	
	13.15	50	25	6	
	9.95	50	25	7	
	4.90	50	25	8	
Bluegills *(Lepomis macrochirus)*	2.86–3.78	soft water	18	7.5	Cairns & Scheier, 1957
	1.93–3.63	soft water	30	7.5	
	10.13–12.50	hard water	18	7.5	
	10.15–12.30	hard water	30	7.5	
Bluegills *(Lepomis macrochirus)*	6.91	10	20	-	Cairns & Scheier, 1959
	7.20	10	20	-	
	7.45	10	20	-	
Fatheads *(Pimephales promelas)*	0.87	20	15	7.5	Pickering & Henderson, 1966
	33.40	360	15	8.2	
Bluegills *(Lepomis macrochirus)*	5.37	20	15	7.5	
	40.90	360	15	8.2	
Goldfish *(Carassius auratus)*	6.44	20	15	7.5	
Guppies *(Poecilia reticulata)*	1.27	20	15	7.5	

RTG-2 cells. However, 96-h exposures to 18 ppm Zn are toxic. The mitotic index is reduced about 72% from the control value of 3.2 (per 1000 cells). Complete mitotic inhibition is achieved after 96 h at 32 ppm Zn. Therefore, based on mitotic indices, cell line RTG-2 is ten times less sensitive than fathead minnow epithelial cells since, with the latter, 1.8 ppm Zn reduces the mitotic index to 50% (Rachlin and Perlmutter, 1968b). Because living rainbow trout are sensitive to Zn in the 0.5–6.0 ppm range (Lloyd, 1960), the RTG-2 cell line from rainbow trout is less sensitive to Zn than are whole fish.

The difficulties of using a cultured cell population to derive toxicity data for whole animals in obvious. Response of one cell type is not a dependable facsimile of the response of the whole organism. In the Rachlin and Perlmutter studies (Rachlin and Perlmutter, 1968b; 1969), one would expect culture line RTG-2 (derived from trout) to respond more like living trout than cultures from fathead minnows. Since the fathead minnow is a fresh-water fish and rainbow trout are anadromous, little rationale exists for using cultures from a fresh-water fish (stenohaline) to derive toxicity data for a euryhaline species. In addition, attached cell lines of a single cell type are suspect because of loss of function, fibroblast overgrowth, unequal exposure to toxicants (with considerable reduction in potential uptake at the attached surface), and a multitude of other problems (Sorensen and Thomas, 1988a; Sorensen, *et al.*, 1988b). Use of cultures of more than one cell type (i.e hepatocyte and fibroblast) increases the probability of prolonging cell function and correlating *in vitro* results with those obtained *in vivo*.

Accumulation and Distribution

Uptake of Zn by fish is altered by exposure time, exposure concentration, fish size, water hardness, length of acclimation to test solutions (hard water and soft water without Zn), feeding level, and trophic level. This is but a partial listing of parameters which change Zn accumulation in the whole body and tissues, but it provides a good initiation point for reviewing a number of studies of Zn uptake in fish.

Accumulation of Zn varies with exposure time. Matthiessen and Brafield (1977) expose sticklebacks (*Gasterosteus aculeatus*) to stable and radioactive Zn in either hard water or soft water for periods up to 400 h. In soft water (0 ppm hardness, as $CaCO_3$), Zn uptake becomes asymptotic at about 24 h at a concentration factor of about 2.8 for whole fish (Fig. 8).

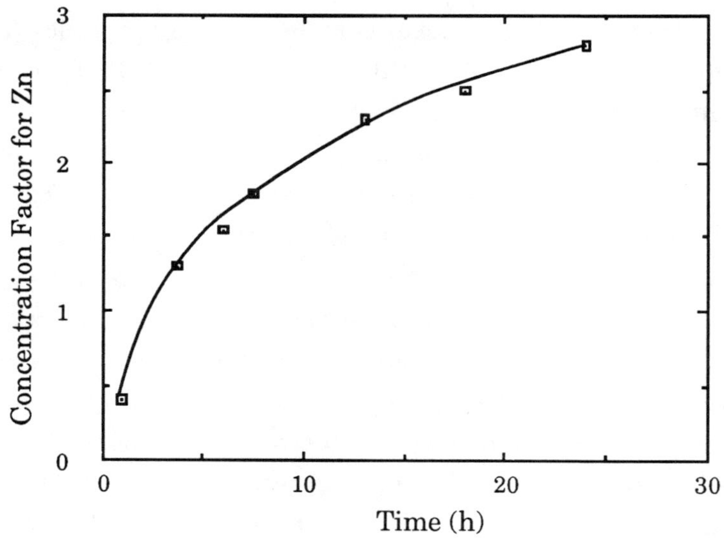

Figure 8. Uptake of Zn in sticklebacks (*Gasterosteus aculeatus*) exposed to 1 ppm stable Zn and 1 μCi Zn-65/l in soft water (0 ppm hardness). Average values for fifteen to thirty fish are given. Uptake is asymptotic at about 24 h at a concentration factor of 2.8. (Modified and redrawn from Matthiessen and Brafield (1977).)

Fish size alters Zn accumulation because metal accumulation is correlated with metabolic rate, and smaller fish have higher metabolism. Results of Matthiessen and Brafield's (1977) study show that Zn-65 uptake increases about five times for 0.1 g fish compared with 1.0 g fish, which have an oxygen uptake difference of five times (hard water exposure). Similarly, for 1.0–2.0 g fish, Zn-65 uptake increases by a factor of 1.6 (from largest to smallest fish) while oxygen uptake also increases by a factor of 1.6. Therefore, Zn uptake is altered by metabolic rate (i.e. size), as expected.

What seems amazing about the Matthiessen and Brafield (1977) study is that sticklebacks in high Ca (hard water) solutions tend to accumulate more Zn-65 than those in low Ca (soft water). This effect is exacerbated two and one-half times if fish are allowed 2 h- rather than 18 h-acclimation to test solutions without Zn (Fig. 9). Moreover, up to three and one-half times higher Zn uptake occurs if fish are exposed in hard water rather than in soft water. This is a puzzling result which seems to conflict with the idea that the presence of Ca reduces membrane permeability to heavy metals. Perhaps the phenomenon shown here is unique to this euryhaline species, since sticklebacks are adapted to exist in solutions of high osmolarity.

Alternatively, and as the authors suggest, perhaps mucus precipitates and sloughs Zn more readily in soft water than in hard water. This might lead to greater Zn accumulation in hard water.

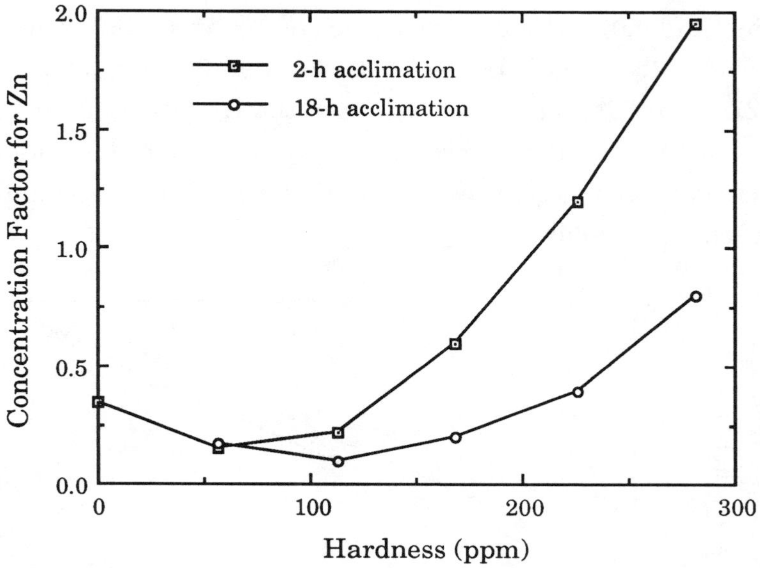

Figure 9. Five-hour Zn-65 uptake for sticklebacks (*Gasterosteus aculeatus*) exposed to 1 ppm stable Zn and 1 μCi Zn-65/l in different hardnesses and following different acclimation times (2 h or 18 h) in the Zn-free solutions. The greater the hardness, the greater the Zn accumulation. Also, the longer fish are allowed to acclimate in the respective test solutions (without Zn), the lower the Zn concentration factor. (Modified and redrawn from Matthiessen and Brafield (1977).)

Feeding level (% ration) alters Zn accumulation. Farmer (1979) exposes juvenile Atlantic salmon *(Salmo salar)* to 0.002 (control), 0.115–1.00 ppm Zn over 21–80 d (12–24 ppm hardness, pH 6.1–7.1, and 10°C) to measure effect of feeding level and voluntary food intake. Initially, exposure to increased Zn (0.420–1.000 ppm) results in decreased appetite; however, voluntary food intake returns to control levels by 40 d. This result suggests that juvenile salmon which survive elevated aqueous Zn levels adapt for at least 3 mo—as indicated by resumption of feeding rate at levels equal to or greater than controls and by the absence of stunting at those feeding levels used in this study.

For salmon exposed 80 d to 0.002 or 0.240 ppm Zn at three feeding levels (satiation, 3.5% ration, and 2.0% ration), exposure to higher levels of

Zn reduces caloric content/g body weight. Increased food intake and decreased caloric content at this exposure level of Zn suggests elevated energy expenditures in these fish. Zinc accumulation is linear at 2.0 and 3.5% rations, but plateaus occur at satiation (5–6% rations) because salmon are growing (Fig. 10). At 2.0–3.5% rations, growth rate is slower; therefore linear growth rates result. However, at satiation feeding, a balance occurs between Zn accumulation and growth such that Zn accumulation plateaus. This result shows the importance of varying biomass of food organisms in long-term metals studies of growing fish. Patterns of metal accumulation are altered at reduced available biomass.

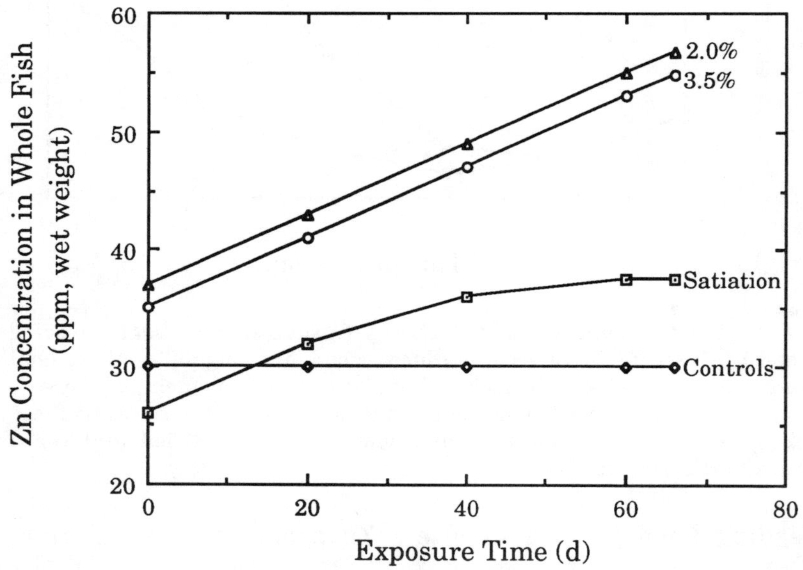

Figure 10. The effect of ration (% dry body weight/d) on Zn accumulation in the whole body of juvenile Atlantic salmon (*Salmo salar*). Zinc is present in water at 0.56 ppm for the upper three curves and 0.002 ppm (control) for the lower curve. Provision of food until satiation (5–6% of ration/d) results in a balance (plateau) between Zn accumulation and growth. However, at 2.0 or 3.5% ration level, growth rate is slower and a linear relationship exists between accumulation and growth rate. Control levels of Zn uptake (shown) correspond to the 2% ration group only, but controls for the other groups are similar. (Modified and redrawn from Farmer *et al.* (1979).)

The feeding level is as important if Zn is present in food chain organisms as it is if Zn is present in water. Baptist and Lewis (1967) expose mummichogs (*Fundulus heteroclitus*) and either croakers (*Micropogon undulatus*) or mojarra (*Eucinostomus* species) as the fourth and third levels, respectively, of a four-step food chain experiment. The upper two levels are phytoplankton and zooplankton (brine shrimp) cultures—prepared fresh daily. A constant level of Zn is provided to the third and fourth levels of the food chain (at 21°C, 24–28 ‰ salinity). Post-larval fish are fed to satiation on brine shrimp, which also are fed to satiation on phytoplankton. In contrast, mummichogs are not fed to satiation.

Higher Zn-65 uptake is observed in post-larval fish—one-hundred fifteen times the initial amount in phytoplankton cultures after 19 d *versus* thirteen times for mummichogs (Fig. 11). Higher metabolic rates for post-larval fish than for mummichogs and feeding levels beneath satiation in mummichogs probably cause the higher Zn uptake in post-larval fish. If mummichogs had consumed eight times the number of post-larval fish (quite likely if feeding had been *ad libitum*), mummichogs would have concentrated about one-thousand times the initial amount in the phytoplankton cultures.

Determination of whether water or food chain uptake is the principle pathway of Zn accumulation in mummichogs is impossible using this study. However, relative comparison of Zn-65 uptake from food-chain or water shows that fish accumulate four or five times more Zn-65 from the food-chain than from water. In contrast, zooplankton accumulate about twice as much Zn-65 from water than from food. Tissue distribution studies of mummichogs show that relatively large amounts of Zn are present in gonads and spleen; that equivalent amounts of Zn are present in the liver, gills, and digestive tract (with contents); and that skeletal muscle accumulates little Zn.

Tissue distribution patterns for Zn in fish are available for stenohaline and euryhaline species. In one study, Mount (1964) exposes bluegills (*Lepomis macrochirus*) for 90 d to levels of Zn up to 5 ppm in the water (Fig. 12). In a chronic exposure such as this, opercular bone accumulates as much Zn as gill tissue (with arch and rakers). However, in acute exposures, gill accumulates 100 times more Zn than bone. Therefore, measurement of gill/bone ratios of Zn provides a means of determining whether fish die from Zn or some other cause.

Field samples of more than one-thousand fish of twenty-two species show almost no overlap between gill/bone ratios of normal fish from the field and fish succumbing to acute exposures of Zn (Mount, 1964). Similar results are reported for two-hundred-three fish of one species in another study

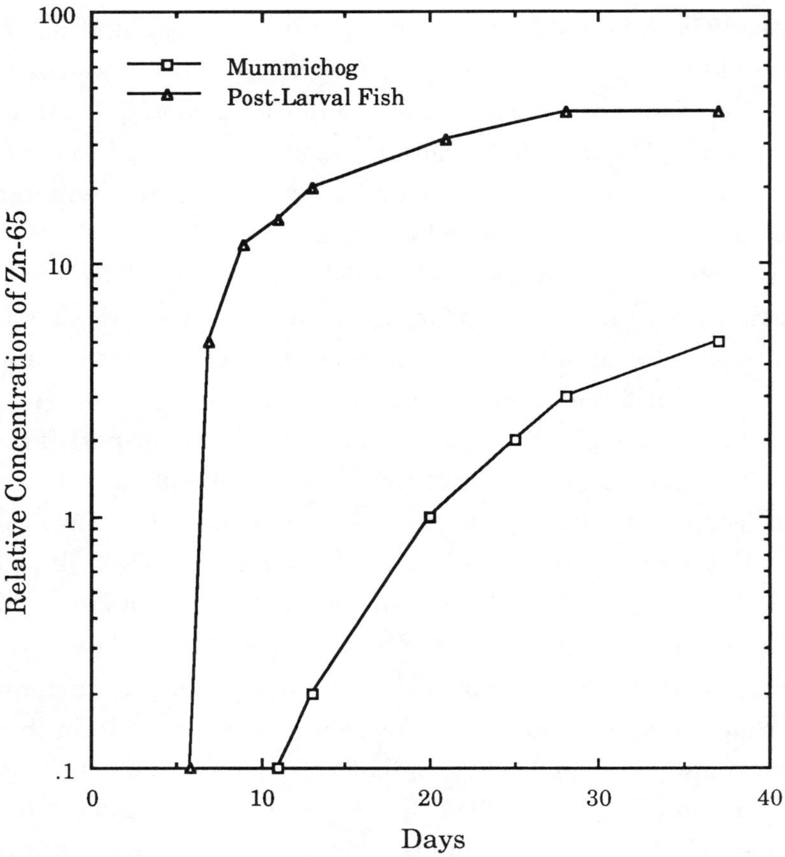

Figure 11. Uptake of Zn-65 through two of four trophic levels of an estuarian food chain using an alternate-day feeding regimen. Values presented are relative to the initial amount of Zn-65 in a phytoplankton culture having a constant level of Zn-65 and a constant value of one throughout the experiment. Although excess phytoplankton is used to supply zooplankton and excess zooplankton is used to supply post-larval fish, only twenty post-larval fish are fed to ten mummichogs on alternate days. Therefore, dietary restrictions for mummichogs resulted in a curve which is almost linear (as with the 2.0 and 3.5% ration curves shown in Fig. 10). (Modified and redrawn from Baptist and Lewis (1967).)

(Cairns, *et al.*, 1971b). Only one species (carp, *Cyprinus carpio*) requires autopsy of more than a few fish for sufficient accuracy of this predictive procedure. For species in which the critical Zn gill/bone ratios are not available, three (and even two) times the control ratio can be used as a conservative estimate of the dividing line between a positive and negative autopsy result. The autopsy method is specific for Zn, can be performed on partially decomposed fish, is not affected by small amounts of other metals, and does not require immediate removal of fish from contaminated water (Mount, 1964).

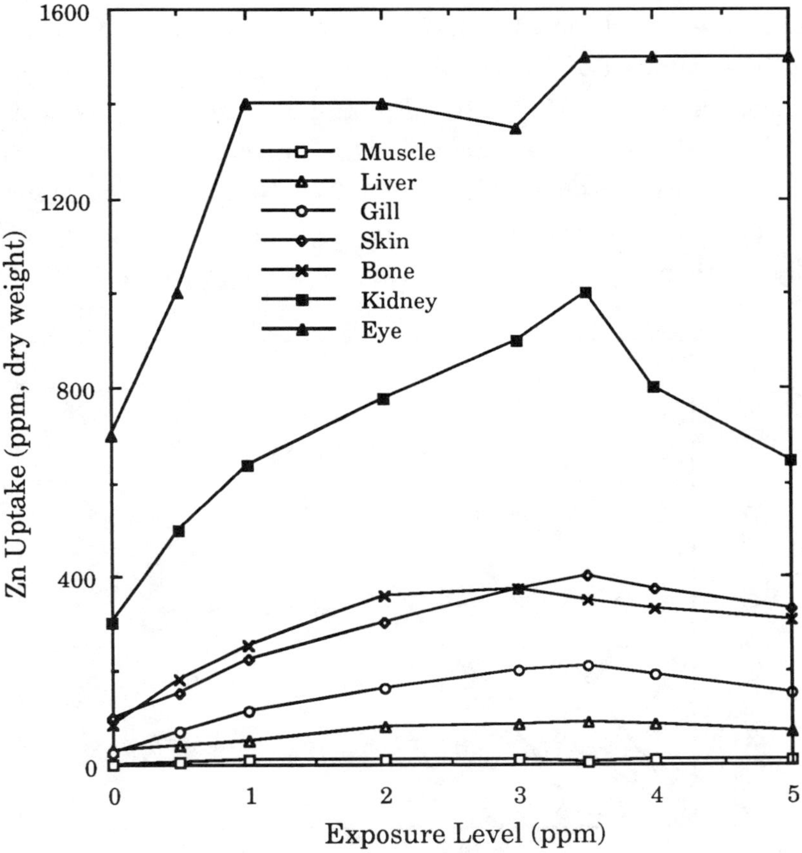

Figure 12. Distribution of Zn in various organs of bluegill (*Lepomis macrochirus*) exposed for 90 d to up to 5 ppm Zn in the water. In this chronic study, gill and bone accumulate about as much Zn; however, in an acute study, gill uptake exceeds bone uptake one-hundred times. (Modified and redrawn from Mount (1964).)

In Mount's (1964) study on bluegills, the eye accumulates extremely high levels of Zn, followed by the kidney, skin(~bone), gill(~gut), liver, and muscle(~spleen) in a chronic exposure (up to 5 ppm for 90 d, Fig. 12). Except for the gastrointestinal tract and the bone, an almost identical order of tissues is observed by Joyner (1971b) in a 96-h exposure of brown bullheads (*Ictalurus nebulosus*) to 0.25–6.00 ppm Zn in water of 74 ppm hardness and pH 7.6. The importance of the kidney, gills, and liver in Zn-processing is obvious from such studies (Joyner, 1961; Mount, 1964).

In one euryhaline species (sticklebacks, *Gasterosteus aculeatus*), the order of tissues concentrating Zn is almost reversed that of fresh-water fish such as bluegills and bullheads (Matthiessen and Brafield, 1977). Sticklebacks exposed for 24 h to 1 ppm stable Zn and 1μCi Zn-65/l show highest Zn accumulation in the gill (of those organs monitored), followed by liver, kidney, and gonad (Fig. 13). After 16-h in 1 ppm stable Zn, stickleback gills, livers, and kidneys have concentration factors (CF) of 5.1, 3.8, and 3.2, respectively (Matthiessen and Brafield, 1977). Gonads, on the contrary, show little or no accumulation of Zn. Although gonads have high Zn levels normally, Zn turnover in these sex organs is low (Pentreath, 1973).

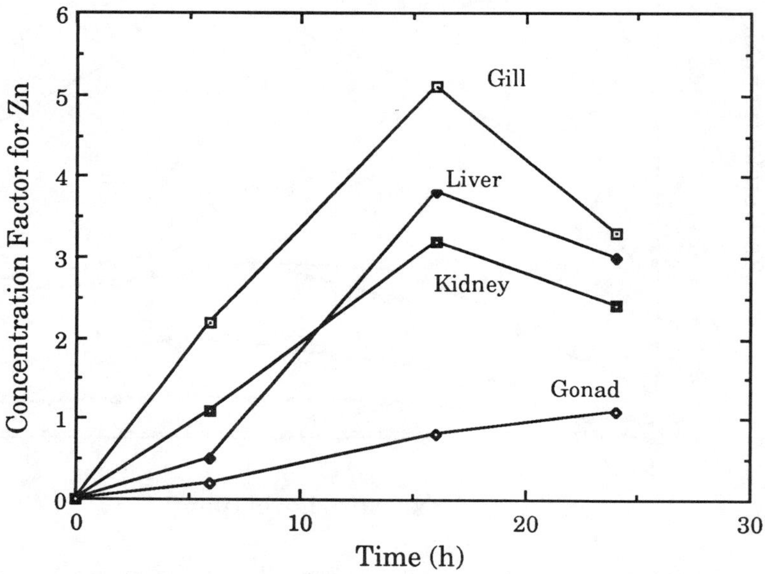

Figure 13. Accumulation of Zn-65 in several organs of sticklebacks (*Gasterosteus aculeatus*) exposed to 1 ppm stable Zn and 1 μCi Zn-65/l for 24 h. After 16 h, gill, liver, and kidney accumulate three to five times the level in water; whereas, gonads do not accumulate appreciable Zn-65. Low turnover of gonadal Zn probably explains this result. Data points represent pooled values for thirty to fifty fish. (Modified and redrawn from Matthiessen and Brafield (1977).)

When fish have been pre-exposed to Zn, CF differ from those in which no pre-exposure occurs. Two such cases involve dogfish (*Scyliorhinus canicula*), an elasmobranch collected from a severely polluted area in the Mediterranean Sea, near Barcelona where wastes from the metropolitan area are poured through the Llobregat River (Crespo, *et al.*, 1979; Flos, *et al.*,

1979). "Control" muscle levels are 57 ppm (dry weight)—about three to seven times higher than eight species of Chondrichthys from unpolluted areas (Windom, *et al.*, 1973). Control pancreas levels of Zn in dogfish average 230 ppm at the beginning of a 25-d exposure to 15 ppm Zn in the water (14–16°C and 35.8–37.2 ‰ salinity). After exposure, 621 ppm is measured in the pancreas—a CF of 26 for the 25-d exposure. Similarly, hepatic levels increase from 47 to 243 ppm Zn (CF, 13); spleen levels increase from 87 to 147 ppm Zn (CF, 4); gill levels increase from 71 to 113 ppm (CF, 3); and skeletal muscle levels increase from 57 to 68 ppm (CF, 0.7). Kidney levels of Zn actually decrease from 121 to 93 ppm. Differential organ binding of Zn by Cd-binding protein (Cd-BP) may account for low renal Zn levels. For example, goldfish injected with 0.0041 ppm Zn show 40% incorporation of Zn in hepatic Cd-BP, but only 1.2% of Zn is bound by Cd-BP in the kidney (Marafante, 1976).

Differences in exposure time, exposure level, and the salinity tolerance of these species prevents precise comparisons between tissue distribution patterns for dogfish, sticklebacks, bullheads, and bluegill. Because some fish (e.g. sticklebacks) are euryhaline and can adapt to eliminate sudden influxes of ions, great differences in Zn processing are to be expected between organs of euryhaline and stenohaline species.

The importance of scales in accumulation, storage, and depuration of Zn has warranted several studies (Sauer and Watabe, 1984; Reader, *et al.*, 1989; Sauer and Watabe, 1989). Unlike Cu and Cd, Zn does not prevent calcification of scales, based on evaluations using alizarin red S staining (Reader, *et al.*, 1989). In fact, Zn actually deposits in circuli (calcified, sequential growth rings on the scale), but not in osteoblasts or uncalcified fibrillary plates (Sauer and Watabe, 1984; ,1989).

In one study, Sauer and Watabe (1984) expose 2-wk-old mummichogs (*Fundulus heteroclitus*) to Zn at high (7.88 ppm Zn), medium (0.81 ppm Zn), or low (0.21 ppm Zn) levels in salt water (25 ‰ salinity) at 20–24°C. After an 8-wk exposure, the Zn level in scales increases linearly with exposure concentration. A maximum of 2130 ppm Zn (dry weight) is accumulated by scales of mummichogs exposed 8-wk to 7.88 ppm Zn. This represents 30% of the Zn body burden. Average uptake values for this group (8 wk and 7.88 ppm exposures) are 19% of the Zn body burden (1336 ppm dry weight). Relative scale weight remains constant during this 8-wk period for fish in all treatments. X-ray microanalysis data indicates that scales from 3-mo-old mummichogs exposed in 7.1 ppm Zn have Zn present in circuli, but not other parts of scales. Zinc-exposure also results in lower Ca/P ratios for circuli (1.28 *versus* 1.63 for controls), as well as higher Zn/Ca and Zn/P ratios than controls.

Interpretation of these results suggests that scales take up Zn actively, rather than passively. Increased percentages of Zn in scales relative to total body burden and the absence of a weight change in scales lead to this conclusion. Only 8–30% of the observed reduction in Ca/P ratios can be accounted for by ion replacement (i.e. Zn^{+2} ions substituting for Ca^{+2} ions in the hydroxyapatite crystal lattice). Therefore, Zn-induced injury at the gill epithelium probably decreases Ca uptake, resulting in less available Ca for scale calcification. In fact, Ca blood levels are lower (91 ppm *versus* 140 ppm) in rainbow trout (*Salmo gairdneri*) held in 40 ppm Zn in water until gill immobilization (Skidmore, 1970). In Skidmore's (1970) study, reduced Ca blood levels are ruled out as a causal factor in death; however, long-term reductions in blood Ca could make Ca less available for scale calcification, as suggested by Sauer and Watabe (Sauer and Watabe, 1984).

In another study, Sauer and Watabe (1989) expose mummichogs (*Fundulus heteroclitus*) to 10 ppm Zn for different periods of time. Up to 746 and 1608 ppm (dry weight) Zn accumulates in the scales in 45 and 90 d, respectively, compared with controls (229 ppm). Energy dispersive X-ray microanalysis of Zn/Ca ratios in the internal calcified scale matrix (rather than the surface of the scale) shows that Zn is deposited within the calcified scale matrix and is not simply absorbed on the scale surface. Number of circuli containing high levels of Zn are related to exposure duration.

Following 45-d exposure to 10 ppm Zn, a 21-d or 49-d depuration reduces scale levels to about 460 ppm Zn. After 49-d depuration in Zn-free water, a 43% reduction occurs in total Zn in scales. Undoubtedly, this is due to removal of Zn from deposition sites, since the scale would have had to double in size in order to add sufficient new skeletal material to the scale to dilute the existent Zn concentration. Dissolution of hydroxyapatite crystals probably allows metal resorption.

Besides Zn, several elements (Mn, Pb, Co, and Sr) are stored in scales; however, some metals (Cd, Cu, Ni, and Al) are not deposited in these integumentary structures (which are calcified tissues often classified with skeletal structures). Differences in the way in which such elements are metabolized allows prediction of results. For example, Cd and Cu (with high affinities for metallothionein) are sequestered in gills, liver, kidney, and other soft tissues. Whereas, Pb, and Sr are metabolized as is Ca; so, skeletal and scale deposition results. Unlike most soft tissues, however, metal in scales and other calcified tissues remains for long periods of time and can provide useful continuous recorders of environmental heavy metal exposure.

Retention

Retention time for Zn in fish exceeds accumulation time (Joyner, 1961; Matthiessen and Brafield, 1977; Willis and Jones, 1977; Newman and Mitz, 1988). For example, Joyner (1971b) exposes juvenile brown bullheads *(Ictalurus nebulosus)* 96 h to 6 ppm Zn in water of 74 ppm hardness (as $CaCO_3$) at pH 7.6. Live counts of fish show that uptake is rapid initially: within 12 h, uptake is near maximum. However, Zn is retained considerably longer. For brown bullheads, only 43% of Zn is lost in 24 h and 54% is lost in 168 h (after 96-h exposure to 6 ppm Zn).

In another retention study, Matthiessen and Brafield (1977) expose a euryhaline species, sticklebacks (*Gasterosteus aculeatus*) to stable and radioactive Zn in either hard water or soft water for periods of up to 400 h. In hard water (282 ppm hardness, as $CaCO_3$), Zn loss occurs at 14–24 h despite the continued exposure to 1 or 4 ppm Zn. Only fish in high Ca solutions show reduced Zn-65 levels while still exposed to Zn. Fish exposed in low Ca solutions do not lose more Zn-65 when placed in Zn-free water. As far as is known, no other species studied thusfar shows a similar reduction of internal Zn burdens despite continued exposure to 1 or 4 ppm Zn. The fact that sticklebacks are euryhaline, adapted to eliminate sudden influxes of ions, may help explain this phenomenon. In the presence of Ca, high internal Zn concentrations cause no toxicity symptoms in sticklebacks, but this is not the case in the absence of Ca. The presence of Ca ions facilitates Zn excretion.

Comparison of results from the Willis and Jones (1977) and the Newman and Mitz (1988) studies show that elimination kinetics depends on whether internalized Zn is in equilibrium with the environment to which fish are exposed. Both studies are conducted with mosquitofish (*Gambusia affinis*). Willis and Jones (1977) use an equilibrium model in which newly-hatched fish are exposed in a uniformly-labeled tank, labeled 3 mo prior and holding 2367 l of water. Mosquitofish are exposed 100 d to 200 ppb stable Zn and 0.1mCi Zn-65/l prior to 120-d depuration studies (Willis and Jones, 1977). However, Newman and Mitz (1988) use mature fish of various sizes (0.016–0.230 g dry weight) and reproductive states. Newman and Mitz (1988) expose fish 77 d to 50 ppb stable Zn and 0.1μCi Zn-65/l prior to up to 31-d depuration studies and assume Zn-65 came only from the water, not from water intake with food during feeding periods or during normal drinking of water.

Willis and Jones (1977) show that fish have reached equilibrium with the environment (~1.2 nCi Zn-65/μg Zn) after 100-d uptake of Zn-65. After uptake, fish are placed in a nonlabeled environment for 120 d to evaluate retention of Zn-65, using live counting techniques. No Zn-65 is detected in components of the unlabeled tank; therefore, the environment does not return excreted Zn-65 to fish under conditions of this experiment. Although no measurements verify this supposition, a 3200% weight increase in fish plus Zn turnover, is said to effectively replace nonlabeled Zn with Zn-65 of the same specific activity. Three mathematical reservoirs of stable Zn are present, each containing 8.72, 4.30, and 91.1% of the element with rate constants of 0.289, 0.051, and 0.00295 d^{-1} and biological half-lives of 2.47, 13.6, and 235 d, respectively. Plots of the logarithm of Zn-65 retention function for mosquitofish (the relative compartmental distribution of Zn) *versus* time show a good fit for actual and theoretical data. Uniformity of Zn-65 labeling is considered essential to determination of compartmental sizes.

For the nonequilibrium model, Newman and Mitz (1988) also describe elimination kinetics for mosquitofish. The median biological half-life for Zn-65 (single component model) is 215 d, with a 468–1816 d range for 95% of equilibrium concentration. Only this slow compartment is identified, supposedly because of the fact that Zn-65 is in water only (rather than food and water, as in the Willis and Jones (1977) study). No consideration is made for intake of Zn-65 when fish are fed or the fact that fresh-water fish drink water. Elimination is determined following 77-d uptake periods at 50 ppb Zn using 31-d depuration times. Rate of Zn elimination increases as fish size decreases showing that small fish have higher metabolic rates. In support of their nonequilibrium model, authors indicate that development of models for size-dependent body burdens of Zn in mosquitofish is invalid when equilibrium time (~2000 d) is longer than the lifetime of fish (~730 d maximum). Nonequilibrium conditions seem to pose the greatest problem in evaluation of elimination kinetics in the Newman and Mitz (1988) study.

In conclusion, retention time for Zn in fish varies but always exceeds accumulation time (Joyner, 1961; Matthiessen and Brafield, 1977; Willis and Jones, 1977; Newman and Mitz, 1988). In bullheads *(Ictalurus nebulosus)* only 43% of Zn is lost in 24 h and 54% is lost in 168 h (after 96-h exposure to 6 ppm Zn) (Joyner, 1961). Matthiessen and Brafield (1977) show that sticklebacks (*Gasterosteus aculeatus*) exposed for 16 h to 1 ppm Zn in high Ca solutions continue to eliminate Zn-65 while still exposed to Zn. This phenomenon may be due to the adaptability of this euryhaline species to sudden ion influx. However, generally, high internal Zn concentrations seem less dangerous to sticklebacks in the presence of Ca. The presence of Ca may accelerate Zn excretion. Willis and Jones (1977) show that mosquitofish which have reached equilibrium with the environment have three mathematical reservoirs of stable Zn. Each pool contains 8.72, 4.30, and 91.1% of the element with rate constants of 0.289, 0.051, and 0.00295 d^{-1} and biological half-lives of 2.47, 13.6, and 235 d, respectively. Uniform Zn-65 labeling seems essential to the evaluation of elimination kinetics.

Biological Effects

Behavioral Changes: A variety of behavioral responses are known for Zn, including avoidance behavior (Jones, 1964; Sprague, 1968), feeding rate changes (Cairns and Loos, 1967), alteration of movement patterns (Cairns and Waller, 1971c; Spoor and Drummond, 1972; Waller and Cairns, 1972), and inability to compensate for torque in a rotating water current (Bengtsson, 1974). Of these behavior changes, avoidance behavior seems to be the most sensitive (5.60 ppb), followed by failure to compensate for alteration of torque (60.00 ppb), movement pattern changes (2.94–3.64 ppm), and feeding rate changes (6.70 ppm). These behavioral changes will be addressed in order of sensitivity.

Sprague (1968) describes an "avoidance apparatus" to document the spontaneous, untrained response of fish when they are presented with a distinct choice between clean water and water containing Zn. He describes a horizontal Plexiglass trough with water flowing into either end and out the center. A sharp vertical boundary at the middle allows the observer to monitor the time spent in either side of the trough. A time-response of 50% is considered no response; whereas, 51–100% shows avoidance, and 0-49% indicates preference.

The typical rainbow trout (*Salmo gairdneri*) can detect and avoid water containing 5.6 ppb Zn (4.3–7.3 ppb, 95% confidence limits) when tested under continuous flow conditions at 13–15 ppm hardness (as $CaCO_3$) and pH 7.2 (Sprague, 1968). Neither water temperature (9.5°C or 17.0°C) nor acclimation level of Zn (3 or 13 ppb) alters this threshold avoidance level. The level of 5.6 ppb Zn is considered extremely low (0.01 times the lethal threshold concentration of 570 ppb). Sprague (1968) reviews results of other workers, who show avoidance levels 0.30–0.45 of the lethal threshold concentration for Zn. A puzzling exception is the Jones (1964) summary showing that sticklebacks and some other fish fail to avoid mildly-lethal Zn concentrations.

Zinc alters the ability of fish to remain upright in a rotating current. Bengtsson (1974) describes an apparatus to observe sublethal effects of Zn on fish. Using the rotatory-flow principle first reported by Lindahl and Schwanbom (1971), individual fish are subjected to a flow of water that is caused to rotate in a narrow tube around the direction of flow. The number of revolutions per min is raised from zero to a point at which the fish is unable to compensate for the torque of rotating flow and is forced to rotate. This point is designated the "critical r.p.m." Prior to testing, fish of three age groups are exposed for up to 150 d to Zn at various concentrations (0.06–0.30 ppm).

When subjected to a rotating flow (1–3 cm/s), Zn-exposed minnows (*Phoxinus phoxinus*) are less able to compensate for the torque and are forced to rotate. Fish less than one year old rotate after 108 d exposure to 0.06 and 0.16 ppm Zn; however, yearlings (exposed 109 d) are more resistant and do not succumb at 0.06 ppm but do rotate at 0.16, 0.33, and 0.78 ppm Zn. A 150-d exposure of yearlings increases Zn tolerance and fish compensate at 0.05–0.20 ppm, but not at 0.30 ppm Zn. For adult fish, an increase in exposure time from 100 d to 270 d reduces sensitivity to Zn. Therefore, older fish and fish exposed longer to Zn are less affected than younger fish and fish exposed for short times. The endpoint of critical rpm (rotational velocity at which fish fail to compensate for torque) is suggested as an early indication of an incipient lethal toxic effect.

Excess Zn alters the rate of feeding. After exposing zebra danios (*Brachydanio rerio*) to various concentrations of Zn, Cairns (1967) measures the time required for fish to consume ten of twenty, 1-mm pieces of tubifex worms. Most Zn-exposed fish require more time to consume ten worms. Doubling Zn exposure levels (3.7 to 6.7 ppm) or increasing exposure times causes more pronounced effects. For example, all except one of the 96-h controls consumes worms in less than 2.5 min; whereas, half of fish exposed 96 h to 6.7 ppm Zn took more than 6.5 min. In contrast, only 30% of fish exposed 96 h to 3.7 ppm Zn require more than 6.5 min.

Despite experiment-induced variation and individual variability, the effect of Zn on feeding rate in zebra danios is to prolong the time required for consumption of half of worms offered to them. Whether olfactory receptors or some other organs are damaged is not determined in this experiment. Even though measurement of the effects of Zn using this feeding behavior is simple, variability is pronounced. As the authors note, "normal" behavior might not be measured in this study because zebra danios generally feed in schools rather than singly (as tested). Also, the bare 2.5 gal aquaria hardly resemble natural habitats.

Zinc exposure results in increased bluegill activity at the surface of aquaria. Measurement of fish movement patterns is designed to monitor Zn levels in the water using in-plant or in-stream monitoring. Despite the "insensitivity" compared with other methods presented here, considerable information has been derived from such monitoring.

Waller (1972) measures movement continuously by light beam interruption during exposure of bluegills (*Lepomis macrochirus*) to 2.93–5.93 ppm Zn at 51 ppm hardness, 19.7°C, and pH 7.1. The apparatus does not interfere with fish movement and allows maintenance of fish for long periods of time. Neither feeding nor time between handling and data collection are detected in the movement patterns. The lowest Zn concentration detected by bluegills in this system is between 2.94 and 3.64 ppm during a 96-h exposure. Turbidity of the water can cause erroneous counts; however, increasing lamp output and alignment of photocells with light beams can increase operational efficiency over a broader turbidity range.

Generally, this method of measuring fish movements provides a reliable way of detecting sublethal Zn levels. However, the method requires more than 6.5 h for detection of toxic levels of Zn (i.e. 10 ppm). For this reason, the method seems to lack the sensitivity of some other methods. Moreover, although fish are likely of detecting Zn at lower levels (5.6 ppb) and within shorter times, the system of fish-photocell-computer is not able to detect Zn as rapidly. Moreover, even if a fish detects a poison immediately, behavior is not necessarily altered in a measurable way at the time of detection.

Perhaps additional means of detecting movement will provide better measures of fish response to Zn for in-plant or in-stream monitoring systems. Spoor (1972) uses a 200-cm, partially shielded, vertical electrode to detect opercular rates, body position, orientation, and locomotor activity. Bluegills are evaluated in this study, but the electrodes are so sensitive that they can detect movements of *Daphnia* (about 1 mm long) at 1 cm from the electrode.

In summary, of the behavioral changes considered here, avoidance behavior of fish is measured with the greatest sensitivity (5.60 ppb), followed by failure to compensate for alteration of torque (60.00 ppb), movement pattern changes (2.94–3.64 ppm), and feeding rate changes (6.70 ppm).

Physiological Alterations: An impressive number of physiological alterations are known to occur in fish exposed to Zn. Over time, increases are noted in ventilation rate, ventilation volume, frequency of coughing, percent coupling (the correspondence between cardiac and ventilatory rhythms) (Skidmore, 1970; Sparks, *et al.*, 1972; ,1975; Hughes and Adeney, 1977). Decreases are noted in the rate of oxygen tension and pH of arterial blood, as well as in oxygen uptake and heart rate (Skidmore, 1970; Sellers, *et al.*, 1975; Hughes and Adeney, 1977).

In an effort to explain the physiological cause of death in rainbow trout (*Salmo gairdneri*), Skidmore (1970) exposes fish to 40 ppm Zn until death (at 15°C, pH 7.1–7.5, and 50 ppm hardness). Changes in blood osmotic concentration and in blood levels of Na, K, Ca, Mg, and Zn are slight. Striking changes, however, are noted in ventilation volume, opercular rate, coughing rate, and QRS wave height (Fig. 14 and 15). Maximum values occur at 80–90% survival time prior to respiratory collapse. In contrast, oxygen utilization, heart rate, and oxygen tension of dorsal aortic blood fall steadily over the same time period. Bradycardia can lead to synchrony of breathing and cardiac rhythms.

Despite equal rates of oxygen uptake over most of the experiment (Fig. 14A), oxygen tension in arterial blood decreases throughout (Fig. 14B). This result suggests a block in the pathway whereby oxygen enters the gill (secondary lamellae, basement membrane, and flanges of pillar cells). Zinc

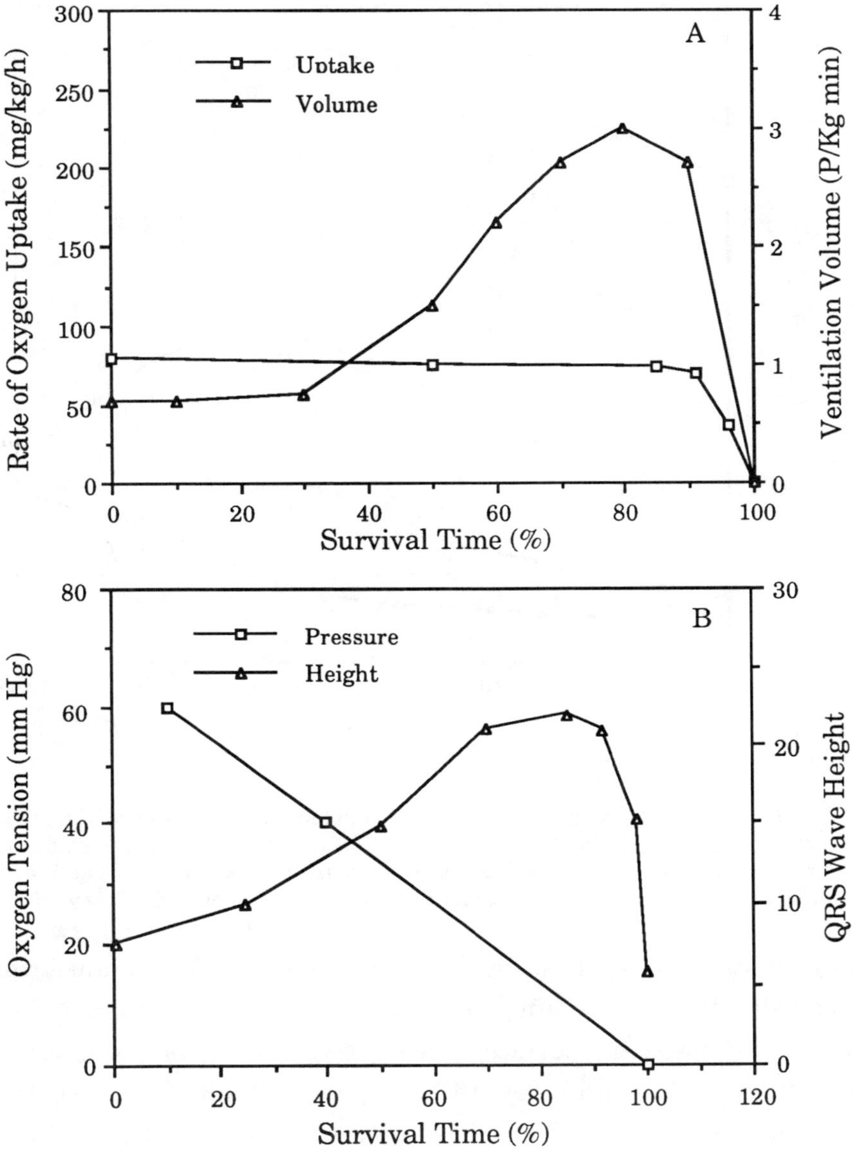

Figure 14. Oxygen uptake rate (A), oxygen tension in the dorsal aorta (B), ventilation volume (A), and QRS wave height (B) in rainbow trout (*Salmo gairdneri*) exposed to 40 ppm Zn in water. Although no change occurs in the rate of oxygen uptake until just before death, oxygen tension of arterial blood continues to fall throughout the experiment. This suggests a block in oxygen uptake at the gill. Ventilation volume and QRS wave height follow the same pattern: increasing steadily until 80–90% of survival time, then decreasing abruptly prior to death. The significance of QRS wave height changes (measured in arbitrary units) is unknown, but results suggest that marked changes occur in heart physiology. (Modified and redrawn from Skidmore (1970).)

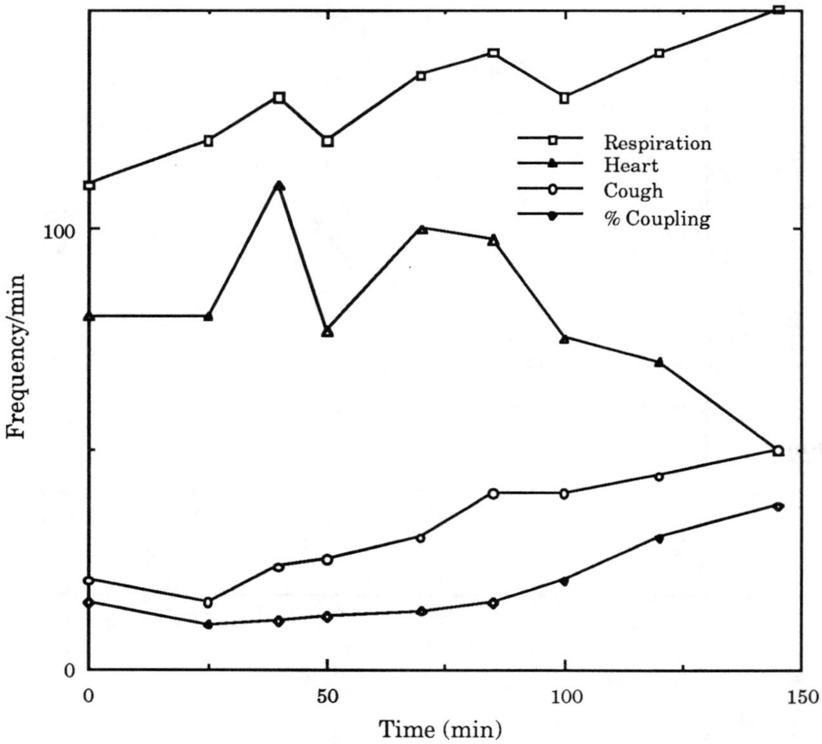

Figure 15. Changes in respiratory and cardiac parameters during exposure of rainbow trout (*Salmo gairdneri*) to 40 ppm Zn in the water. Ventilation and coughing frequency increase, but heart rate decreases. Percent coupling (the relationship between cardiac and respiratory rhythms) increases with time. (Modified and redrawn from Hughes and Adeney (1977).)

injures gills structurally, altering the ability of bluegills to transfer oxygen and decreasing the oxygen diffusion capacity by increasing barrier thickness and decreasing functional surface area for oxygen transfer (Skidmore, 1970; Hughes, 1973; Hughes and Perry, 1976; Hughes and Adeney, 1977). Death of trout probably is caused by Zn-induced decreased permeability of the gills to oxygen (but not to a number of cations). Tissue hypoxia would occur when maximum ventilation no longer supplies oxygen requirements of rainbow trout.

Hughes and Adeney (1977) also expose rainbow trout (*Salmo gairdneri*) to 40 ppm Zn. Increased respiratory frequency and decreased heart frequency result (Fig. 15), as is the case for fish exposed to hypoxic conditions (Randall and Shelton, 1963). The relationship between cardiac and ventilatory rhythms (% coupling) is increased in trout having low coupling prior to exposure. The rate of coughing also increases during Zn exposure (at 40 ppm) and during a 4-h exposure of trout to 10 ppm Zn, but to a lesser extent in the latter. Measurement of cardiac and ventilatory frequencies prior to and following Zn exposure, reveals less marked bradycardia and less pronounced increases in ventilatory frequency after exposure than prior to exposure. Again, Zn probably interferes with oxygen uptake at the gills.

Sellers and associates (1975) cannulate rainbow trout (*Salmo gairdneri*) of 226 g body weight in order to monitor opercular and buccal pressure changes (ventilation amplitude, ventilation rate, and coughing frequency) during exposure to Zn singly and in combination with Cu. An intermediate concentration of Zn (0.69 ppm Zn) elicits greater responses (in ventilation amplitude, ventilation rate, and coughing frequency) than higher concentrations. As might be expected, higher Zn levels cause more variation in ventilation pressure changes than lower levels. Singly or in combination with Cu, Zn causes increased coughing frequency. For Zn-Cu combinations, synergism is observed (see Chapter IX). Coughing frequency at 0.04 ppm Cu and 0.66 ppm Zn combined is about one and one-half times greater than the sum of the responses to the single toxicants.

Noteworthy is the marked individual variability in ventilatory activity in the study by Sellers *et al.* (1975). The general irritation of buccal and possibly gill membranes results in aberrant movement. Amplitude of ventilation may be low and ventilation rate may be high for one subject (or for one concentration) but not for a second subject (or concentration). Unfortunately, Sellers *et al.* (1975) do not report total respiratory activity for individual fish using ventilation rate and amplitude date (i.e. volume of water breathed per unit time). Perhaps the individual variation would be reduced in such comparisons.

Arterial oxygen tension and pH are also monitored and provide useful data (Sellers, *et al.*, 1975). In 21-h exposures to 1.43 ppm Zn (the LC50 at 51–68 ppm hardness) 66% decreases in arterial oxygen tension result. Some block in the respiratory pathway may cause this decrease in oxygen tension. Perhaps the increased water to blood diffusion distance (Skidmore and Tovell, 1972) plays a major role. Zinc (1.43 ppm for 21 h) causes pH of arterial blood to decrease—an indication of lactic acid and/or carbon dioxide accumulation (Burton, *et al.*, 1972b).

The ventilatory changes which occur during exposure of fish to Zn are fairly reliable measures of the level of Zn in water. In fact, breathing rate changes can be used to detect Zn in in-plant and in-stream monitoring systems—recommended to safeguard surface waters. Sparks and associates (Sparks, *et al.*, 1972) design such a system to measure breathing rate using analog-to-digital converters, mini-computers, and teleprinters. Recordings (15-min) are taken at half-hour intervals at specific periods throughout the day. Bluegills (*Lepomis macrochirus*) can "detect" 2.55–8.70 ppm using this method. Bluegills detect 2.55, 4.16, and 5.22 ppm Zn within 8, 11, and 4 h, respectively, if detection is based on two fish showing the response simultaneously.

Because increased breathing amplitude is observed in addition to the increases in ventilation frequency, the better in-plant or in-stream monitoring systems might be designed to measure both changes simultaneously. Heath (1972) publishes a review of six methods of measuring ventilation changes and indicates that the respiratory pressure recording method yields the most information with the least amount of detrimental effect on the fish. This method requires catheterization (through holes in the clythrum and snout), and connection of catheters to pressure transducers. Respiratory volume can be obtained in addition to respiratory frequency. Only frequency is obtained for the other five methods (visual observation, implanted monopolar electrode, dual external electrodes, opercular wires, and oral membrane system). Such a method would allow measurement of two vital physiological parameters as fish are exposed to Zn, which is a common component of many types of pollution.

Histopathology: Tissue injury is caused by Zn-poisoning of a number of species of fish, including the common guppy (*Poecilia reticulata*), grass carp (*Ctenopharyngodon idellus*), common carp (*Cyprinus carpio*), and the stickleback (*Gasterosteus aculeatus*). In addition to causing growth inhibition and retardation of sexual maturity in the common guppy (Crandall and Goodnight, 1963), excess Zn causes injury to gills, kidney, liver, and skeletal muscle in the guppy and several other species (Crandall and Goodnight, 1963; Matthiessen and Brafield, 1973; Wong, *et al.*, 1977).

Crandall and Goodnight (1963) use a 95-d exposure of the common guppy (*Poecilia reticulata*) to 1.15 or 2.30 ppm Zn at 80 ppm hardness, pH 7.9–8.2, and 25–27°C. Notable changes include histopathological damage, growth inhibition, and retardation of sexual maturity in newly-hatched fish. Renal tubular dilation, tubular casts, reduced mesenteric fat, and decreased renal peritubular lymphoid tissue are present. Gonadal development is retarded, and hepatocyte damage (including the appearance of "ring" nuclei in parenchymal cells and vacuolation) are noted. Skeletal muscles show small diameter fibers and degenerative changes. Progressively more severe damage is noted as Zn level and exposure time are increased. Zinc-exposed fish are severely stunted—possibly due to poor food consumption.

Wong and associates (1977) report that grass carp *(Ctenopharyngodon idellus)* and common carp *(Cyprinus carpio)* show unequal sensitivity to Zn. Grass carp (3–5 cm in length) are more susceptible to Zn (in water of 195 ppm hardness, 22–23°C, and pH 6.2–7.1). Gill and liver tissues show a particulate deposit, but no effort is made to identify deposits. Since polarization of particles is not checked, it is not known whether these deposits are metallic. Liver cells for carp exposed to 25 ppm Zn show coagulation necrosis—a condition in which irreversible dilatation occurs and protoplasmic viscosity increases, as noted by increased staining affinity. The presence of necrotic, degenerating hepatocytes is considered indicative of considerable hepatic injury. The LC50 value of Zn is not calculated but data suggest that the value approximates near-zero levels.

Skidmore (1970) observes that most gill damage occurs between 75–90% of survival time in rainbow trout (*Salmo gairdneri*). In a subsequent study, Skidmore and Tovell (1972) report histopathological injury at different points in survival time. Fingerling rainbow trout (*Salmo gairdneri*) are exposed to 40 ppm Zn and timed for one of three responses: surfacing (189 min and 89% of survival time), overturn (201 min and 94% of survival time), or immobilization (213 min and 100% of survival time or death). The first sign of gill damage is a partial detachment of chloride cells from the underlying epithelium. After 60% of survival time elapses, half of secondary lamallae are curled and epithelium on the concave (but not convex) side is lifted. A 20% reduction occurs in the maximum distance of the lamellae margin from the filament surface. At overturn, half of secondary lamallae are curled such that a 50% reduction occurs in the length of the secondary lamellae. Central blood spaces are partially occluded, but proximal and marginal margins are enlarged. The subepithelial space enlarges and forms on both lamellar surfaces. Epithelium is still attached normally along the marginal channel. At immobilization, additional curling occurs with only lamellar margins remaining attached. Central blood spaces are completely occluded and the subepithelial space is further enlarged. Removal of Zn-exposed fish at the surfacing stage and transfer to Zn-free water results in overlapping, adhering secondary lamellae, which parallel the filaments. After 7 d, although some sections appear normal, control lamellar structure is not reestablished. Therefore, the Zn-induced reactions appear typical of an acute inflammatory reaction resulting from direct contact between the gill epithelium and Zn.

Matthiessen and Brafield (1973) find that levels of 0.5–1.0 ppm Zn in water kill mature sticklebacks in 1–3 d and injure gill secondary lamellae. Epithelial cells detach, adjacent epithelia coalesces, vacuolation occurs, and nuclei and mitochondria swell. Despite such damage, recovery occurs for many acutely poisoned fish if these are moved to Zn-free hard water. Epithelia regenerate, accompanied by a temporary appearance of chloride cells on the secondary lamellae. Over 700 h exposures to 2–6 ppm Zn in hard water (282 ppm, as $CaCO_3$) are not toxic; however a metabolite (possibly lipid) accumulates and membrane-bounded vesicles are formed. Active chloride cells appear on the secondary lamallae. Fish exposed to 1 ppm Zn for 16 h show vacuolation and epithelial detachment with coalescing adjacent secondary lamellae; but in 5 d, recovery occurs if fish are placed in Zn-free water. After 9-d recovery periods, secondary lamellae appear to be completely normal morphologically.

In control sticklebacks, water to blood distance in the gill normally averages 966 nm; whereas, abnormal gills from Zn-poisoned fish average 2110 nm. Detachment of epithelial cells from basal lamina causes the two-fold increase in distance. In hard water tests of Zn-induced morphological changes, the appearance of large numbers of chloride cells over secondary lamallae increases the mean thickness of the water to blood pathway to 2745 (about three times the distance in controls). This proliferation of chloride cells occurs when fish are exposed about 100 h at 2–6 ppm Zn. These cells may be produced as a response to influx of Zn ions. And although such a great increase in water to blood distance would seem to make gaseous exchange more difficult, chloride cells appear to be involved in Zn excretion and recovery from Zn-induced damage.

Histochemical studies using cultured cells allow determination of intracellular constituents which react with Zn. Studzinski (1965) exposes cultured HeLa cells to 20 μM Zn at pH 6.0–6.5 and 37°C. Zinc is retained by the nucleolus. Basicity of the Zn-binding protein is revealed by acid extraction. Circumscribed areas of the nucleus have what are considered exceptionally high concentrations of stainable amino and imidazole groups. Extraction procedures reveal that proteolytic enzymes rapidly and completely remove substance(s) responsible for Zn binding in the nucleolus. Primary amino acid groups (especially the more reactive) are found to be essential for Zn binding; and the conclusion is reached that the unusually reactive proteins could be nascent polypeptides produced by the nucleolus but not yet linked to RNA or other cellular constituents. Proteins rich in lysine appear likely candidates, since high concentrations of unbound, accessible, basic nitrogenous groups are present in lysine. Imidazole groups of histidine residues may play a part in Zn retention. Zinc binding in fish cells could be expected to be similar to that of these cultured human cells.

To summarize, Zn-exposure of both euryhaline and fresh-water species of fish results in histopathological alterations of gill, liver, kidney, and skeletal muscle. Damage becomes progressively more severe as Zn level and exposure time are increased. Morphological evaluation of the gill shows that Zn-induced injuries seem to be repaired completely after a 9-d recovery period. Severe stunting of Zn-exposed fish may be due to poor food consumption, suggesting secondary effects of inanition and/or prolonged Zn-induced stress. Therefore, most histopathological changes induced by Zn (and also by other metals) are believed to be nonspecific.

<u>Enzymatic Alterations</u>: Shepart and Simkiss (1978) collect the 35,000 g fraction rich in Ca^{++}-activated ATPase from gill filaments of unexposed roach (*Rutilus rutilus*), a fresh-water fish. A 30-min incubation of this fraction with Zn (50μM) inhibits Ca^{++}-ATPase activity to about 40% of the initial level. Considerably greater inhibition is obtained by adjusting incubation procedures. In comparison, Cu and Hg cause 50% inhibition of Ca^{++}-ATPase at <5 μM. In fact, 8-wk exposures of roach *in vivo* to 5 or 10 ppb Cu cause slightly decreased (nonsignificant) Ca^{++}-ATPase activity per mg protein per hour, but significantly increased protein content to a maximum of 78%. Moreover Ca^{++}-ATPase activity per g gill filaments increased at the 10 ppb, 8-wk exposure to Cu. Thus, Cu apparently induces production of extra

membrane proteins, although not all extra protein is likely to be Ca^{++}-ATPase. A corresponding increase in the number of enzyme units would be necessary to maintain the overall Ca^{++}-ATPase activity in the 35,000 g pellet. Therefore, both Zn and Cu inhibit Ca^{++}-ATPase activity.

Watson and Beamish (1980) expose rainbow trout (*Salmo gairdneri*) to a maximum of 2 ppm Zn. Gill ATPase activity for Mg^{++}-, Ca^{++}-, Na^{+}-K^{+}-, and Na^{+}-NH_4^{+}-ATPase is augmented. However, only minute changes are noted at ambient temperature (11°C) compared with optimal enzyme temperature (37°C). Moreover, changes are noted at 11°C for only 0.29 ppm Zn-exposed trout (30 d) and not in trout exposed 30 d to 0.99 or 1.98 ppm Zn in water. Neither serum osmolarity nor electrolyte concentrations (for serum Cl^{-}, K^{+}, Ca^{++}, Mg^{++}, and Zn^{++}) are affected by 30-d exposures to about 2 ppm Zn. No significant differences occur in brachial microsomal protein with Zn exposure. Ultimately, augmentation of gill ATPase activities might reduce survival due to Zn-impairment since ATPase activity would be expected to be higher after exposure to Zn-polluted surface waters. Moreover, poorer growth rate might result from the metabolic cost of energy expenditure required to maintain an increased rate of ATP hydrolysis.

Anadromous fish (e.g. rainbow trout) and catadromous fish (e.g. American eel) would be expected to respond differently than fresh-water fish to this minimal Zn challenge (Watson and Beamish, 1980). Perhaps a more measurable result would be evident since anadromous fish can adapt to hyperosmotic solutions with less stress than fresh-water fish. Salt water fish expel salts (such as Zn) from the gills; whereas, fresh-water fish actively transport salts into the body *via* the gills. Therefore, it is likely that ATPase activity would differ markedly in fresh-water fish challenged with Zn—much more so than in salt water fish.

Protective Mechanisms: Eddy and Fraser (1982) find that the presence of 2–20 ppm Zn in aqueous solutions alters mucous production in 20-50 g rainbow trout *(Salmo gairdneri)* exposed in fresh water (Na, 0.22 mM; Cl, 0.3 mM; Ca, 0.1 mM) or in seawater (32 ‰). Rate of production of sialic acid (a measure of relative mucous production) is normally low for rainbow trout exposed in fresh water (1–2 nM/h/g), corresponding to a mucous production of 6 mg/h/g body weight. Placing trout in 20 ppm Zn or in seawater for 6 h, increases sialic acid production rates to about 80 nM/h/g, corresponding to 370 mg mucus/h/g. This represents a sixty-fold increase over the normal rate.

Normally, therefore, the coat of mucus on rainbow trout is replaced at a slow rate. However, when trout are exposed to stressors such as heavy metals or salt water, production rate of mucus is rapid. This research assumes that mucus collected is identical to mucus secreted (despite the established pattern of hydration after secretion), that composition of mucus (i.e. sialic acid content) is identical for both gills and body surface (very likely to be true), and that production rates are identical for gills and body surface (true for some species but not for others).

Summary

Zinc causes mortality, growth retardation, histopathological alterations, respiratory and cardiac changes, inhibition of spawning, and a multitude of additional detrimental effects which threaten survival of fish. From 0.87 to 40.90 ppm Zn is reported as the 96-h LC50 level for several species (Table 1). Variability is caused by differences in salinity tolerance, water hardness, temperature, pH, the presence of other elements, species and additional factors. Moreover, use of various behavioral paradigms yields behavioral changes in the 5.6 ppb to 6.7 ppm Zn concentration range. Avoidance behavior is altered at 5.6 ppb Zn. Fish are unable to compensate for torque alterations at 60.00 ppb Zn. Movement patterns change at 2.94–3.64 ppb Zn, and feeding rate changes at 6.70 ppm. Zinc causes nonspecific histopathological alterations in a number of species. Gill, liver, kidney, and skeletal muscle are damaged. The first sign of gill damage is detachment of chloride cells from underlying epithelium. The subepithelial space enlarges because of detachment of epithelial cells from basal lamina. Water to blood distance can more than double, making gaseous exchange more difficult. Therefore, even though Zn is an essential element serving many useful functions in fish, Zn is extremely toxic to fish when present at elevated levels.

References

Baptist, J.P., and C.W. Lewis, Transfer of Zn-65 and Cr-51 through an estuarine food chain, Symp. Radioecol., CONF-670503, 2nd Nat'l. Symp., May 15-17, Ann Arbor, Michigan, (1967) 420.

Barrington, E.J.W., The Chemical Basis of Physiological Regulation, Scott, Foresman and Co., Glenview, Ill., (1968) 1.

Bengtsson, B.-E., The effect of zinc on the ability of the minnow, *Phoxinus phoxinus* L., to compensate for torque in a rotating water current, Bull. Environ. Cont. Toxicol., 12 (1974) 654.

Beverton, R.J.H., and S.J. Holt, Review of the life span and mortality rates of fish in nature and their relation to growth and other physiological characteristics, Ciba Foundation Colloquia on Aging, 5 (1959) 142.

Brungs, W.A., Chronic toxicity of zinc to the fathead minnow, *Pimephales promelas* Rafinesque, Trans. Am. Fish. Soc., 98 (1969) 272.

Burton, D.T., E.L. Morgan, and J. Cairns, Jr., Mortality curves of bluegills (*Lepomis macrochirus* Rafinesque) simultaneously exposed to temperature and zinc stress, Trans. Am. Fish. Soc., 101 (1972a) 435.

Burton, D.T., A.H. Jones, and J. Cairns Jr., Acute zinc toxicity to rainbow trout (*Salmo gairdneri*): Confirmation of the hypothesis that death is related to tissue hypoxia., J. Fish. Res. Bd. Can., 29 (1972b) 1463.

Cairns, J., Jr., and A. Scheier, The effects of temperature and hardness of water upon the toxicity of zinc to the common bluegill (*Lepomis macrochirus* Raf.), Not. Nat. Acad. Sci., Phil., 299 (1957) 1.

Cairns, J., Jr.,, and A. Scheier, The effects of periodic low oxygen upon the toxicity of various chemicals to aquatic organisms, 12th Industrial Waste Conf. (1957), Lafayette, Indiana, Purdue Univ., (1958) 165.

Cairns, J., Jr., and A. Scheier, The relationship of bluegill sunfish body size to tolerance for some common chemicals, Proceedings 13th Indust. Waste Conf., Purdue Univ. Engn. Bull., (1959) 243.

Cairns, J., Jr., and J.J. Loos, Changed feeding rate of *Brachydanio rerio* (Hamilton-Buchanan) resulting from exposure to sublethal concentrations of zinc, potassium dichromate, and alkyl benzene sulfonate detergent, Proc. Penn. Acad. Sci., 40 (1967) 47.

Cairns, J., Jr., T.K. Bahns, D.T. Burton, K.L. Kickson, R.E. Sparks, and W.T. Waller, The effects of pH, solubility and temperature upon the acute toxicity of zinc to the bluegill sunfish (*Lepomis macrochirus* Raf.), Trans. Kan. Acad. Sci., 74 (1971a) 81.

Cairns, J., Jr., N.H. Cromer, T.K. Bahns, and W.T. Waller, A confirmation of Mount's autopsy technique for zinc-caused fish mortality, Water Res. Bull., 7 (1971b) 956.

Cairns, J., Jr., and W.T. Waller, The use of fish movement patterns to monitor zinc, Water Pollut. Cont. Res. Ser., Project #18050EDP12/71, (1971c) 1.

Cairns, J., Jr., A.G. Heath, and B.C. Parker, The effects of temperature upon the toxicity of chemicals to aquatic organisms, Hydrobiologia, 47 (1975) 135.

Chapman, G.A., Toxicities of cadmium, copper, and zinc to four juvenile stages of chinook salmon and steelhead, Trans. Am. Fish. Soc., 107 (1978) 841.

Crandall, C.A., and C.J. Goodnight, The effects of sublethal concentrations of several toxicants to the common guppy, *Lebistes reticulatus*, Trans. Am. Microsc. Soc., 82 (1963) 59.

Crespo, S., R. Flos, F. Balasch, and G. Alonso, Zinc in the gills of the dogfish (*Scyliorhinus canicula* L.) related to experimental aquatic zinc pollution, Comp. Biochem. Physiol., 63C (1979) 261.

Cusimano, R.F., D.F. Brakke, and G.A. Chapman, Effects of pH on the toxicities of cadmium, copper, and zinc to steelhead trout (*Salmo gairdneri*), Can. J. Fish. Aquat. Sci., 43 (1986) 1497.

Eddy, F.B., and J.E. Frasher, Sialic acid and mucus production in rainbow trout (*Salmo gairdneri* Richardson) in response to zinc and seawater, Comp. Biochem. Physiol., 73C (1982) 357.

Everall, N.C., N.A.A. MacFarlane, and R.W. Sedgwick, The interactions of water hardness and pH with the acute toxicity of zinc to the brown trout, *Salmo trutta* L., J. Fish Biol., 35 (1989) 27.

Farmer, G.J., O. Ashfield, and H.S. Samant, Effects of zinc on juvenile Atlantic salmon *Salmo salar*: Acute toxicity, food intake, growth, and bioaccumulation, Envir. Pollut., 19 (1979) 103.

Flos, R., A. Caritat, and F. Balasch, Zinc content in organs of dogfish (*Scyliorhinus canicula* L.) subject to sublethal experimental aquatic zinc pollution, Comp. Biochem. Physiol., 64C (1979) 77.

Goodman, J.R., Toxicity of zinc for rainbow trout (*Salmo gairdneri*), Calif. Fish Game, 37 (1951) 191.

Goodyear, C.P., and C.E. Boyd, Elemental composition of largemouth bass (*Micropterus salmoides*), Trans. Amer. Fish. Soc., 3 (1972) 545.

Hamilton, S.J., P.M. Mehrle, and J.J. Jones, Evaluation of metallothionein as a biological indicator of stress from cadmium in brook trout, Trans. Am. Fish. Soc., 116 (1987) 551.

Heath, A.G., A critical comparison of methods for measuring fish respiratory movements, Wat. Res., 6 (1972) 1.

Hughes, G.M., Respiratory responses to hypoxia in fish, Amer. Zool., 13 (1973) 475.

Hughes, G.M., and S.F. Perry, Morphometric study of trout gills: A light-microscopic method suitable for the evaluation of pollutant action, J. Exp. Biol., 64 (1976) 447.

Hughes, G.M., and R.J. Adeney, The effects of Zn on the cardiac and ventilatory rhythms of rainbow trout (*Salmo gairdneri*, Richardson) and their responses to environmental hypoxia, Water Res., 11 (1977) 1069.

Jones, J.R.E., The metals as salts, Fish and River Pollution, Butterworth & Company, London, (1964) 66.

Joyner, T., Exchange of zinc with environmental solutions by the brown bullhead, Trans. Am. Fish Soc., 90 (1961) 444.

Judy, R.D., Jr., and P.H. Davis, Effects of calcium addition as $Ca(NO_3)_2$ on zinc toxicity to fathead minnows, *Pimephales promelas*, Rafinesque, Bull. Envir. Cont. Toxicol., 22 (1979) 88.

Krumholz, L.A., Reproduction in the western mosquitofish, *Gambusia affinis affinis* (Baird & Girard), and its use in mosquito control, Ecol. Mono., 18 (1948) 1.

Lindahl, P.E., and E. Schwanbom, A method for detection and quantitative estimation of sublethal poisoning in living fish, Oikos, 22 (1971) 210.

Lloyd, R., The toxicity of zinc sulfate to rainbow trout, Ann. Appl. Biol., 48 (1960) 84.

Marafante, E., Binding of mercury and zinc to cadmium-binding protein in liver and kidney of goldfish (*Carassius auratus* L.), Experientia, 15 (1976) 149.

Matthiessen, P., and A.E. Brafield, The effect of dissolved zinc on the gills of the stickleback *Gasterosteus aculeatus* (L.), J. Fish Biol., 5 (1973) 607.

Matthiessen, P., and A.E. Brafield, Uptake and Loss of dissolved zinc by the stickleback *Gasterosteus aculeatus* L., J. Fish Biol., 10 (1977) 399.

McDonald, D.G., The effects of H^+upon the gills of freshwater fish, Can. J. Zool., 61 (1983) 691.

Mount, D.I., An autopsy technique for zinc-caused fish mortality, Trans. Am. Fish. Soc., 93 (1964) 174.

Mount, D.I., The effect of total hardness and pH on acute toxicity of zinc to fish, Air Water Pollut. Int. J., 10 (1966) 49.

Mount, D.I., Chronic toxicity of copper to fathead minnows (*Pimephales promelas,* Rafinesque), Wat. Res., 2 (1968) 215.

Murphy, B.R., G.J. Atchison, and A.W. McIntosh, Cadmium and zinc content of fish from an industrially contaminated lake, J. Fish Biol., 13 (1978) 327.

Newman, M.C., and S.V. Mitz, Size dependence of zinc elimination and uptake from water by mosquitofish *Gambusia affinis* (Baird and Girard), Aq. Toxicol., 12 (1988) 17.

Odum, H.T., Ecological potential and analogue circuits for the ecosystem, Amer. Sci., 48 (1960) 1.

Pagenkoph, G.K., Gill surface interaction model for trace-metal toxicity to fishes: Role of complexation, pH, and water hardness, Envir. Sci. Technol., 17 (1983) 342.

Pentreath, R.J., The accumulation and retention of ^{65}Zn and ^{54}Mn by the plaice, *Pleuronectes platessa* L., J. Exp. Mar. Biol. Ecol., 12 (1973) 1.

Pickering, Q.H., and W.N. Vigor, The acute toxicity of zinc to eggs and fry of the fathead minnow, Prog. Fish.-Cult., 27 (1965) 153.

Pickering, Q.H., and C. Henderson, The acute toxicity of some heavy metals to different species of warmwater fishes, Air Water Pollut. Int. J., 10 (1966) 453.

Pickering, Q.H., Some effects of dissolved oxygen concentrations upon the toxicity of zinc to the bluegill, *Lepomis macrochirus*, Raf., Water Res., 2 (1968) 187.

Pippy, J.H.C., and G.M. Hare, Relationship of river pollution to bacterial infection in salmon (*Salmo salar*) and suckers (*Catostomus commersoni*), Trans. Amer. Fish. Soc., 98 (1969) 685.

Prosser, C.L., and F.A. Brown Jr., Comparative Animal Physiology, 2nd ed., Philadelphia, (1966) 688.

Rachlin, J.W., and A. Perlmutter, Response of an inbred strain of platyfish and the fathead minnow to zinc, Prog. Fish Cult., 30 (1968a) 203.

Rachlin, J.W., and A. Perlmutter, Fish cells in culture for study of aquatic toxicants, Wat. Res., 2 (1968b) 409.

Rachlin, J.W., and A. Perlmutter, Response of rainbow trout cells in culture to selected concentrations of zinc sulfate, Prog. Fish-Cult., 31 (1969) 94.

Randall, D.J., and G. Shelton, The effects of changes in environmental gas concentration on the breathing rate of a teleost fish, Comp. Biochem. Physiol., 9 (1963) 229.

Reader, J.P., N.C. Everall, M.D.J. Sayer, and Morris, The effects of eight trace metals in acid soft water on survival, mineral uptake and skeletal calcium deposition in yolk-sac fry of brown trout, *Salmo trutta* L., J. Fish Biol., 35 (1989) 187.

Rehwoldt, R., G. Bida, and B. Nerrie, Acute toxicity of copper, nickel, and zinc ions to some Hudson River fish species, Bull. Environ. Cont. Toxicol., 6 (1971) 445.

Roch, M., J.A. McCarter, A.T. Matheson, M.J.R. Clark, and R.W. Olafson, Hepatic metallothionein in rainbow trout (*Salmo gairdneri*) as an indication of metal pollution in the Campbell River System, Can. J. Fish. Aquat. Sci., 39 (1982) 1596.

Roch, M., and J.A. McCarter, Hepatic metallothionein production and resistance to heavy metals by rainbow trout (*Salmo gairdneri*)—I. Exposed to an artificial mixture of zinc, copper, and cadmium, Comp. Biochem. Physiol., 77C (1984a) 71.

Roch, M., and J.A. McCarter, Hepatic metallothionein production and resistance to heavy metals by rainbow trout (*Salmo gairdneri*)—II. Held in a series of contaminated lakes, Comp. Biochem. Physiol., 77C (1984b) 77.

Sauer, G.R., and N. Watabe, Zinc uptake and its effect on calcification in the scales of the mummichog, *Fundulus heteroclitus*, Aq. Toxicol., 5 (1984) 51.

Sauer, G.R., and N. Watabe, Temporal and metal-specific patterns in the accumulation of heavy metals by the scales of *Fundulus heteroclitus*, Aq. Toxicol., 14 (1989) 233.

Sellers, C.M., A.G. Heath, and M.L. Bass, The effect of sublethal concentrations of copper and zinc on ventilatory activity. Blood oxygen and pH in rainbow trout (*Salmo gairdneri*), Wat. Res., 9 (1975) 401.

Shepart, K., and K. Simkiss, The effects of heavy metal ions on Ca^{2+} ATPase extracted from fish gills, Comp Biochem. Physiol., 61B (1978) 69.

Skidmore, J.F., Oxygen uptake by zebrafish (*Brachydanio rerio*) of different ages in relation to zinc sulfate resistance, J. Fish. Res. Bd. Can., 24 (1967) 1253.

Skidmore, J.F., Respiration and osmoregulation in rainbow trout with gills damaged by zinc sulfate, J. Exp. Biol., 52 (1970) 481.

Skidmore, J.F., and P.W.A. Tovell, Toxic effects of zinc sulfate on the gills of rainbow trout, Water Res., 6 (1972) 217.

Solbé, J.F.d.L.G., The toxicity of zinc sulphate to rainbow trout in very hard water, Wat. Res., 8 (1974) 389.

Sorensen, E.M.B., and T.L. Bauer, Hematological dyscrasia in teleosts chronically exposed to selenium-laden effluent, Arch. Environ. Contam. Toxicol., 12 (1983a) 135.

Sorensen, E.M.B., J.S. Bell, and C.W. Harlan, Histopathological changes in selenium-exposed fish, Amer. J. Forensic Med. Pathol., 4 (1983b) 111.

Sorensen, E.M.B., and T.L. Bauer, A correlation between selenium accumulation in sunfish and changes in condition factor and organ weight, Environ. Pollut., 34 (1984a) 357.

Sorensen, E.M.B., and T.L. Bauer, Planimetric analysis of redear sunfish (*Lepomis microlophus*) hepatopancreas following selenium exposure, Environ. Toxicol. Chem., 3 (1984b) 159.

Sorensen, E.M.B., and P. Thomas, Selenium accumulation, reproductive status, and histopathological changes in environmentally exposed redear sunfish, Arch. Toxicol., 61 (1988a) 324.

Sorensen, E.M.B., T.L. Bauer, and M.G. Krause, Multielemental analysis of the hepatopancreas of selenium-exposed sunfish, Toxicologist, 8 (1988b) 108.

Sparks, R.E., J. Cairns Jr., and A.G. Heath, The use of bluegill breathing rates to detect zinc, Water Res., 6 (1972) 895.

Spoor, W.A., and R.A. Drummond, An electrode for detecting movement in gradient tanks, Trans. Am. Fish. Soc., 101 (1972) 714.

Sprague, J.B., Lethal concentrations of copper and zinc for young Atlantic salmon, J. Fish. Res. Bd. Can., 21 (1964) 17.

Sprague, J.B., Avoidance reactions of rainbow trout to zinc sulphate solutions, Wat. Res., 2 (1968) 367.

Studzinski, G.P., Selective binding of zinc by basic proteins of the Hela cell nucleolus, J. Histochem. Cytochem., 13 (1965) 365.

Wallace, W., The mortality of plaice, Nature, 115 (1925) 337.

Waller, W.T., and J. Cairns Jr., The use of fish movement patterns to monitor zinc in water, Water Res., 6 (1972) 257.

Watson, T.A., and F.W.H. Beamish, Effects of zinc on brachial ATPase activity in vivo in rainbow trout, *Salmo gairdneri*, Comp. Biochem. Physiol., 66C (1980) 77.

Williams, S.R., Nutrition and diet therapy, C. V. Mosby Company, St. Louis, (1969) 686.

Willis, J.N., and N.Y. Jones, The use of uniform labeling with zinc-65 to measure stable zinc turnover in the mosquito fish, *Gambusia affinis*—I. Retention, Health Physics, 32 (1977) 381.

Windom, H., R. Stickney, R. Smith, B. White, and F. Taylor, Arsenic, cadmium, copper, mercury and zinc in some species of North Atlantic finfish, J. Fish. Res. Bd. Can., 32 (1973) 275.

Wong, M.H., K.C. Luk, and K.Y. Choi, The effects of zinc and copper salts on *Cyprinus carpio* and *Ctenopharyngodon idellus*, Acta Anat., 99 (1977) 450.

Zitko, V., and W.G. Carson, A mechanism of the effects of water hardness on the lethality of heavy metals to fish, Chemosphere, 5 (1976) 299.

CHAPTER VI
CADMIUM

Introduction

Cadmium (Cd) is a metallic, transition element with two electrons in the outermost shell and a penultimate shell filled with eighteen electrons. As a member of the Group IIB triad in the periodic classification of elements, Cd shares properties with Zn and Hg. Like Zn, Cd forms ionic halides, but has been predicted to be incapable of environmental methylation due to instability of monoalkyl derivatives (Moore and Ramamoorthy, 1984). In contrast, Hg is capable of environmental methylation, forms covalent halides, and tends to form a large number of organometallic compounds which are stable in air and water. Also, Cd and Zn are softer, more electropositive, and have lower melting points than neighboring transition group metals.

Similarities and differences also exist in the metabolism of elements of the Group IIB triad. Metallothionein and other Cd-binding proteins (Cd-BP) hold and/or transport Cd, Zn, and Hg within the body (Marafante, 1976).

Cadmium, like Zn and Hg, accumulates in gill, liver, and kidney—presumably through strong binding with cysteine residues of metallothionein or components of other Cd-binding proteins (Marafante, 1976; Beattie and Pascoe, 1979; Noel-Lambot, *et al.*, 1978; Reichert, *et al.*, 1979; Roberts, *et al.*, 1979; Kito, *et al.*, 1982; Roch, *et al.*, 1982; Roch and McCarter, 1984a; 1984b; Woodworth, *et al.*, 1983; Wofford and Thomas, 1984). The high affinity of Cd for sulfhydryl groups and the ability of Cd to impart moderate covalency in bonds, results in increased lipid solubility, bioaccumulation, and toxicity. All three Group IIB elements can disrupt or terminate enzymatic activity (Jackim, *et al.*, 1970; Christensen, 1971/72; Vallee and Ulmer, 1972; Watson and Benson, 1987). Cadmium can displace Zn from high molecular weight proteins (Bryan and Hildalgo, 1976; Frazier and Puglese, 1978; Noel-Lambot, *et al.*, 1978), and the same carrier proteins probably shuttle Cd and Zn across plasma membranes (Huang, *et al.*, 1980). However, uptake mechanisms are different for Cd and Hg (Gerson and Shaikh, 1984). Therefore, on the basis of what is commonly known today, Cd is more similar to Zn than to Hg chemically and metabolically.

As a nondegradable cumulative pollutant, Cd is considered capable of altering aquatic trophic levels for centuries. The ubiquity of Cd and the early recognition of Cd toxicity to fish, combined with the relative simplicity of Cd analysis, has led to a multiplicity of possible studies. Environmental research compares Cd uptake and distribution under natural circumstances for fish species occupying different environments and various trophic levels. Laboratory studies provide causal information—interrelating water hardness, pH, alkalinity, salinity, temperature, parasitism, and other factors with Cd toxicity. Enzymatic, physiological, hematologic, reproductive, and histopathological changes are correlated with the severity of Cd exposure. This review begins with environmental uptake and distribution studies and integrates some of the expanding literature reporting Cd poisoning of fish.

Environmental Exposures

Most published environmental studies involving Cd exposures of fish focus on uptake and tissue distribution rather than on causal effects because of the difficulty in establishing definitive correlates. Published environmental research indicates that whole-body Cd levels in fish can range from 0.2 to 8.0 ppm on a dry weight basis (Table 1). Metal concentration varies with tissue handling and exposure level. For example, Vinikour *et al.* (1980) obtain among the lowest whole-body Cd levels in fish—partly as a result of the removal of gastrointestinal tract contents from Fox River fish. Moreover, their study reports the absence of any major industrial effluent; the only reported effluents are those from recreational boating and a sewage treatment plant. Undoubtedly, these sample handling techniques and low exposure levels result in the low 0.2 ppm whole-body Cd levels of Fox River fish. In contrast, discharge from an electroplating factory and aqueous metal loading of 17.3 ppm dissolved Cd and 30.3 ppm suspended Cd gave whole-body Cd levels forty-fold higher for the Palestine Lake fish (Murphy, *et al.*, 1978).

Table 1. Mean whole-body concentrations of cadmium (ppm, dry weight) in various fish species. Sample size is reported as n.

Species	Location	[Cd] (n)	Reference
Redbreast sunfish	Back Creek, Virginia	0.40 (30)	Ney & Van Hassel, 1983
Rock bass		0.38 (14)	
Northern hog sucker		0.40 (12)	
White sucker		0.44 (26)	
Blacknose dace		0.76 (19)	
Fantail darter		0.60 (19)	
Yellow perch	Fox River, Illinois	0.20 (40)	Vinikour *et al.*,1980
Bluegill		0.20 (61)	
Black crappie		0.20 (26)	
Black bullhead		0.30 (10)	
Cichlid	Lake Nakuru, Kenya, Africa	0.26 (3)	Greichus *et al.*,1978
Bluegill	Palestine Lake, Indiana	3.12 (35)	Murphy *et al.*, 1978
Redear sunfish		8.00 (10)	
Warmouth		1.43 (3)	
Orangespot sunfish		6.74 (1)	
Largemouth bass		1.11 (33)	
Golden shiner		0.66 (3)	
White sucker		1.99 (3)	
Brown bullhead		2.83 (3)	
Bowfin		0.06 (1)	
Bluegill	Palestine Lake, Indiana	3.40 (35)	Atchison *et al.*, 1977

Generally, intra- and inter-species comparisons of tissue levels Cd assume (among other things) that exposure time, growth rate, and bioavailability are the same. Needless to say, such is not the case. Bendell-Young *et al.* (1986), for example, find that five of six populations of white suckers (*Catostomus commersoni*) have different growth rates which account for measured differences in Cd concentrations. Moreover, acid levels alter the growth rate of fish—some fish grow faster while others grow more slowly (Almer, *et al.*, 1974; Schofield, 1976; Ryan and Harvey, 1977; 1980). Faster-growing, acid-stressed fish may have greater susceptibility to the toxic effects of Cd (and other pollutants) than slow-growing, acid stressed fish (Bendell-Young, *et al.*, 1986). Much the same phenomenon is true of rapidly growing larval forms—increased susceptibility to metals is noted compared with the more slowly metabolizing adult forms (Eaton, 1974).

Despite intra- and inter-species specific variability in tissue loading of Cd, levels in skeletal muscle tend to fluctuate less than those in the liver (Table 2). As an example, Cd concentration in skeletal muscle from fish from an Egyptian study (El Nabawi, *et al.*, 1987) and a Lake Powell U.S. study (Bussey, *et al.*, 1976) averaged 0.018±0.005 ppm (28% error) and 0.018±0.008 ppm (44% error), respectively, while the hepatic concentrations of fish from the two studies were 0.046±0.042 ppm (91% error) and 0.377±0.350 ppm (93% error), respectively. Some workers consider muscle Cd concentrations too unreliable for comparative purposes, largely because the low accumulated levels are at the limit of detection (Bendell-Young, *et al.*, 1986). Lower Cd levels in muscle may result from elevated concentrations of cystine and methionine compared with other proteins (Beveridge, 1947). Absence of sulfhydryl groups in these sulfur-rich amino acids probably plays a role in decreasing Cd binding in skeletal muscle.

Despite variability in Cd accumulation after chronic exposure, levels of hepatic Cd generally provide one of the best indicators of severity of exposure in fish (Rowe and Massaro, 1974; Smith, *et al.*, 1976) and in humans (Friberg, *et al.*, 1974). However, after acute exposures, gill levels of Cd provide a more suitable monitor since little hepatic or renal Cd uptake occurs during acute exposures (Mount and Stephen, 1967). Exclusive use of renal levels as an indicator of Cd poisoning (at least in mammalian studies), has been problematic since Cd is excreted with proteins in cases of severe Cd-induced renal damage (Friberg, *et al.*, 1974).

Table 2. Mean liver and skeletal muscle concentrations of cadmium (ppm, dry weight) in several fish species. Sample size is reported as n.

Location (Reference)	[Cd] liver (n)	[Cd] muscle (n)	Fish Species
Arizona	0.532 (12)	0.007 (12)	Largemouth bass
(Bussey *et al.*, 1976)	0.126 (12)	0.018 (12)	Black crappie
	0.797 (12)	0.026 (12)	Walleye
	0.054 (12)	0.021 (12)	Rainbow trout
Artic (Bohn & McElroy, 1976)	0.68 (7)	<0.5 (7)	Artic cod
Gulf of Mexico	---	0.12 (5)	Flatfish
(Horowitz & Presley, 1977)	---	0.10 (5)	Long-spined porgy
	---	0.15 (2)	Rough scad
	---	0.09 (4)	Sea robin
	0.89 (1)	0.11 (2)	Sand seatrout
	---	0.14 (1)	Black-ear bass
	---	0.10 (1)	Atlantic croaker
	---	0.08 (1)	Wenchman
Canada (Bohn and Fallis, 1978)	4.10 (61)	1.40 (3)	Shorthorn sculpins
Canada	3.3(23)	---	Northern Pike
(McFarlane & Frazin, 1980)	10.1 (17)	---	White sucker
Ontario (Bendell-Young *et al.*, 1986)	4.7 (14)	<0.0006 (14)	White sucker
Egypt	0.035 (16)	0.023 (5)	Pandora
(El Nabawi *et al.*, 1987)	0.024 (16)	0.018 (5)	European barracuda
	0.018 (14)	0.012 (5)	Gurnard
	0.108 (10)	0.023 (4)	Rabbitfish

Compared to marine species, fresh-water fish are particularly vulnerable to Cd exposure. When fresh-water species are exposed to Cd, species-specific differences in behavior and diet can mean the difference between life and death. The degree of association of fish with sediments seems to play a critical role in Cd accumulation. For example, higher whole-body levels of Cd are observed in benthic (bottom-dwelling) than in pelagic (free-swimming) species of fish (Ney and Van Hassel, 1983). Blacknose dace and fantail darters (benthic species) accumulate 0.60 to 0.76 ppm compared with redbreast sunfish and rock bass which accumulate 0.38 to 0.40 ppm (Table 1). Therefore, behavioral differences in sediment contact alter Cd uptake.

Diet influences Cd accumulation in fish. Comparison of Cd levels in redear sunfish (a snail-eater) with those of bluegill shows almost three-fold higher Cd concentration in redear than in bluegill sunfish (Murphy, *et al.*, 1978) (Table 1). Carnivorous largemouth bass (*Micropterus salmoides*) and black crappie (*Pomoxis nigromaculatus*) contain among the lowest whole-body trace metal concentrations, while the more omnivorous warmouth (*Chaenobryttus gulosus*) and bluegill (*Lepomis macrochirus*) generally contain higher concentrations (Murphy, *et al.*, 1978). Needless to say, dietary preference is considered one of the primary factors in determining Cd accumulation in fish exposed environmentally. Nonetheless, trace metal concentrations in fish are apparently related to factors other than food habits and trophic status (Wiener and Giesy, 1979). This is shown by significant differences in trace metal concentration in pickerel (*Esox niger*) and largemouth bass (*Micropterus salmoides*), which occupy the highest trophic level and have similar dietary preferences. Therefore, environmental studies generally show that measurement of hepatic Cd levels is preferable to measurement of skeletal muscle Cd levels and that marked intra- and inter-species specific differences exist in Cd accumulation in feral fish.

Toxicity

Christensen and colleagues (1977) observe glucose, and ion changes in plasma of brook trout (*Salvelinus fontinalis*) exposed to 0.06–6.35 ppb aqueous Cd for 2 or 8 wk. These results are used to set no-effect or safe levels at 6.35 and 0.50 ppb Cd for 2 and 8 wk, respectively. Maximum allowable water concentrations for domestic water supplies (10 ppb) are compared with those for fresh-water aquatic life (0.40 ppb), as well as safe limits set by other workers (0.04 to 6.35 ppb) (Christensen, *et al.*, 1977). Surprisingly, bioassays of the alevin-fry stage are approximately equivalent to those obtained from year-long studies with all life stages. Safe levels for some species in the alevin-fry life stage may be as low as 0.04 ppb and comparable to safe levels for adults exposed in soft water and/or in chronic assays (Christensen, *et al.*, 1977). Insensitivity of the alevin life stage to metals is marked (see Fig. 6 in Chapter V).

Therefore, laboratory studies indicate that Cd is extremely toxic to fish. Lethal concentrations for 50% of the sample population (LC50) vary from about 2 to 55 ppm at 96 h (Table 3). Changes in water quality (e.g. salinity, hardness, pH, alkalinity, temperature, and organic fraction) and species differences impact Cd toxicity (Brown, 1968; Eisler, 1971; Giesy, *et al.*, 1977;

Pascoe and Cram, 1977a; Pascoe and Mattey, 1977b; McCarty, *et al.*, 1978; Pascoe, *et al.*, 1986).

Table 3. Approximate water hardness (as ppm $CaCO_3$) compared with 96-h lethal concentrations for 50% of the sample group (LC50) for aqueous exposures.

Species	Hardness (ppm)	LC50 (ppm)	Reference
Goldfish	20	2.1	McCarty *et al.*, 1978
Stickleback	115	6.5	Pascoe & Cram, 1977
Striped killifish	--	21.0	Eisler, 1971
Stickleback	121	23.0	Pascoe & Mattey, 1977
Goldfish	140	46.8	McCarty *et al.*, 1978
Fathead minnow	202	31.0	Pickering & Gast, 1972
Mummichog	--	50.0	Eisler, 1971
Sheepshead minnow	--	55.0	Eisler, 1971

Acute Cd toxicity and salinity are inversely related (Fig. 1). As salinity is increased from 5 to 35 ppt, 192-h LC50 values for mummichogs decrease about two to three times (Eisler, 1971). Lower salinity in an estuary occurs at ebb tide; therefore, at ebb tide, the effect of metals would be expected to be greater than at high tide.

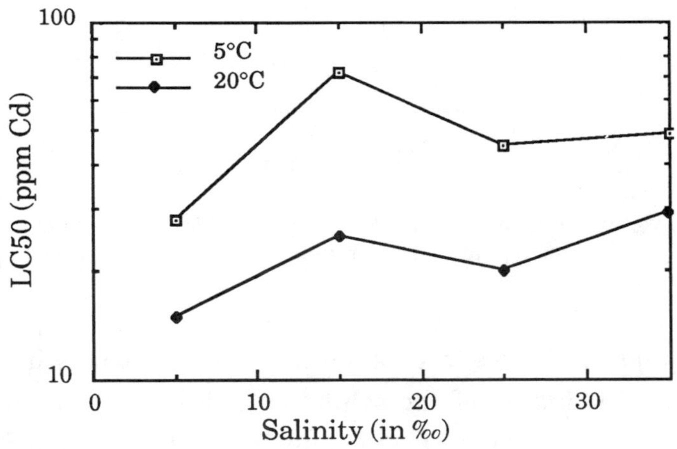

Figure 1. Salinity and temperature alter 96-h LC50 values for mummichogs (*Fundulus heteroclitus*). Increasing salinity increases survival; whereas, increasing temperature decreases survival of fish exposed to Cd. Sixty fish were used to compute values at each point. (Adapted from Eisler (1971).)

The marked effects of water hardness on Cd toxicity are common knowledge (Fig. 2 and 3) (Brown, 1968; McCarty, *et al.*, 1978; Pascoe, *et al.*,

1986). When fish are exposed in soft water, cumulative mortality curves are of a simple sigmoidal type (McCarty, *et al.*, 1978). However, in hard water, mortality curves are biphasic or concentration dependent in the 10 to 100 ppm range, but concentration independent below 10 ppm (Fig. 3). Biphasic mortality curves suggest that Cd toxicity in hard water results from more than one mechanism (Pascoe and Cram, 1977a; Pascoe and Mattey, 1977b; McCarty, *et al.*, 1978). McCarty *et al.* (1978) postulate that (at higher Cd levels) some component of lethality may stem from concomitant pH effects, and not from Cd toxic action(s) *per se*. Despite analytical problems, Carroll *et al.* (1979) report Ca-induced protection against Cd toxicity in the absence of the Cd precipitation observed in the McCarty *et al.* (1978) study.

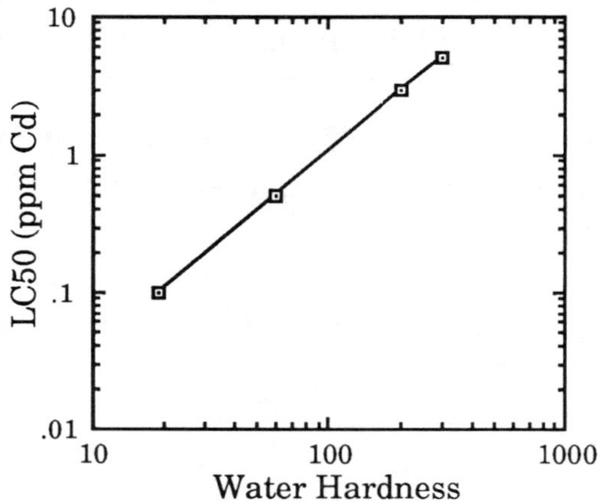

Figure 2. Total water hardness (ppm, as $CaCO_3$) affects 48-h LC50 values in rainbow trout (*Salmo gairdneri*). Increasing hardness increases survival of fish exposed to Cd. (Adapted from Brown (1968).)

Understanding of the effects of hardness on Cd toxicity in fish seems to rest on an understanding of the solubility of Cd in water during toxicity testing. In 96-h tests of six metals to killifish (*Fundulus heteroclitus*) only Cd (at 27 ppm) remains in solution, whereas up to 79.1% and 84.4% losses (due primarily to solubility changes) are observed for Pb and Cu, respectively (Jackim, *et al.*, 1970). In addition, the Pascoe *et al.* (1986) and the Carroll *et al.* (1979) studies show stable Cd concentrations at the lower aqueous levels of Cd—levels less likely to result in formation of Cd precipitates under certain conditions.

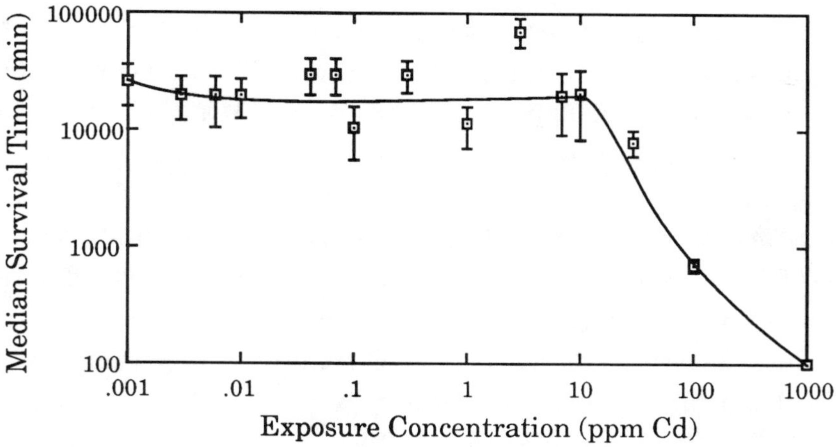

Figure 3. Biphasic median survival curve for sticklebacks (*Gasterosteus aculeatus*) exposed to various Cd concentrations in hard water. Survival is independent of Cd concentration below 10 ppm, but dependent on Cd concentration in the 10 to 100 ppm Cd exposure range. In soft water, survival is similar to that in the 10–100 ppm Cd concentration range. (Adapted from Pascoe and Cram (1977a) and Pascoe and Mattey (1977b).)

However, some workers note formation of insoluble precipitates of Cd, as well, during exposures of fish (Pickering and Henderson, 1966; Eaton, 1974; McCarty, *et al.*, 1978). McCarty and colleagues (1978) document 10% declines in aqueous Cd concentrations within 96 h in hard water and stable concentrations in soft water. Apparently, use of water hardened according to the USEPA (1975) formula can result in a 35% reduction in Cd aqueous concentrations after 1 h (Kinkade and Erdman, 1975). Pascoe and co-workers (1986) report 30% and 48% reductions in aqueous Cd concentrations at a water hardness of 100 ppm (as $CaCO_3$) after 24 h, as well as 85% and 96% reductions in aqueous levels of Cd in water hardened to 300 ppm (as $CaCO_3$) using the USEPA (1975) formulations. Pickering and Henderson (1966) record cloudiness and layering of precipitate over test chambers in hard water solutions during 48-h toxicity testing. Eaton (1974) notes that the solubility limit of Cd is exceeded in chronic and acute toxicity tests. Since Cd concentration can decrease with time (McCarty, *et al.*, 1978), use of USEPA (1975) testing solutions or similar formulations precludes definitive evaluation of dose-response effects.

The changes in Cd concentrations with hardness in some studies results in problems in use of application factors for Cd in fish. Application-factor ranges are proposed to establish acceptable toxicant limits for fish (Mount and Stephen, 1967). Maximum acceptable toxicant concentration (MATC) range values are divided by median lethal threshold concentrations (MLTC). For Cd, values of 31 and 80 ppb MATC, as well as MLTC values of 20.4 ppb (more than twice the rejected level for Cd in drinking water of humans (Flick, *et al.*, 1971)), yield an application factor range of 0.0015–0.0039 for bluegills (Eaton, 1974). Pickering and Gast (1972) obtain application factor ranges of 0.005–0.008 ppb for Cd by using 96-h TL50 acute values. The difference in the application-factor range is probably due to precipitation of Cd in the hard water solutions used by Eaton (1974). As mentioned, when solubility limits are exceeded, aqueous Cd concentrations vary. Therefore, for some studies, application factors can be different in soft water than in hard water since the toxicity of Cd is greater in soft water. This is probably due in part to the fact that Cd solubility limits are less likely to be exceeded during soft water testing. Higher ionic levels of Cd result. Rather than the use of application factors, Eaton (1974) suggests the use of 30–60 d exposures of eggs and larvae to more accurately establish the chronic toxicity of Cd. Certainly eggs and larvae represent the most sensitive of life stages and what might be called the "bottlenecks" in species survival.

Many other parameters can alter Cd toxicity in fish. Among a few of the best studied are pH (Cusimano, *et al.*, 1986), temperature (Eisler, 1971; Giesy, *et al.*, 1977; Roch and Maly, 1979), and parasitism (Pascoe and Cram, 1977a; Pascoe and Mattey, 1977b). As pH decreases from 4.7 to 5.7 to 7.0, 96-h LC50 values decrease from 28 ppb to 0.7 ppb to <0.5 ppb; at pH 5.7 and 7.0, the absence of Ca ion results in extreme Cd toxicity (Cusimano, *et al.*, 1986). Ameliorating effects of low pH may be due to a) chelation of Cd by mucous secretions which have been stimulated by increased aqueous levels of hydrogen (H) ion, b) Cd speciation changes with subsequent alteration of toxicity, c) competitive inhibition of Cd^{+2} ion by H^{+} ion at gill plasma membrane sites, d) proton-induced steric modification of binding sites (Cusimano, *et al.*, 1986). In short, lower aqueous pH increases fish survival during Cd exposures.

Increased temperatures apparently increase Cd toxicity by increasing metabolic rate of affected species. Cadmium exposed, cold-acclimated (6°C) rainbow trout (*Salmo gairdneri*) survive almost three times longer (140 h) than warm-acclimated (12°C or 18°C) trout (Roch and Maly, 1979). Mummichogs (*Fundulus heteroclitus*) also are about three times less susceptible to Cd at 5°C than at 20°C (Fig. 1) (Eisler, 1971). Mosquitofish (*Gambusia affinis*) show 96-h LC50 values of 2.6 ppm, 1.5 ppm, and 1.3 ppm Cd at 28.0°C, 30.7°C, and 30.2°C, respectively (Giesy, *et al.*, 1977). For rainbow trout (*Salmo gairdneri*) exposed to a mixture of Zn:Cu:Cd (400:20:1), LC50 values decrease from approximately 800 ppb Zn in the mixture to 350 ppm Zn (hardness: 25 ppm as $CaCO_3$) as temperature increases from 5 to 22°C. A Q_{10} of 1.7 is obtained for rainbow trout in this study. Therefore, lower temperatures prolong teleost survival during Cd exposures.

The importance of species differences is illustrated by 96-h toxicity tests which indicate that goldfish (*Carassius auratus*) are more resistant to Cd than are fathead minnows (*Pimephales promelas*) (Pickering and Gast, 1972; McCarty, *et al.*, 1978). Moreover, when striped killifish (*Fundulus majalis*), mummichogs (*Fundulus heteroclitus*), and sheepshead minnows (*Cyprinodon variegatus*) are exposed under the same conditions, killifish succumb at a 96-h LC-50 value of 21 ppm; whereas, mummichogs and sheepshead minnows have 96-h LC-50 values of 50 and 55 ppm, respectively (Eisler, 1971) (Table 3). Relative toxicant resistance of the straw-colored minnow (*Notropis blennius*), blunt-nosed minnow (*Pimephales notatus*), and goldfish (*Carassius auratus*) can vary from 1.0 to 3.0 (Powers, 1920). The outlier-prone nature of dose-response effects and metal uptake in fish exposed to Cd and other metals remains a mystery.

The presence of plerocercoids of a cestode in the three-spined stickleback (*Gasterosteus aculeatus*) reduces median survival time during Cd exposure by about times compared with non-parasitized fish (Pascoe and Cram, 1977a). Generally, the presence of these infective metacestodes (as well as other parasites) is thought to reduce the resistance of fish to other stresses. Weakened, parasitized fish have lower condition factors, retarded sexual maturation, decreased organ weights, reduced hematocrits, and delayed oocyte maturation (Arme and Owen, 1967; Pennycuick, 1971). Such changes may reflect nutrient competition between host and parasite (Walkey and Meakins, 1970). Parasitized fish are more susceptible to Cd, as well as other environmental stressors. Since many parasites are internal, reference toxicants can be used to estimate the condition of fish at the time of testing. This is a logical method of insuring that test organisms are healthy prior to toxicity tests (Davis and Hoos, 1975; Hall, *et al.*, 1986).

Exact mechanisms causing death in acute Cd poisoning of fish have long puzzled scientists. Early workers report that fish simply suffocate in their own mucus after exposure to certain heavy metals or to some acids. Copious amounts of mucus are produced when some species of fish are exposed to Pb, Cd, Cu, Hg, Cu, sulfuric acid, or other acids and metals (Carpenter, 1927; Westfall, 1945; Mount and Stephen, 1967). Normal movement of gill filaments is restricted or rendered impossible in sticklebacks (*Gasterosteus aculeatus*) exposed to several metals or acids. Passage of water through the gill chamber is so restricted that much of the water taken in is expelled through the mouth instead of through the opercular apertures, often with such violence that the fish is propelled backwards (Jones, 1939). However, apparently, Cd-exposed coho salmon (*Oncorhynchus kisutch*) do not show these sneeze-like reactions nor do they appear to "fight" the toxicant, but simply are narcotized before death (Schreck and Lorz, 1978). Therefore, in contrast to sticklebacks, salmon respond rather placidly.

Excess mucous production in response to metals or acids is diagnosed as coagulation film anoxia. Oxygen deficiency is thought to occur as a result of the orange-brown precipitate of Cd with the mucus which covers gill filaments (Carpenter, 1927). Progressively increasing deposition of dead cellular material on the gills is thought to led to asphyxiation in some species. In fact, the presence of Cd accelerates mucus production to the extent that mucus turnover becomes an important excretory mechanism for Cd in fish (Varanasi and Markey, 1978). Repeated convulsive, frenzied swimming movements (Eaton, 1974; Pascoe and Mattey, 1977b) also suggest behavioral changes designed to dislodge mucus from the gills. For about half a century, coagulation film anoxia was the only accepted cause of death of fish poisoned by specific metals and acids.

In chronic exposures, other mechanisms seem to be involved. Cearley and Coleman (1974) describe erratic, uncoordinated swimming movements, muscle spasms and convulsions, followed by loss of equilibrium, with periods of quiescence and paralysis. In many instances, muscle spasms precede death and cause uncontrolled swimming, such that fish swim upside down and spiral (Cearley and Coleman, 1974; Pascoe and Mattey, 1977b). Such abnormal behavior suggests more than simple dislodging of mucus in some species. At death, the fins of largemouth bass (*Micropterus salmoides*) and bluegill (*Lepomis macrochirus*) are fully spread, the branchiostegals and opercula are expanded, and the body is arched laterally in the area between the base of the pectoral fins and the middle of the dorsal fin. Cadmium-induced lateral distortions (displacing the longitudinal axis as much as 80° from normal) may cause this arching of the body as well as the spinal cord lesions found by Bengtsson *et al.* (1975). Events preceding and following death suggest that the nervous and/or muscular systems are sites of damage. Inhibition of acetylcholinesterase with resultant death by paralysis of respiratory muscles and/or depression of the respiratory center are possible causes of death (Cearley and Coleman, 1974).

Muscular and/or neuronal abnormalities have led scientists to suspect hypocalcemia as an underlying cause. Low plasma Ca is capable of causing such abnormalities. In fact, Roch and Maly (1979) show that Cd-induced mortality is correlated with the reduction in plasma levels of Ca to 3.0 meq/l. Rate of mortality (the reciprocal of median survival times) is correlated with the rate of decline in plasma Ca, indicating that the decline of Ca concentrations in plasma may be the direct cause of mortality.

Other workers believe that Cd-induced death is caused by destruction of respiratory function. Basically, necrotic damage may disrupt gaseous exchange and salt/water balance (Cusimano, *et al.*, 1986). Marked accumulation of Cd in bluegill (*Lepomis macrochirus*) gill tissue (Mount and Stephen, 1967), direct cellular injury of the gills of mummichogs (*Fundulus heteroclitus*) (Gardner and Yevich, 1970), and a 15% decrease in heart rate in tautogs (*Tautoga onitis*) suggest direct Cd-induced injury of gills (Eisler, 1971) and subsequent alteration of respiratory function. Therefore, for some species, necrotic damage of gill tissues and hypocalcemia may be more critical than excess mucous production leading to asphyxiation. Species-specific differences in the response of fish to Cd and multiple damage to several tissues could complicate the identification of a single mechanism explaining the cause of death. Most likely, several factors play a role in causing death in fish exposed to Cd—especially in acute *versus* chronic exposures.

In conclusion, the toxicity of Cd in fish is altered by water quality parameters including salinity, water hardness, pH, alkalinity, and temperature of exposure solutions. Parasitism can exacerbate the effects of Cd, and species-specific differences in response to Cd challenge are important. Application factors are used to establish acceptable toxicant concentration ranges (0.0015–0.0080 ppb Cd) depending on water quality, species, and life stage. The cause of death in fish poisoned by Cd appears to involve a combination of factors such as mucous production and asphyxiation, inhibition of acetylcholinesterase, depression of the respiratory center, hypocalcemia, and/or gill injury. On a comparative basis, the toxicity of Cd in some species of fish is twenty-five times less than that of Hg and approximately equal to the toxicity of Pb (Carpenter, 1927; Jones, 1939).

Accumulation and Tissue Distribution

The mode of metal uptake in fish is ascribed to dietary sources alone, water alone, or total environmental exposure (Lucas, *et al.*, 1970; Atchison, *et al.*, 1977; Murphy, *et al.*, 1978). However, a complex interaction of many factors seems to be involved for numerous ecosystems (Giesy, *et al.*, 1977; Wiener and Giesy, 1979). Numerous factors play a role in determining the level of accumulated Cd—including Cd source, exposure level, distance from contamination source and the presence of other ions (especially Ca and Zn).

The two major pathways of incorporation of Cd for fish are absorption across the gill surface and across the intestinal mucosa. However, the gastrointestinal route of Cd uptake may become more important than the gill route as aquatic ecosystems become progressively more contaminated with Cd. For example, of eleven species of fish from an industrially-polluted lake, carnivorous black crappie (*Pomoxis nigromaculatus*) and largemouth bass (*Micropterus salmoides*) contain low whole-body levels; whereas, more omnivorous warmouth (*Chaenobryttus gulosus*) and bluegill (*Lepomis macrochirus*) contain high levels (Murphy, *et al.*, 1978). Even higher levels are found in redears (*Lepomis microlophus*) which consume a diet consisting of a considerable quantity of snails, which live in Cd-rich sediments (800 ppm Cd, dry weight). Therefore, these data show that dietary uptake can provide a major pathway of Cd incorporation in some ecosystems.

In contrast, McFarlane and Franzin (1980) evaluate hepatic Cd concentrations in northern pike (*Esox lucius*) and white sucker (*Catostomus commersoni*) in five less polluted lakes (sediments contain a maximum of 16 ppm Cd, dry weight). When adjusted for age, mean hepatic Cd concentrations do not reflect the difference in mean levels of Cd in gut contents (6.5 and 32.0 ppm, dry weight, for pike and suckers). Absence of a correlation between hepatic Cd levels and Cd levels of gut contents suggests that water rather than dietary uptake is a major route of Cd uptake in species exposed in less contaminated environments.

Despite the Cd-concentrating capability of fish, many species are extremely sensitive to Cd. Even though some fish can accumulate Cd to levels ten to one-thousand times higher than the level in ambient water (Fleischer, *et al.*, 1974), lethality can occur at whole-body concentrations below 1 ppm (Pascoe and Mattey, 1977b). Laboratory studies show that aqueous exposure level is important in determining the level of Cd accumulated by the whole body. During 29 d exposures to 0.1, 1.0, or 10.0 ppm Cd, accumulation of Cd in the whole-body of rainbow trout increases with reported values at 4.2, 8.7, and 47.0 ppm (dry weight), respectively (Pascoe, *et al.*, 1986). Pascoe and Mattey (1977b) observe whole-body Cd residues averaging 50.6 ppm (wet weight) in sticklebacks (*Gasterosteus aculeatus*) surviving a median time of 6.9 h. Interestingly, whole-body Cd levels for laboratory-exposed fish can exceed those of feral fish (Atchison, *et al.*, 1977; Greichus, *et al.*, 1978; Murphy, *et al.*, 1978; Vinikour, *et al.*, 1980; Ney and Van Hassel, 1983).

To measure how much Cd is accumulated with respect to aqueous exposure concentrations, concentration factors (CF) are used. Cadmium uptake in the whole-body (or in organs) is divided by aqueous concentrations of Cd (Wiener and Giesy, 1979; Woodworth and Pascoe, 1983). For bluegills (*Lepomis macrochirus*), CF can vary from 71 to 240 for the whole-body (Atchison, *et al.*, 1977; Wiener and Giesy, 1979). Woodworth and Pascoe (1983) obtain lower whole fish CF values of 6.5 (at 1 ppm Cd aqueous exposures) and 19.6 (at 0.1 ppm Cd aqueous exposures) in sticklebacks (*Gasterosteus aculeatus*). The reason CF values are higher in the Woodworth and Pascoe (1983) study than in the Wiener and Giesy (1979) study seems to be a difference in exposure times. Woodworth and Pascoe expose sticklebacks 170 h; whereas, Wiener and Giesy report CF values for bluegills exposed about 500 d to a lifetime. Theoretically, the most stable CF values would be those obtained at equilibrium time, when the concentration of Cd in the body (or tissue) is stable with respect to aqueous concentrations. Therefore, the higher CF values reported by Atchison *et al.* (1977) and Wiener and Giesy (1979) are probably more representative of chronic that acute Cd exposures.

Concentration factors vary with exposure levels (Fig. 4). For Cd, CF in three-spined sticklebacks (*Gasterosteus aculeatus*) exposed for up to 79 d are inversely related to exposure levels. For example, values of CF range from 0.51 at a 100 ppm aqueous concentration to 511 at a 0.001 ppm exposure concentration (Pascoe and Mattey, 1977b). Ultimately, this relationship is probably due to a finite limit in the ability of fish to accumulate Cd (Eisler, 1974). However, in 96-h tests of mummichogs, the inverse relationship between CF and exposure level is not evident: whole-body CF ranges from 0.17 to 0.3 for exposure concentrations up to 75 ppm (Eisler, 1971; 1974). The shorter exposure time (96 h) for mummichogs probably causes this result; and again, the more stable CF values appear to be those obtained at equilibrium time for any given species.

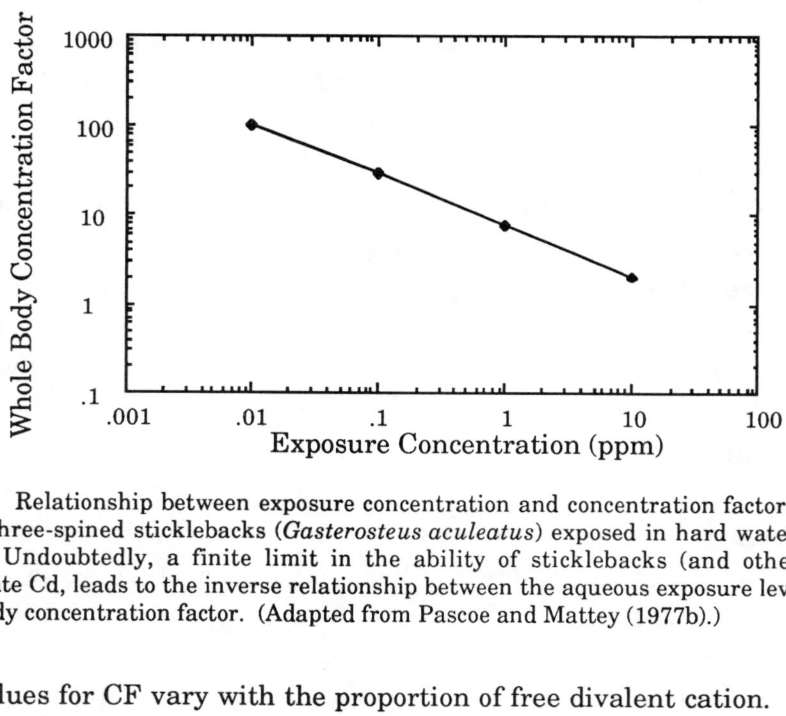

Figure 4. Relationship between exposure concentration and concentration factors in whole fish for three-spined sticklebacks (*Gasterosteus aculeatus*) exposed in hard water (Y=0.76-0.61X). Undoubtedly, a finite limit in the ability of sticklebacks (and other fish) to accumulate Cd, leads to the inverse relationship between the aqueous exposure level and the whole body concentration factor. (Adapted from Pascoe and Mattey (1977b).)

Values for CF vary with the proportion of free divalent cation. Ionic Cd is thought to be more available for direct uptake through plasma membranes of gills of fish than bound Cd (Holcombe, *et al.*, 1976; Merlini and Pozzi, 1977a; Merlini and Pozzi, 1977b; Ramamoorthy and Blumhagen, 1984). Therefore, a higher CF for Cd probably indicates that proportionately more free divalent cation is present in exposure solutions. This is more likely to be the case in acidic, organic-rich ecosystems rather than in low organic, hard water systems. Little bound Cd exists when pH is low and organic carbon concentration is high (Giesy and Wiener, 1978; Wiener and Giesy, 1979); therefore, CF values would be expected to be higher because these conditions enhance Cd uptake in fish. In contrast, the presence of organic matter is known to decrease uptake of Hg and Zn by rainbow trout (*Salmo gairdneri*) (Ramamoorthy and Blumhagen, 1984).

The use of CF values for Cd presumes (and justifiably so) the absence of homeostatic regulation. In homeostatic regulation, concentrations of trace elements are regulated and independent of the concentration of that element in the water (Murphy, *et al.*, 1978; Wiener and Giesy, 1979). This can be the case for essential elements such as Cu, Mn, and Zn—although alternate theories exist (see Chapter V). Generally, concentrations of Cd in fish tissues are highly variable and seemingly independent of the strict physiological control evident with the essential elements. This is evidenced by the absence of a correlation between Cd concentration and fish length, fish weight, organ weight, or condition factors of fish (Lovett, *et al.*, 1972; Havre, *et al.*, 1973; Mathis and Kevern, 1975; Giesy, *et al.*, 1977; Murphy, *et al.*, 1978; Wiener and Giesy, 1979).

Whole-body uptake patterns are different for survivors than for fish which succumb to Cd: generally, survivors accumulate less than fish which die from Cd poisoning. Mummichogs surviving 264 h accumulate 100 ppm or less (ash weight) compared with dead fish which accumulate from about 100 to 7000 ppm (Eisler, 1971). Dead mummichogs accumulate more Cd than living fish: fifty-three times more in 24 h tests and eighty-nine times more in 48 h tests.

Early in an exposure (prior to 3 h), Cd is probably loosely bound to cellular secretions at the body surface and especially the gills (Mount and Stephen, 1967; Eisler, 1974; Varanasi and Markey, 1978). Unusually elevated levels of Cd are found in the gill tissues in acute exposures (Mount and Stephen, 1967). Cadmium loss by sloughing of dead cellular material and mucus from the gills and body surface accelerates excretion (Varanasi and Markey, 1978). This mechanism seems to have adaptive significance in increasing teleost survival, especially during bolus, point source contamination.

In acute exposures of bluegills (*Lepomis macrochirus*), fish dying from Cd poisoning may accumulate 150 ppm or more (dry weight) in the gill compared with living fish which accumulate 130 ppm or less in the gills (Mount and Stephen, 1967). Gills are thought to accumulate Cd to an upper limit (about 130 ppm) prior to respiratory failure and death (Mount and Stephen, 1967). However, absence of overlap in gill uptake data between living and dead fish may be due to Mount and Stephen's use of gill sample composites, which reduce variability. In chronic exposures, gills accumulate much less than 130 ppm prior to death of fish. In fact, following 11-mo exposures to 0.031–2.140 ppm aqueous Cd, bluegills (*Lepomis macrochirus*) die of Cd poisoning with mean gill levels of 21–80 ppm (Eaton, 1974).

Initially, it might seem that the Eaton (1974) data contradict those of Mount and Stephen (1967). However, the gill threshold level of 130 ppm Cd is proposed for an acute rather than chronic exposure. Short duration tests (90 d) characterize Mount and Stephen's (1967) work compared with the 9.5-mo study of Eaton (1974). Mount and Stephen make no pretense of using the threshold level for chronic studies. Therefore, in acute toxicity testing, gill Cd levels become an excellent monitor of the cause of death. In fact, some workers consider the gills of fresh-water fish the most important organ for evaluation of Cd accumulation from water.

Besides the gill, the liver, kidney, and/or gastrointestinal tract tissues seem to be the three organs of greatest importance in Cd accumulation (Mount and Stephen, 1967; Eisler, 1974; McFarlane and Franzin, 1980; Ney and Van Hassel, 1983; Bendell-Young, *et al.*, 1986). After 38-wk exposures to 3.4 ppb aqueous levels of Cd, the kidney, liver, and gill accumulate Cd in a time- and dose-dependent fashion in first- and second-generations of brook trout (*Salvelinus fontinalis*) (Benoit, *et al.*, 1976). Following this exposure, the kidney, liver, and gill accumulate the greatest amount of Cd—approximately 50, 10, and 7 ppm (dry weight), respectively. Aqueous exposures of largemouth bass (*Micropterus salmoides*) show a direct relationship between Cd accumulation and exposure time or exposure concentration in pooled internal organs (liver, kidney, spleen, and digestive system) (Fig. 5) (Cearley and Coleman, 1974). After 4-mo exposures to 0.08 ppm Cd, for example, the pooled internal organs accumulate about 0.05 ppm (ash weight). Unfortunately, the pooling of internal organs limits understanding of Cd distribution to the organs which detoxify and remove Cd from fish since inter-species specific variability in Cd uptake is pronounced.

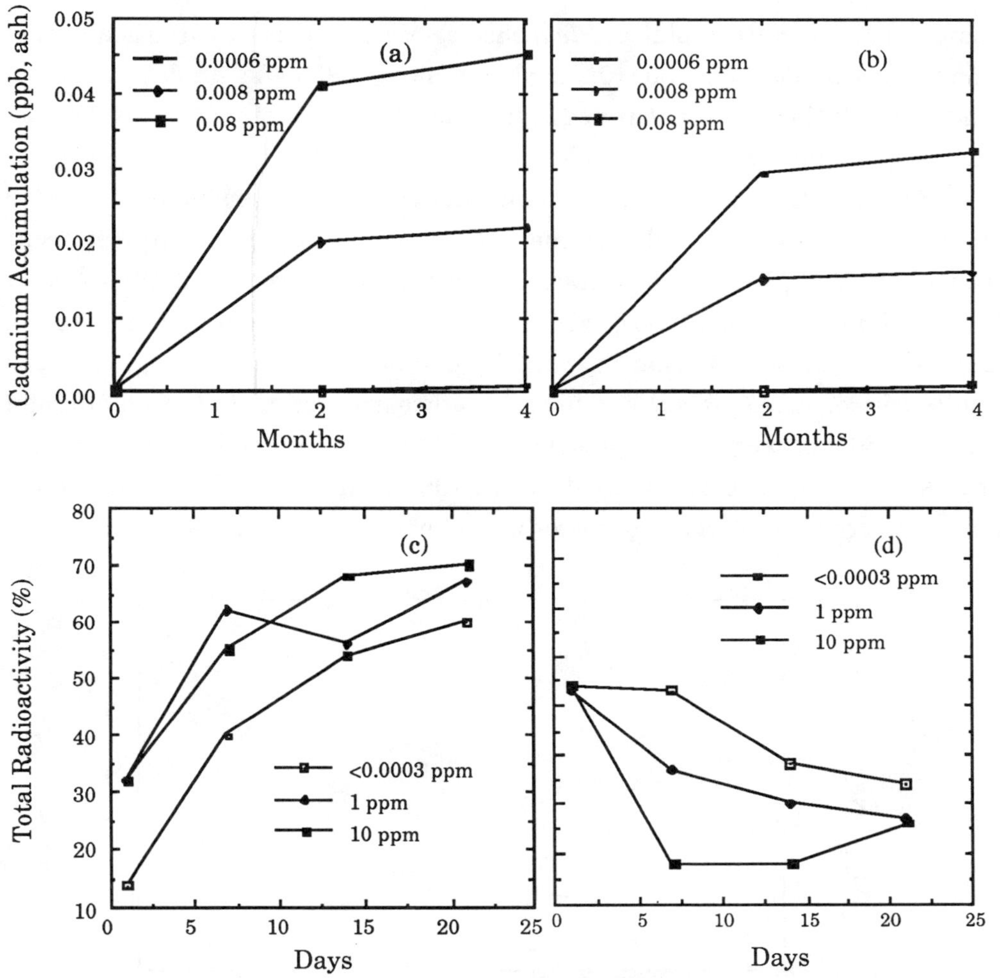

Figure 5. Cadmium accumulation in the viscera (a,c) and gills (b,d) of largemouth bass (a,b) and mummichogs (c,d) exposed for variable times at different Cd concentrations. For viscera, Cd uptake increases with increasing exposure level and exposure time, except at the highest aqueous exposure concentrations (i.e. 10 ppm). For gills, direct (b) or inverse (d) relationships exist. (Modified from Eisler (1974) and Cearley and Coleman (1974).)

Most Cd in chronic exposures of fish is associated with the liver or the kidney. Intraperitoneal injection of Cd (about 50 µl of 125 µg/ml) in sticklebacks results in 75% visceral deposition with over half of this amount in the liver (Woodworth and Pascoe, 1983). Gills and kidneys contain only 6.0 and 4.5%, respectively. Similarly, the liver is shown to contain the highest level of Cd following aqueous exposures of sticklebacks (*Gasterosteus aculeatus*) exposed 170 h to 0.1 or 1.0 ppm Cd (Woodworth and Pascoe, 1983) and bluegills (*Lepomis macrochirus*) exposed 11 mo to 0.031–2.140 ppm Cd

(Eaton, 1974). Mature plaice (*Pleuronectes platessa*) given three sequential Cd intraperitoneal injections (0.2 mg/kg on day 1, 0.6 mg/kg on day 4, and 2.0 mg/kg on day 7) have eight times higher hepatic levels of Cd on day 12 (151 ppm, wet weight) than renal levels (Overnell, *et al.*, 1977).

However, Cd levels in the posterior kidney are comparable to or greater than hepatic levels in coho salmon (Reichert, *et al.*, 1979), brown trout (Gottofrey, *et al.*, 1988), brook trout (Sangalang and Freeman, 1979), rainbow trout (Thomas, *et al.*, 1985), white suckers (Bendell-Young, *et al.*, 1986), channel catfish (Fig. 6) (Smith, *et al.*, 1976), and white catfish (Fig. 7) (Rowe and Massaro, 1974). Results of teleost studies are similar to those of human studies. Some cases show higher hepatic than renal uptake of Cd; whereas, others show the reverse (Bonnell, *et al.*, 1960). Therefore, liver and kidney are considered the critical organs for Cd detoxification and removal.

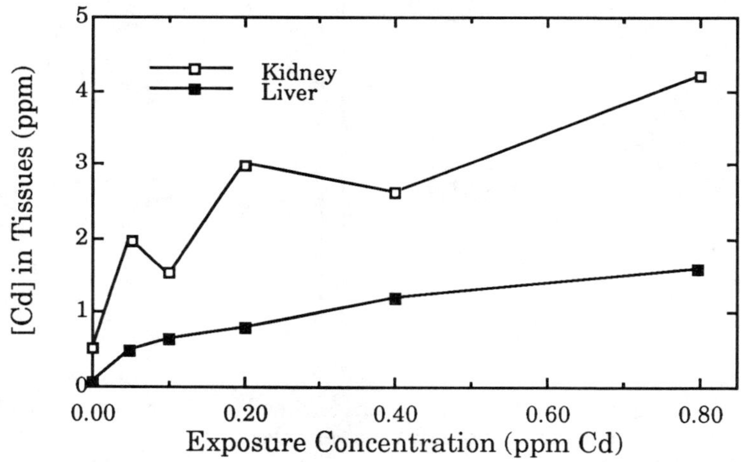

Figure 6. Average concentrations of Cd in hepatic and renal tissues of channel catfish (*Ictalurus punctatus*) exposed 3 wk to aqueous solutions up to 0.8 ppm Cd. Renal levels are higher than hepatic levels. (Modified from Smith *et al.* (1976).)

More than one year is suggested as the biological half-life for hepatic and renal Cd. Haux and Larsson (1984) expose rainbow trout (*Salmo gairdneri*) juveniles 18 or 30 wk to 10 or 100 ppb Cd and allow them to recover 25 or 57 wk in clean water. The absence of Cd elimination from the liver after fish are exposed to the lower dose (10 ppb), even following 57 wk of recovery, is noteworthy and attests the refractory nature of Cd storage in teleosts. Furthermore, following 30 wk of exposure to Cd and 57 wk of recovery in clean water, a positive correlation exists between renal Cd and hepatic Cd (r, 0.90).

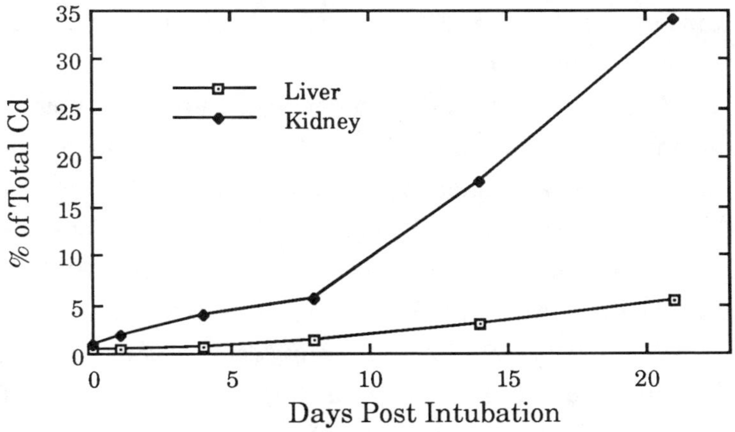

Figure 7. Percentage of Cd distributing to the liver and kidney of white catfish (*Ictalurus catus*) following a single intragastric dose of 0.2 mg/kg. Fish regurgitated from 39 to 56% of the dose; therefore, values represent the percentage of the remaining dose in the respective organs. Renal levels exceed hepatic levels at all time points. Each point is the average of data from two fish. (From data generated by Rowe and Massaro (1974).)

Concentration factors for the liver and kidney of channel catfish (*Ictalurus punctatus*) show the same inverse relationship as for the whole body of other fish (Smith, *et al.*, 1976). For example, liver CF after 21-d exposures was 12 at an exposure concentration of 0.05 ppm compared with a CF of 2 at an exposure concentration of 0.80 ppm. Kidney CF is two to four times that of the liver in one study. Greater renal than hepatic accumulation is found for both intragastric and aqueous administration of Cd in fish.

Distance from the contamination source and the level of Ca in the water affect the level of hepatic Cd in fish collected from the environment (McFarlane and Franzin, 1980). Four-fold higher levels of Ca in lake water (i.e. 39 ppm) decreased Cd hepatic levels about three to six times. Lower Cd exposure levels and elevated water hardness (higher Ca levels) seem to produce elevated hepatic, rather than renal, levels of Cd. However, no difference in accumulation of Cd was found in whole bodies of live or dead fish which could be related to the hardness of test solutions during 29-d exposures to 0.1, 1.0, or 10.0 ppm Cd (Pascoe, *et al.*, 1986). Mount and

Stephan (1967) also show that water hardness does not affect gill or hepatic uptake of Cd in 90-d testing.

Whole-body autoradiography of brown trout (*Salmo trutta*) exposed in water containing 1 ppb Cd-109 shows strong labeling of the gills, trunk kidney, olfactory apparatus, liver, and pyloric cecum (in that order) (Gottofrey, *et al.*, 1988). Axonal transport *via* the olfactory nerve is thought to explain strong accumulation of Cd by the olfactory apparatus (olfactory rosette, olfactory nerve, and olfactory bulb of the brain (Tjalve, *et al.*, 1986)). In contrast, skeletal muscle is not labeled. Unlike Hg, Cd does not appear to accumulate in fish flesh (Eisler, 1974). Bussey *et al.* (1976), Bohn and McElroy (1976), Horowitz and Presley (1977), Bohn and Fallis (1978), McFarlane and Franzin (1980), Bendell-Young *et al.* (1986), and El Nabawi *et al.* (1987) find that Cd concentrations are higher in livers than in skeletal muscle and that muscle levels range from 0.007–1.400 ppm, dry weight.

Therefore, gill Cd concentration can be used to monitor acute exposures; hepatic, renal, and gastrointestinal Cd levels tend to be useful in chronic exposures. The liver and kidney are critical organs in detoxification of Cd as evidenced by marked accumulation in these organs. The kidney continues to accumulate Cd after exposure ceases, probably as a result of the redistribution of Cd from large stores in the liver. In long-term exposures, hepatic and renal Cd concentrations provide useful information because of this binding and mobilization by metallothionein. However, in mammalian studies, renal Cd levels can decrease in cases of severe poisoning because of Cd is released with urinary proteins (Friberg, *et al.*, 1974). This factor, as well as increased variability in accumulation patterns in the kidney compared with the liver (Smith, *et al.*, 1976), tend to spotlight the liver as the preferred organ in cases of Cd intoxication in fish. Generally, however, differences in tissue distribution of Cd and lack of information regarding the chronology of Cd intoxication in most environmental studies, necessitate metal analysis of several tissues.

At equilibrium, uptake rates of Cd approximate elimination rates, and Cd concentrations in fish are considered to be in balance with levels of Cd in the ecosystem. For largemouth bass (*Micropterus salmoides*) and bluegill (*Lepomis macrochirus*), an equilibrium time of 2 mo exists—based on the absence of significant additional whole-body accumulation after the second month of exposure to aqueous solutions of 0.0005 to 0.8500 ppm Cd (Cearley and Coleman, 1974). Mount and Stephen (1967) find 30–60 d equilibrium times for bluegills exposed subacutely, based on the absence of gill and liver uptake after that time. Moreover, about 4-mo equilibrium times are observed for first- and second-generation brook trout (*Salvelinus fontinalis*) exposed to 0.9, 1.7, or 3.4 ppm Cd (Benoit, *et al.*, 1976). Therefore, depending on conditions of exposure and the species, 1- to 4-mo equilibrium times are reported for Cd, with death presumably following as a consequence of the failure of elimination mechanisms and/or stimulation of uptake mechanisms.

Elimination

Established depuration rates for Cd indicate that fish remove Cd rapidly (Eisler, 1974). Mummichogs (*Fundulus heteroclitus*) loose 90% of whole-body levels of Cd in 180 d following 25 d exposures to 0.0003, 1, or 10 ppm Cd (Eisler, 1974). Despite the thirty-three-thousand-fold difference in exposure levels for mummichogs, the depuration rate does not change over the 180-d depuration interval. Ultimately, the complete (or near complete) clearance of Cd from the body probably results in the log-normal, outlier-prone distributions for Cd in the whole body of five species of fish (Giesy, *et al.*, 1977). In fact, the existence of a constant elimination rate tending toward total elimination of Cd, rather than elimination to a prescribed concentration, would result in the observed positively skewed distributions for Cd in five species of fish (Giesy, *et al.*, 1977).

Clearance of Cd from specific tissues can lead to the marked reduction in the body burden of Cd (Eisler, 1974; Benoit, *et al.*, 1976; Reichert, *et al.*, 1979). In 12 wk, brook trout (*Salvelinus fontinalis*), previously exposed 105 wk to 3.4 ppm Cd, loose 79% of Cd from gills; so that, levels approach those measured in control fish (0.20±0.06 ppm) (Benoit, *et al.*, 1976). Although hepatic Cd concentrations in brook trout are about the same following depuration, kidney levels increase slightly (Benoit, *et al.*, 1976). Livers of coho salmon (*Oncorhynchus kisutch*) show little change in Cd concentration during a 37-d depuration period [following 15 d exposure to 150 ppb in seawater (28 ‰)] (Reichert, *et al.*, 1979). The kidney Cd levels, however, double over the 37 d depuration to 0.8 ppm (wet weight). Therefore, during depuration of Cd from fish, renal levels generally increase while hepatic levels decrease or remain constant.

Mucous turnover provides an important means of reducing the body burden of Cd in fish. Coho salmon (*Oncorhynchus kisutch*) exposed to 150 ppb for 14-d, concentrate about 180 ppb Cd (wet weight) in mucus, compared with about 5 ppb Cd after 37-d of depuration (Varanasi and Markey, 1978). Loss of Cd *via* the mucus, as well as substantial mucous production (13.1 mg/g fish weight) makes mucous turnover an important process for Cd excretion in salmon. Following intravenous injections, four to five times more Cd than Pb is discharged from mucus (Varanasi and Markey, 1978). Interestingly, in the same study, 51% of accumulated Cd is retained in the scales of coho salmon following a 37-d depuration, preceded by 14-d exposures to 150 ppb water-borne Cd (Varanasi and Markey, 1978). Therefore, scales may serve as a temporary storage site for Cd during depuration, as mucus continues to release Cd.

The hepatic-biliary excretory route is also important in eliminating Cd from fish. Mummichogs (*Fundulus heteroclitus*) show markedly elevated levels of Cd in the gallbladder (Eisler, 1974). During the 180-d depuration study (following 21-d exposures to aqueous 10 ppm), the gallbadder contains 17.7% of total body levels of Cd in mummichogs (*Fundulus heteroclitus*). Cadmium probably is packaged by the liver and released into the intestine with bile pigments, to begin gastrointestinal redistribution and/or release of Cd in feces, as is the case for other heavy metal toxicants in fish (Grahl, *et al.*, 1985; Sorensen, *et al.*, 1979) and other vertebrates (Friberg, *et al.*, 1974; Klaassen and Kotsonis, 1977; Webb and Cain, 1982). The importance of the liver in the detoxification process seems due to both Cd uptake in the liver (Eaton, 1974; Benoit, *et al.*, 1976; Reichert, *et al.*, 1979; Sangalang and Freeman, 1979; Woodworth and Pascoe, 1983; Bendell-Young, *et al.*, 1986; Gottofrey, *et al.*, 1988) and binding of Cd by metallothionein and other hepatic proteins in teleosts (Marafante, 1976; Noel-Lambot, *et al.*, 1978; Beattie and Pascoe, 1979; Reichert, *et al.*, 1979; Kito, *et al.*, 1982; Roch, *et al.*, 1982; Woodworth, *et al.*, 1983; Roch and McCarter, 1984a; 1984b; Wofford and Thomas, 1984; Thomas, *et al.*, 1985; Olsson and Haux, 1986; Hamilton, *et al.*, 1987).

Translocation of Cd from the gastrointestinal tract to the liver is shown in the 180-d depuration study of mummichogs (*Fundulus heteroclitus*) previously exposed 21 d to aqueous levels less than 0.3 ppb Cd (Eisler, 1974). Gut levels decrease from 36% to 2% while hepatic levels increase from 8% to

66% in mummichogs. Therefore, although some studies suggest sequestration (rather than elimination) as a major detoxification mechanism (Giesy, *et al.*, 1977), a combination of sequestration and elimination seems to characterize the process of detoxification and removal of Cd from fish.

Biological Effects

Vertebral Alterations: Bengtsson *et al.* (1975) report spinal deformities (vertebral fractures) in mature minnows (*Phoxinus phoxinus*) exposed for 70 d to aqueous Cd as low as 7.5 ppb (0.00019 of the 96-h LC50 level). Exposure to 960 ppb for 70 d results in the maximum number of injuries (67%) (Fig. 8). Seventy-nine percent of fractures occur in the caudal region, especially in the second to seventh caudal vertebrae (64%), which are surrounded by the largest muscle mass. Fractures seem to be related to muscular convulsions which are observed frequently, especially in fish exposed to the highest aqueous levels of Cd. Cadmium-induced prolongation of skeletal muscle contractions, longitudinal contraction of the whole body, and vertebral overloading probably collapses vertebrae. No evidence of osteomalacia (such as that occurring in Itai-Itai disease in humans) is observed in these minnows.

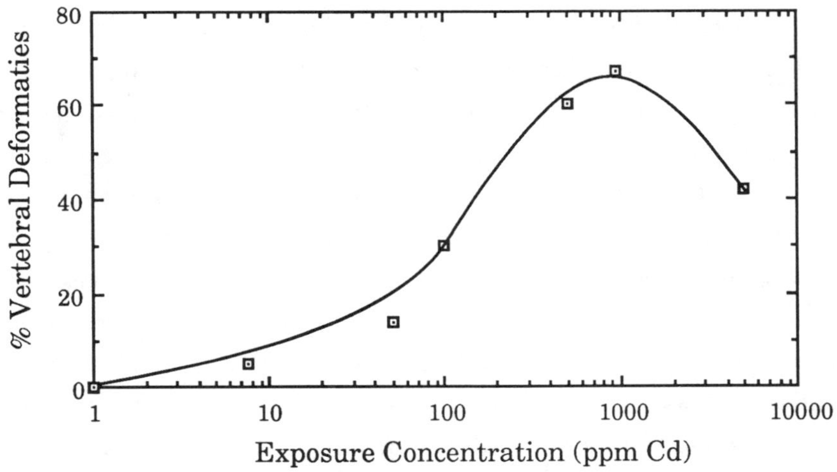

Figure 8. Percentage of surviving minnows (*Phoxinus phoxinus*) with vertebral fractures after 70-d exposures to various concentrations of aqueous Cd. (Adapted from Bengtsson *et al.* (1975).)

Weak vertebrae, low collagen and Ca levels in the bone, and decreases in bone density are associated with elevated levels of tissue Cd and decreased levels of tissue Zn. Dwyer *et al.* (1988) reports altered mechanical and biochemical properties in bone of longear sunfish (*Lepomis megalotis*) collected from a stream (Big River) contaminated with Pb, Cd, and Zn. Skeletal muscle concentrations of Pb and Cd are elevated at three sites downstream from a tailings break. Biological availability of Pb, Cd, and Zn is confirmed (Schmitt, *et al.*, 1984). Despite lower exposure levels of Cd than Pb and Zn, Cd seems to be affecting fish from Big River. Particularly noteworthy is the fact that contaminated sunfish from Washington State Park have lower levels of Zn in skeletal muscle (compared with other contaminated areas), as well as the weakest vertebrae (18% decrease compared with controls), the lowest P concentrations in bone, the highest proline:hydroxyproline ratio (13% increase compared with controls), the second-lowest bone concentration of collagen and Ca (5.3% and 22.7% decreases, respectively), and decreases in bone density (10% compared with controls). Skeletal muscle tissue concentrations of Cd are correlated with mechanical stress, elastic limit, and modulus of elasticity. Unfortunately, Cd levels in bone are not published in this study.

The flounder is not as susceptible to vertebral deformities as are *L. megalotis* and *P. phoxinus*. Larsson and co-workers (1981) expose male and female flounders (*Platichthys flesus*) 4–9 wk to 5–500 ppb Cd (0.00025–0.02100 of the 96-h LC50 value) in brackish water (7–7.5 ‰). No demineralization, fractures, or any other vertebral deformities are noted. Unfortunately, the flounder does not show vertebral damage despite the presence of ionic changes in the plasma. For comparison, a 10-wk exposure of fresh-water minnows (*Phoxinus phoxinus*) to 500 ppb Cd causes spinal deformations in 60% of specimens (Bengtsson, *et al.*, 1975); whereas, a similar exposure (9 wk at 500 ppb Cd) does not result in skeletal deformities in flounder (Larsson, *et al.*, 1981). Besides being less susceptible to Cd poisoning because it is a saltwater species, the flounder probably has an acellular bone tissue which is thought to be largely unaffected by a disturbed Ca balance (Larsson, *et al.*, 1981).

Vertebral abnormalities are not unique to Cd-exposed fish, but are known to occur in fish exposed to high temperatures or to pesticides. Hubbs (1959) reports 100% and 63% vertebral deformities in two collections of mosquitofish (*Gambusia affinis*) at the Texas Natural History Collection; 74 and 48%, respectively, are strongly malformed. Constant high temperature (34.5–35.5°C) at Graham Ranch Warm Spring (Big Bend National Park, Texas) and "bathtub spring" (near Coahuila, Mexico) probably cause the serpentine vertebral columns, (with scarcely any straight segment), malformed fins, and distorted scale patterns. Therefore, meristic characters are considered to be thermally plastic during ontogeny (Hubbs, 1959).

Pollutants other than Cd cause vertebral deformities. McCann and Jasper (1972) report hemorrhaging and fracturing of the caudal vertebrae in bluegill fingerlings (*Lepomis macrochirus*) exposed to six different pesticides 4-8 h after treatment at concentrations above and below LC50 values for each formulation. Kinking of tails (20° laterally from normal), cracked vertebrae (centra, spines, and arches), intrusion of muscle and bone into the lumen of the centrum occurs. Aktron® (0-[2-chloro-1-(2,5-dichlorophenyl) vinyl] 0,0-diethyl phosphorothioate) causes hemorrhaging in up to 85% of bluegills tested at higher concentrations (1.23–1.42 ppm) and in 10–15% of bluegills tested at lower concentrations (0.29–0.38 ppm).

Normally, in unpolluted/unheated fish populations, the incidence of vertebral abnormalities is low. Gill and Fisk (1966) find that 3% of about twenty-thousand salmon have vertebral abnormalities based on radiographic analysis of collections taken from 1958 to 1962 from Washington to Alaska, as well as from Japan and Russia. Fusions, distortions, arch formation, duplicate arch elements on single vertebra, and extra hemal processes are noted in what are considered normal populations of sockeye, pink, and chum salmon. Therefore, in these three salmon species (and probably in other fish as well), less than 5% of fish can be expected to show vertebral abnormalities under normal conditions. This low level is a striking contrast to the 67% and 85% vertebral abnormalities in fish exposed to about 1 ppm Cd for 70 d or to about 1 ppm Aktron ® for 4–8 h, respectively. The fact that the second to seventh (or eight) caudal vertebrae are the location of abnormalities from either Cd, Aktron ®, or electroshocking (Spencer, 1967) suggests that chemically- or electrically-induced prolongation of action potentials might cause the high percentage of vertebral abnormalities. Whether thermal elevation during ontogeny induces vertebral deformities by this mechanism remains to be determined.

Respiratory Changes: Morgan and Kuhn (1974) observe an increased opercular rhythm of largemouth bass (*Micropterus salmoides*) from 35 oscillations per min to a maximum of 90 oscillations per min after exposure to 0.1–1.0 ppm aqueous Cd (Fig. 9). Maximum opercular frequency is observed in bass after 2.0–3.5 d following exposure of bass to Cd, compared to 0.5–1.0 d following exposure of bass to Cu. Use of the respiratory response of fish (along with seralogic and histopathologic studies) is recommended for monitoring of industrial releases; however, one obvious problem in the use of respiratory responses as a monitor of Cd release is that fish species respond differently to the same Cd challenge. For example, after exposure of sticklebacks (*Gasterosteus aculeatus*) to Cd, the normal movement of gill filaments is so restricted by excess mucus that much of the water taken in is expelled through the mouth instead of through the opercular apertures, often with such violence that the fish is propelled backwards (Jones, 1939). Delays caused by these sneeze-like responses reduce the continuity of measurements and reduce measurement time. In contrast, Cd-exposed coho salmon (*Oncorhynchus kisutch*) are simply narcotized before death (Schreck and Lorz, 1978). Therefore, either placid or violent respiratory responses hinder measurement of opercular oscillation. Undoubtedly, some species are better suited for measurement of Cd-induced changes in respiratory rate. Species responding at the center (rather than the extremes) of the continuum can be expected to provide good correlation between Cd challenge and respiratory rate.

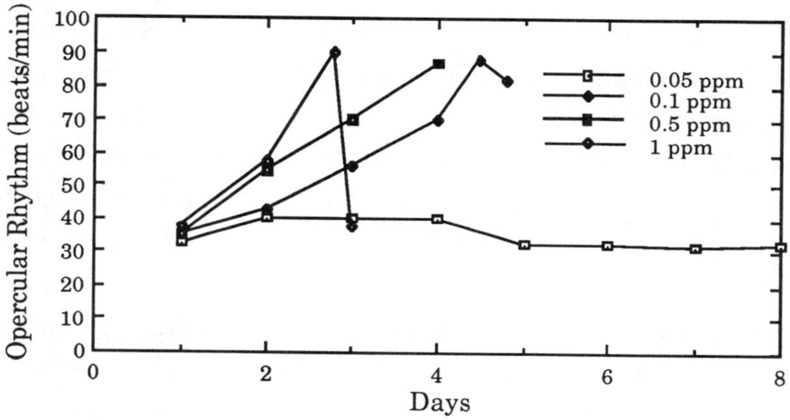

Figure 9. Daily average opercular frequencies of largemouth bass (*Micropterus salmoides*) exposed to different concentrations of aqueous Cd. (Adapted from Morgan and Kuhn (1974).)

Histopathology: Gardner and Yevich (1970) report histopathological changes in a euryhaline killifish (*Fundulus heteroclitus*, the common mummichog) exposed up to 48 h to 50 ppm Cd in the water. Gastrointestinal, renal, and gill tissue changes occur in 1, 12, and 20 h, respectively. Intestinal mucosa shows increased mucous cell activity, epithelial swelling, and release of a mucoid-cellular exudate. Kidney proximal tubules are degenerate, with casts and nuclear debris. Hypertrophy of gill filaments, increases in numbers of mitotic figures, and hyperplasia of epithelial surfaces occurs. Neither collecting tubules nor glomeruli of killifish are affected by Cd. Temperature, pH, and salinity affect the severity and the sequence of appearance of lesions (Gardner and Yevich, 1969). Interestingly, adverse histopathological effects are similar for Cd and Cu, but recovery from tissue injury does not occur following exposure of largemouth bass (*Micropterus salmoides*) to 0.1–1.0 ppm aqueous Cd (Morgan and Kühn, 1974).

Following 4-wk aqueous exposures to 100 ppb Cd, rainbow trout (*Salmo gairdneri*) show renal Cd levels at 1.7 ppm (dry weight) and hyaline droplet degeneration in 80% of proximal tubules, as well as dilated lumina, reduced brush borders, rod- and ring-shaped mitochondria, and proliferation of lysosome-like bodies and apical vacuoles (Forlin, *et al.*, 1986). Comparable aqueous Cd exposures result in 1.1 ppm (dry weight) Cd in the liver and hypertrophied Golgi complexes, detached ribosomes, dilated/fragmented rough endoplasmic reticulum, fragmented mitochondria, glycogen depletion, and a tendency toward fibrosis (Forlin, *et al.*, 1986). Tafanelli and Summerfelt (1975) also find that livers of goldfish are less detrimentally affected than kidneys following 10 mg/kg intraperitoneal injections. Histopathological alterations of the kidney may reduce tubular reabsorption of Ca and produce hypocalcemia (Roch and Maly, 1979; Larsson, *et al.*, 1981; Haux and Larsson, 1984).

Mammalian studies show similar histopathological effects. Bonnell *et al.* (1960) use adult albino rats for administration of intraperitoneal injections of 0.75 or 1 mg/kg Cd, three times weekly for periods up to 1 yr. Proteinuria, urinary casts, and hematuria are found regularly. Renal lesions occur after about 1 mo when renal Cd levels are less than 50 ppm (whether this value is based on wet weight or dry weight is not reported) and resemble those of chronic pyelonephritis due to Cd poisoning. Male rats are more severely affected than are females. Lesions are characterized by tubular atrophy and dilatation, glomerular crowding, degenerative glomerular lesions, interstitial fibrosis, and subcapsular lymphocytic infiltration. Therefore, Cd-induced tissue injury is similar for mammals and fish, with most severe changes in the kidney, liver, gill, and intestine.

Reproductive Effects: Sangalang and O'Halloran (1972) report histopathological and hormonal changes in testicles of mature brook trout (*Salvelinus fontinalis*) exposed in vivo to 25 ppb Cd for 24 h or 10 ppb for about 21 d. Extensive hemorrhagic necrosis, disintegration of lobule-boundary cells (comparable to Leydig cells—the site of steroid production), and distended or collapsed blood vessels result. Unlike goldfish (*Carassius auratus*) which show damage to the entire seminiferous tubules (Tafanelli and Summerfelt, 1975), brook trout show no evidence of damage of primordial germ cells (Sangalang and O'Halloran, 1972). In contrast to both goldfish and brook trout; however, killifish testicles are not affected by Cd (Gardner and Yevich, 1970). In brook trout, weak sudanophil staining indicates that synthesis and/or release of lipid material (steroids) in lobule-boundary cells is disturbed. In vitro tests show that Cd inhibits the conversion of exogenous pregnenolone to free steroids and reduces 11-ketotesterone synthesis from pregnenolone. Therefore, low levels of Cd probably cause subtle, direct effects on steroid enzyme systems of male brook trout—effects adversely altering the reproductive capability. As noted, apparently much variation exists in the sensitivity of the male reproductive system to Cd.

Eaton (1974) observes that an aqueous exposure level of 80 ppb Cd (but not 31 ppb Cd) causes adverse effects in bluegills (*Lepomis macrochirus*) based on a chronic, 11-mo bioassay. Adult bluegills reproduce at even 2140 ppb Cd, but 60-d embryo survival times and mean embryo weights decrease as a function of the exposure level. Therefore, sensitivity of bluegills may be similar to that of fathead minnows (*Pimephales promelas*) tested under similar conditions (Pickering and Gast, 1972). For fathead minnows, 37 ppb was considered the safe level and 57 ppb was considered unsafe.

Severe congenital deformities are found in larvae of adult bluegill (*Lepomis macrochirus*) which had exposed 9 mo to 0.08–2.14 ppm aqueous Cd (Eaton, 1974). Pericardial and abdominal edema, dorsal lordosis (more than 90° from normal), shortened and deformed caudal fin and peduncle, delayed yolk sorption, and microcephalia occur. Larvae from adults exposed to 0.08 ppm or higher aqueous levels move abnormally, and those from adults exposed to 2.14 ppm are capable of only weak, quivering motions. Low blood volume and clotting are characteristic of larvae from adults exposed to 0.08 ppm or higher exposure levels.

Hematology: Gill and Pant (1986) expose adult ornamental barbs (*Puntius conchonius*), a fresh-water fish from northern India, to 0.63 ppm Cd for up to 12 wk. Erythrocyte abnormalities occur and include chromatin condensation at the nuclear periphery, increased interchromatin spaces in erythrocyte nuclei, nuclear "puffs" (probably chromatin leakage from erythrocyte nuclei), hypochromia, cytoplasmic erythrocyte stippling, cell membrane deterioration, and schistocytosis (the accumulation of fragmented erythrocytes, such as that observed in human cases of hemolytic anemias). Chronic exposure of this species results in a persistent anemia (Gill and Pant, 1985). Disturbed RNA synthesis probably causes these erythrocyte changes in *Puntius conchonius* after Cd-poisoning, as observed in isolated mammalian hepatocytes (Puvion and Lange, 1980; Puvion-Dutilleul and Puvion, 1981).

Sjobeck *et al.* (1984) report data from a collection of female perch (*Perca fluviatilis*) from Cd-contaminated sites downstream from a storage battery factory operable about 76 yr but closed since 1976. Fish from Station 3 show a weak anemic response (9% decrease in hematocrit, 12% decrease in hemoglobin, and 21% decrease in the number of erythrocytes compared with controls), lymphocytosis (70%), slight thrombocytosis (34%), slight neutropenia (34%), and increased hepatic Cd levels (470%). In yet another species (i.e. mummichogs), eosinophilia (45%) is noted after 48 h exposure to 50 ppm Cd in the water; eosinophils and thrombocytes show nuclear pleomorphism and/or vacuolation and plasma membrane changes (Gardner and Yevich, 1970). Therefore, both laboratory-exposed and feral fish show hematological changes after Cd exposure.

Roberts *et al.* (1979) expose rainbow trout (*Salmo gairdneri*) and brown trout (*Salmo trutta*) to 9 ppb aqueous Cd for 2 wk or 3 mo. Despite Cd accumulation in the gills and liver, hematocrit does not change. Haux and Larsson (1984) expose rainbow trout (*Salmo gairdneri*) juveniles 18 or 30 wk to 10 or 100 ppb Cd and allow them to recover 25 or 57 wk in clean water. Anemia develops within 18 wk, does not improve after 30 wk, but disappears within 6 mo when fish are moved into clean water. Roberts *et al.* (1979) expose trout in water with a total hardness of 260–270 ppm and Haux and Larsson expose trout at 7 ‰ salinity. Therefore, Cd seems to be less detrimental under these conditions, as indicated by the absence of changes in hematocrit and by the fairly rapid recovery from anemia. This contrasts with the anemia observed in fresh-water barbs exposed in water of 46 ppm hardness (Gill and Pant, 1985; 1986) and environmental exposures of fish in areas of low hardness (40–50 ppm as $CaCO_3$) (Sjobeck, *et al.*, 1984).

Plasma Changes: Larsson and co-workers (1981) expose male and female flounders (*Platichthys flesus*) 4–9 wk to 5–500 ppb Cd (0.00025–0.02100 of the 96-h LC50 values) in brackish water (7–7.5 ‰). Although no differences in plasma levels of Na and Cl occur, hypocalcemia, hypokalemia, hypermagnesemia, and hyperphosphatemia are reported. Ion disturbances are thought to be caused by tissue injury of the gills, kidneys, and intestine, which are in involved in ion exchange between the organism and the surrounding water. Unfortunately, flounders show ionic disturbances of plasma without showing vertebral damage.

Giles (1984) reports significantly lower plasma Na, K, Ca, and Cl, as well as elevated Mg concentrations in rainbow trout (*Salmo gairdneri*) exposed to 3.6–6.4 ppm aqueous Cd for up to 178 d. At the 6.4 ppb exposure level, protein, Na, and osmolarity are elevated and Ca and Mg levels are reduced. Contrary to the Larsson *et al.* (1981) supposition, the results of Giles (1984) on flounders indicate that electrolyte imbalances may not result from renal impairment in all species. In rainbow trout, hyperglycemic response probably accounts for average EOF (fraction of plasma osmolarity accounted for by the five measured electrolytes) of 64.2% in control and Cd-treated trout. Small plasma electrolyte changes, the maintenance of water permeability relationships, and the apparent competence of kidney function suggest that Cd induces electrolyte redistribution from extracellular to intracellular spaces in rainbow trout.

Haux and Larsson (1984) expose rainbow trout (*Salmo gairdneri*) juveniles 18 or 30 wk to 10 or 100 ppb Cd and allow them to recover 25 or 57 wk in clean water. Hypocalcemia, hypermagnesemia, and hyperglycemia develop within 18 wk. Although disturbed ion balance disappears within 6 to 13 mo of recovery in clean water, hyperglycemia persists up to 57 wk after Cd exposure ceases. The interference of Cd with glucose homeostasis (possibly interfering with insulin production or release from beta cells) may be due to persistence of Cd in the liver (and associated exocrine/endocrine pancreas).

Christensen and colleagues (1977) observe glucose, and ion changes in plasma of brook trout (*Salvelinus fontinalis*) exposed to 0.06–6.35 ppb aqueous Cd for 2 or 8 wk. Plasma chloride increases 7% after 8 wk exposures to 3.40 ppb Cd, but not after 2- or 8-wk exposures to 6.35 ppb. Plasma glucose decreases 12% after 8 wk exposures to the same Cd levels. These results are used to set no-effect or safe levels at 6.35 and 0.50 ppb Cd for 2 and 8 wk in brook trout, respectively.

Sjobeck and colleagues (1984) note changes in ion and lactate levels in perch (*Perca fluviatilis*) females collected from Cd-contaminated sites downstream from a storage battery factory operable about 76 yr but closed since 1976. Females from Station 3 do not show hyperglycemia or hypocalcemia, but do show decreased blood lactate levels (66%), hypermagnesemia (12%), and increased hepatic Cd levels (470%). At two other Cd-contaminated stations, decreased muscle and liver glycogen levels are observed, as well as hyperlactemia, hypokalemia, and 632–727% increases in hepatic Cd. The authors suggest that the weaker response of fish from these sites (than from laboratory studies) is due to adaptation to Cd in the field; however, some species of fish are known to avoid low concentrations of metals when given opportunity to do so (Jones, 1964). Therefore, whether the weaker response is indicative of lower exposure levels or adaptation to Cd may be difficult to distinguish in field studies.

Sastry and Subhadra (1982) observe elevated plasma glucose and lactate in a fresh-water catfish (*Heteropneustes fossilis*) exposed to 260 ppb Cd for 15, 30, or 60 d. After 30 d, fish are hyperglycemic (by 62%) and hyperlactemic (by 13%), but 60 d marks a return to normal blood glucose and lactate. Liver and muscle glycogen are decreased (by 33%) at 30 d, but increased at 60 d to 443% for liver and 125% for muscle. Transient hyperglycemia might be due to Cd accumulation in beta cells of the pancreatic islets and a subsequent, brief reduction in insulin production and/or release, as suggested by Haux and Larsson (1984) for rainbow trout. Cadmium-induced stress could increase catecholamines and corticosteroids to increase blood lactate briefly due to increased breakdown of muscle glycogen. In general, activities of lactate dehydrogenase, pyruvate dehydrogenase, and succinate dehydrogenase increase in liver, kidney, intestine, ovary, gills, and skeletal muscle of catfish exposed to 260 ppb Cd for 15 or 30 d (Sastry and Subhadra, 1982). Cadmium also induces increased muscular activity in catfish at 15 and 30 d, as indicated by decreased muscle glycogen, increased blood lactate, and enhanced activity of the three dehydrogenases. Transient changes in carbohydrate metabolism apparently are stress-related in fresh-water catfish exposed to Cd.

As observed for perch and fresh-water catfish in the two previous studies, Lowe-Jinde and Niimi (1984) observe depletion of liver and muscle glycogen in rainbow trout (*Salmo gairdneri*) exposed to 4, 12, or 36 ppb Cd for 1 to 84 d. Even 1 d post-exposure, levels are depleted to about one-third of control levels. Glycogen depletion averages about three times less than control levels the first 3 d of Cd exposure. Liver size increases the 56th and 84th day following aqueous exposure to 4 ppb Cd. Slight increases in carbohydrate metabolism (indicated by increased blood lactate) are noted 7 and 14 d following exposure to 12 ppb; the increase (34%) on the 21 d to 12 ppb Cd is statistically significant. Increases in liver size and carbohydrate metabolism probably occur as fish adapt to maintain homeostasis under Cd stress.

Bilinski and Jones (1974) observe 50% inhibition of the oxidation of lactate by gills of rainbow trout (*Salmo gairdneri*) after exposure of trout 24 h to 1.12 ppm Cd in soft water (4 ppm as $CaCO_3$). At this time, the average Cd level in gills is 12.8 ppm (wet weight). A 20-min exposure of gill filaments in vitro to 10 ppm Cd results in 22–29% inhibition of lactate oxidation. Therefore, oxidative activity is decreased by Cd at higher concentrations in vitro than in vivo, possibly due to slower penetration of Cd into cells of in vitro preparations of gill filaments or to differences in the external milieu of the intact gill and the gill preparations. Disruption of enzyme systems and/or cellular organization could explain the inhibitory action of Cd.

In summary, exposure of fish to Cd causes alteration of plasma levels of ions, glucose, and lactate, among other substances. Hypermagnesemia and/or hypokalemia can occur in flounder, rainbow trout, and perch (Larsson, *et al.*, 1981; Giles, 1984; Haux and Larsson, 1984; Sjobeck, *et al.*, 1984). Hypocalcemia occurs in flounder and rainbow trout (Larsson, *et al.*, 1981; Giles, 1984; Haux and Larsson, 1984). Low or high plasma Cl (Christensen, *et al.*, 1977; Giles, 1984), low plasma Na (Giles, 1984), and hyperphosphatemia (Larsson, *et al.*, 1981) also are found in flounder and rainbow trout. Hyperglycemia persists in rainbow trout (Haux and Larsson, 1984), but not in fresh-water catfish (Sastry and Subhadra, 1982). Blood electrolytes may be altered by stress as a consequence of increased adrenaline levels which alter water permeability at gill membranes (Mazeaud, *et al.*, 1977). In saltwater fish, water loss leads to weight loss and hemoconcentration of plasma, followed by secondary compensatory electrolyte shifts. However, fresh-water fish respond differently, imbibing water for a weight gain and hemodilution prior to a secondary electrolyte shift. Understandably, the salt tolerance of the species involved plays a major role in the effect which Cd has on plasma electrolyte balance.

As mentioned, brook trout show hypoglycemia in response to 1.65–3.40 ppb Cd (but not to 6.35 ppb) after 8-wk exposures (Christensen, *et al.*, 1977). Alterations in blood levels of glucose (and lactate) probably result from a generalized stress reaction akin to that of mammals (Mazeaud, *et al.*, 1977). Chromaffin cells release catecholamines as a response to hypothalamic stimulation; catecholamines initially elevate blood glucose and lactate as liver and muscle glycogen are broken down. In the stress response, elevation in blood glucose and lactate (alarm reaction) is followed by stabilization (resistance), and subsequently by decreased levels (exhaustion) (James and Wigham, 1986). Therefore, levels of glucose and lactate can be increased, decreased, or stable depending on the time of blood sampling. Obviously transient changes such as these require prolonged monitoring at brief intervals; otherwise, alterations in blood chemistry could be overlooked.

Fish seem to respond to stress in much the same fashion as mammals. Mazeaud and co-workers (1977) find catecholamine release in response to various types of stress (handling, salinity changes, hypoxia, and thermal shock). Coho salmon (*Oncorhynchus kisutch*), sockeye (*O. nerka*), and chinook (*O. tschawytscha*) release more circulating catecholamines, mainly adrenaline, as a response to several types of stress. Coho and chinook females respond faster to stress than do males. Secondary effects result, such as a pronounced increase in blood glucose and either an increase or decrease in the plasma level of free fatty acid, depending on the species. Metabolic disturbances are of relatively long duration (several days) despite a more rapid clearance of circulating catecholamines. Catecholamine action may explain stress-induced water imbibition of fish in fresh-water and dehydration of fish in seawater.

Mazeaud *et al.* (1977) note the wide range of biological variation in quantitative hormonal response to stress in fish. Two studies attest the variation possible when fish are exposed to Cd. James and Wigham (1986) show time-dependent changes in plasma cortisol levels of rainbow trout (*Salmo gairdneri*) exposed to 0.05 and 0.1 ppm aqueous Cd. These changes are thought to reflect a general response to continuing stress (i.e. alarm, resistance, and exhaustion). On the contrary, Schreck and Lorz (1978) do not find cortisol or ion changes in the plasma of juvenile coho salmon (*Oncorhynchus kisutch*) exposed to aqueous Cd. Neither a cortisol elevation, nor a change in serum Cl occurs. Therefore, the practicality of using cortisol levels as a general index of environmental quality or fish health seems limited to species having a high stress response (i.e. rainbow trout) rather than coho salmon. These two studies (Schreck and Lorz, 1978; James and Wigham, 1986) attest the range of biological variation in quantitative hormonal response to stress in fish. As done for chickens and turkeys (Siegal, 1971; Brown and Nestor, 1973), genetic selection for lines of low or high stress response might be possible for fish prior to additional testing.

Enzymatic Effects: Cadmium-induced changes in enzyme activity are measured in liver (Jackim, *et al.*, 1970; Jackim, 1973; Roberts, *et al.*, 1979), gill (Roberts, *et al.*, 1979; Watson and Benson, 1987), and plasma (Christensen, 1971/72; Christensen, *et al.*, 1977; Roberts, *et al.*, 1979). Most studies show that changes in enzyme activity are not correlated with the severity of Cd exposure and that in vivo and in vitro results are not correlated. For example, in vitro and in vivo studies using killifish (*Fundulus heteroclitus*) liver enzymes (Jackim, *et al.*, 1970) show 19, 17, and 11% inhibition of acid phosphatase, catalase, and xanthine oxidase, respectively, following 96-h exposure of fish to 27 ppm Cd. However, in vitro tests of enzyme inhibition after direct exposure of a liver homogenate to 11.2 ppm Cd (24 s to 25 min) gives 80 and 6% inhibition of alkaline phosphatase and catalase, respectively, but no change in activities of acid phosphatase or xanthine oxidase. A 96-h in vitro exposure of RNAase to 11.2 ppm Cd results in a 50% decrease in activity after a transitory stimulation at 24 h (about 4% increase in activity). Cellular (or tissue) barriers may explain lack of correlation between in vivo and in vitro data: specific cations could be excluded from active sites because of membrane damage. Jackim and co-workers do not design their studies to differentiate symptomatic changes from causal changes. Nonetheless, they believe that changes in enzyme activity are of sufficient magnitude and consistency to cause death of fish exposed to Cd.

Roberts *et al.* (1979) expose rainbow trout (*Salmo gairdneri*) and brown trout (*Salmo trutta*) to 9 ppb aqueous Cd for 2 wk or 3 mo. Despite Cd accumulation in the gills and liver, no change occurs in the activities of ten enzymes in those tissues, or in plasma and erythrocytes. No serious alteration is noted for rainbow trout gill enzymes despite accumulation of 17.7 ppm Cd (dry weight)—a level much greater than the 15 ppb which results in 50% mortality within 21 d. No derangement of the activity of metabolically important enzymes is noted in tissues of either species, consistent with other studies showing only marginal changes in activities of a variety of liver enzymes (Jackim, *et al.*, 1970; Jackim, 1973).

Following exposure of mummichogs (*Fundulus heteroclitus*) to 10 ppm for 4 or 14 d, 10% and 7% increases, respectively, occur in the activity of ALA-D (∂-aminolevulininate dehydratase) in the liver (Jackim, 1973). Induction of some liver enzymes such as ALA-D is considered a common toxicological phenomenon (Jackim, *et al.*, 1970; Jackim, 1973). Zinc and silver also increase ALA-D activity; whereas, Pb decreases ALA-D activity (Jackim, 1973). Because ALA-D catalyzes the formation of porphobilinogen from aminolevulinic acid, increased ALA-D activity (as a result of Cd exposure) could result in formation of more hemoglobin in Cd-exposed fish. Such is not the case, as numerous workers report (Gardner and Yevich, 1970; Haux and Larsson, 1984; Sjobeck, *et al.*, 1984; Gill and Pant, 1985).

Watson and Benson (1987) compare gill ATPase in bluegill sunfish (*Lepomis macrochirus*), fathead minnows (*Pimephales promelas*), and golden shiners (*Notemigonus crysoleucas*) exposed to Cd. Bluegills show significant stimulation of Na/K-ATPase after 48 h aqueous exposures to 1 ppb Cd, but no effect after 48 h exposures to 10 or 100 ppb Cd. In contrast, fathead minnow gill ATPase is not affected by 48 h exposures to 1 ppb Cd, but shows significant inhibition after exposure to 10 or 100 ppb Cd in the water. Gill ATPase of shiners is not affected by Cd, possibly because of an acquired tolerance to Cd (due to higher background levels as determined by accumulated concentrations of Cd in gills of control fish—0.52 ppb, dry weight).

Explanation of these differences among bluegills, fathead minnows, and shiners is difficult. Gill ATPase is involved in active transport of electrolytes across the gill of teleosts (Epstein and Katz, 1967; Pfeiler and Kirschner, 1972). Therefore, increased stress from Cd exposure might cause sufficient adrenaline release to stimulate ATPase with concurrent permeability changes in the gill (Mazeaud, *et al.*, 1977). However, the established differences in quantitative hormonal response in different fish lends credibility to the thought that any measured differences in ATPase activity (Watson and Benson, 1987) might be overshadowed by variation in ATPase activity in high-stress and low-stress species or individuals. Therefore, shiners may not be affected by Cd (in part) because of a lower stress response than bluegills and fathead minnows. In any case, measurement of plasma adrenaline and gill electrolyte transport (along with measurement of gill ATPase activities) might provide new dimensions in understanding species-specific responses of fish to Cd.

In contrast to the previously mentioned studies, Christensen and colleagues (Christensen, 1971/72; Christensen, *et al.*, 1977) show good correlation between inhibition of enzyme activity and the severity of Cd exposure in vitro. In one of these studies, plasma from white suckers (*Catostomus commersoni*) is incubated for 6–16 min with 0–2000 ppm Cd to measure the activity of GOT (now referred to AST, aspartate aminotransferase) and LDH (lactate dehydrogenase) (Christensen, 1971/72). Exposing enzymes present in the plasma of suckers to Cd concentrations from 0 to 2000 ppm causes progressive inhibition of GOT (about 60% in 16 min) and of LDH (about 30% in 6 min). Therefore, Cd apparently inactivates both GOT and LDH upon contact. In comparative in vivo studies, exposure of fish to Cd causes release of GOT and LDH from injured tissues, thereby providing a measure of tissue injury. In such an in vivo study, Christensen *et al.* (1977) expose brook trout (*Salvelinus fontinalis*) to 0.06–6.35 ppb aqueous Cd for 2 or 8 wk. Plasma lactic dehydrogenase increases 54% after 8 wk exposures to 1.65 ppb Cd, but not to 6.35 ppb Cd. Obviously, more study is needed to establish reasons for such differences in release of transferase or dehydrogenase enzymes into plasma after teleost Cd exposure to different concentrations.

Cadmium blood levels in fish are ordinarily low compared with that found after the 2000 ppm tests used by Christensen (1971/72). Fish blood contains 28±20 ppm Cd following exposure of coho salmon to 150 ppb for 15 d (Reichert, *et al.*, 1979). Therefore little detrimental effect would be expected for either GOT or LDH released from injured tissues into the bloodstream. At about 28 ppm Cd, GOT and LDH inhibition would be less than 5% from data presented by Christensen (1971/72). In contrast, 1000 times this Cd blood level is required to produce the 30% and 60% inhibition previously mentioned. In any case, measurement of LDH and GOT (or other serum enzymes) from fish poisoned with Cd can be expected to produce underestimates since Cd present in plasma can inactivate enzymes such as GOT and LDH. The extent of the underestimate depends on the length of exposure of the serum enzymes to Cd and the sensitivity of the enzyme to Cd. Although the exact involvement of enzymes in heavy metal toxicity is obscure, enzyme inhibition may be a key factor for some enzymes but not for others. For Cd, toxicity may be function of its electron-attracting properties and ability to combine with electron-rich functional groups such as sulfhydryl groups in enzymes such as GOT, LDH, and others.

In conclusion, for most enzymes tested, the use of changes in enzyme activity as a monitor of Cd-poisoning of fish seems limited. Moreover, changes seem to be of insufficient magnitude and consistency to be an immediate causal factor in death of fish. Inconsistency is observed in the direction of the response in vivo and in vitro. Inhibition of enzyme activity (Jackim, *et al.*, 1970; Christensen, 1971/72; Watson and Benson, 1987), enzymatic induction (Jackim, 1973; Christensen, *et al.*, 1977; Watson and Benson, 1987), and no change in activity (Jackim, *et al.*, 1970; Christensen, *et al.*, 1977; Watson and Benson, 1987) are recorded in the literature. Except for the measurement of GOT and LDH in vitro, no clear trends are evident in studies discussed here for fish poisoned with Cd. One major problem with the measurement of enzymatic activity is the frequent use of the mammalian thermal assay regimen of 37°C, rather than the optimal temperature of the poikilotherm species (sometimes as low as 11°C). This can lead to inconsistent patterns of enzyme activity, as determined in studies of the effects of 37°C *versus* 11°C on the activity of ATPase in rainbow trout (Watson and Beamish, 1980).

Protective Mechanisms: As is the case with mammals, fish possess an armamentarium of mechanisms to counteract the detrimental effects of poisons. Some of these mechanisms probably are altered as a result of meeting the challenge of processing a poison such as Cd (e.g. Cd-BP, metallothionein, and ascorbic acid). Other protective systems apparently are altered because of the direct interaction between Cd and enzymes in these systems. As is the case for GOT and LDH, such enzymes probably are altered because they make contact with Cd (e.g. glutathione peroxidase and the cytochrome P-450 system). Cadmium may bind protein groups which promote the catalytic function.

Teleosts form Cd-binding proteins which have storage and detoxification mechanisms similar to those of other vertebrates and which may be responsible for increased resistance of fish to Cd following pretreatment with low levels of Cd. Metallothionein and Cd-binding proteins are considered important in fish exposed to Cd (Overnell, *et al.*, 1977; Beattie and Pascoe, 1979; Reichert, *et al.*, 1979; Brown, *et al.*, 1984; Thomas, *et al.*, 1985; Hamilton, *et al.*, 1987). Intracellular binding of Cd by these proteins can lead to lengthy biological half-lives, as mentioned. Ascorbic acid may limit Cd accumulation and intestinal uptake of Cd (Sahagian, *et al.*, 1967; Fox, 1975; Thomas, *et al.* 1982). Cadmium causes decreases in the level of ascorbic acid in mullet. And because phase I and II reactions are basically enzymatic, Cd can inhibit these reactions by altering biotransformation enzymes (Reddy, *et al.*, 1981; Forlin, *et al.*, 1986). These and other protective mechanisms are altered during Cd poisoning in fish.

Proteins which bind Cd are of considerable interest. Ion-exchange chromatography of trout tissues on DEAE-cellulose after gel filtration on Sephadex G-75 resolves two fractions of Cd-BP and two fractions of (Zn, Cu)-metallothionein (Thomas, *et al.*, 1985). These proteins are produced by rainbow trout (*Salmo gairdneri*) pre-exposed to 9 ppb aqueous Cd for 10 mo prior to exposure to 54 ppb Cd for 2 mo—a dose which produces 70% mortality without pre-exposure. Shorter pre-exposures to 9 ppb Cd produce less tolerance. More than 99% of accumulated Cd is present in liver, kidney, and gills. All three organs contain two isomorphic forms of metallothionein having 22% cysteine but very little phenylalanine, tyrosine, or histidine. Roughly twice as much Cu as Zn is present, and 7 g atoms of metal is bound per mole of protein. These features are consistent with values obtained by others for isoforms of metallothionein. Cadmium is unable to displace Zn (or Cu) from endogenous metallothionein, except in vitro when absurdly elevated Cd concentrations (1000 ppb) are introduced by dialysis.

As indicated, rainbow trout produce two Cd-binding proteins (Cd-BP) in addition to metallothionein isoproteins (Thomas, *et al.*, 1985). Normally, these Cd-BP do not bind Zn. However, pretreatment of trout with Zn before Cd exposure causes Cd distribution between Cd-BP and metallothionein in liver, kidney, and gills. Zinc pretreatment does not protect trout against Cd poisoning. Perhaps Cd is bound to an apothionein form as the nascent protein is being newly synthesized (due to Zn stimulation); whereas, under the conditions of this study, Cd is (of itself) unable to induce synthesis of new apothionein. Therefore, induction of protein synthesis by Zn during pre-exposures can explain sequestration of Cd by apo-metallothionein during Cd post-exposure and the presence of Cd in both Cd-BP and metallothionein.

Unfortunately, Hamilton *et al.* (1987) measure only metallothionein from brook trout (*Salvelinus fontinalis*) exposed 30 d to 3.6–60.6 ppb Cd in the water. These workers designate Cd not bound by metallothionein as "free Cd" rather than evaluating Cd-BP directly. As expected from the Thomas *et al.* (1985) study, metallothionein is not saturated with Cd before the appearance of pathological effects (Hamilton *et al.* 1987). Rather, dose-response relationships are established between free Cd and three parameters—mortality, Cd exposure concentration, and whole-body residues. Evaluation of the results of the Hamilton *et al.* (1987) study indicates that correlations are equally good or better between these three parameters and Cd in the 100,000 g supernatant as for free Cd (both probably primarily Cd-BP). Overnell *et al.* (1977) also observes that most hepatic Cd in plaice liver is associated with the 100,000 g supernatant after subcellular fractionation. In their study, mature plaice (*Pleuronectes platessa*) are given three, consecutive intraperitoneal injections of Cd (0.2 mg/kg on day 1, 0.6 mg/kg on day 4, and 2.0 mg/kg on day 7) prior to sacrifice on day 12. Carbohydrate is present in the Cd-binding glycoprotein recovered from the 100,000g supernatant. Ratios of thiol groups to Cd are 2.6 to 1. The plaice Cd-binding glycoprotein is similar to the low molecular weight Cu-BP isolated from the rat (Winge, *et al.*, 1975).

Reichert and colleagues (1979) administer Cd intravenously to coho salmon (*Oncorhynchus kisutch*) and find Cd bound on a 8,900 molecular weight protein (Cd-BP fraction) and high molecular weight proteins (>13,000 molecular weight fraction). The highest proportion of Cd found in the Cd-BP fraction is in the liver. Of total cytosolic hepatic Cd, 79, 45, and 84% are found 3, 24, and 48 h post-injection. This compares with 9, 23, and 24% Cd in the Cd-BP fraction of the kidney at the same times. The proportion of Cd bound to high molecular weight structures in the liver and kidney cytosol reaches a maximum at 24 h and then decreases; whereas, the proportion of Cd in Cd-BP is a minimum at 24 h and increases at 48 h. A 2-wk pre-exposure of salmon to 150 ppb Cd in the water prior to intravenous injection, results in marked increases in total cytosolic Cd. This effect is most pronounced in the kidney and gills; two-fold minimum increases are noted, compared with fish not pre-exposed to Cd. Induction of Cd-BP in liver and kidney (and possibly the gills) of coho salmon occurs as rapidly following Cd challenge as in mammals.

Beattie and Pascoe (1979) dose rainbow trout (*Salmo gairdneri*) intraperitoneally, giving a 3 mg/kg Cd dose over a 4-d period prior to sacrifice on day 7. Gel filtration using Sephadex G-75 results in association of Cd with a 6,700 dalton hepatic Cd-BP, similar to mammalian metallothionein (6,000–10,000 daltons). Minimal absorbance at 280 nm indicates that the Cd-BP is deficient in aromatic amino acids. About 6.4 μg of Cd is associated with each mg of hepatic protein. Certainly these studies give adequate evidence of the ubiquity of Cd-BP and metallothionein in fish and the importance of these proteins in cases of Cd exposure.

When goldfish (*Carassius auratus*) are injected intraperitoneally with a mixture of 30 nM each of Cd (0.0033 ppm), Zn (0.0041 ppm), and Hg (0.0081 ppm), all three metals are incorporated into Cd-BP (Cd-binding protein) in the liver and kidney (Marafante, 1976). About 25% of Cd is bound to Cd-BP in the liver with less than 2% bound to higher molecular weight proteins in that organ. This compares with 5.2% bound in Cd-BP in the kidney (with 0.9% bound to higher molecular weight proteins). Of the total Hg dose administered, 1.2 and 0.5% are bound in the liver and kidney, respectively. Moreover, only 3.20% and 0.03% of Zn is bound in the respective organs. Of the total hepatic Zn in the liver, Cd-BP incorporates 40%; whereas, the kidney binds 1.2% of Zn. Of the Hg in the liver and kidney, Cd-BP incorporates 17 and 12%, respectively. Therefore, Cd-BP represents an important binding site for elements of the Group IIB triad. *De novo* synthesis of Cd-BP in response to Cd challenge and incorporation of Hg, Zn, and Cd in Cd-BP (and metallothionein) could explain some interactions in metabolism of these metals in fish exposed to mixtures of metals in polluted waters.

Metallothionein concentrations in the livers of rainbow trout (*Salmo gairdneri*) from Zn, Cu, and Cd contaminated lakes show good correlation with the degree of aqueous metal contamination (Roch and McCarter, 1984b). However, after 4 wk of exposure, no enhancement of tolerance is apparent in rainbow trout. Recommended safe levels of 50 ppb Zn, 2.5 ppb Cu, and <0.5 ppb Cd for combined metals provide a 75% or more margin of safety for rainbow trout exposed under conditions of this study. This level corresponds to metallothionein concentrations of 100 nM/g liver, a minimal biological response in these trout.

Whereas, Cd bound by metallothioneins and other proteins can be said to represent "detoxified Cd; " Cd associated with higher molecular weight proteins (>20,000 daltons) can be said to represent the potential for toxicity. Brown and co-workers (1984) expose scorpionfish (*Scorpaena guttata*) 96-h to 25 or 50 ppm Cd (0.4 and 0.8 LC50, respectively) for evaluation of what they call "detoxification/toxification profiles" for the liver, intestine, gill, and kidney. Cadmium concentration in the "metallothionein pool" (3,000–20,000 daltons) represents detoxified Cd; whereas, Cd concentration in the "enzyme pool" (>20,000 daltons) represents the potential for toxicity. As indicated by the high proportion of Cd in the enzyme pool (relative to the metallothionein pool), kidney tissue from Cd-exposed fish has the least ability to detoxify Cd and is considered the most sensitive tissue. Liver from Cd-exposed scorpionfish with more than 98% of Cd in the metallothionein pool, has the greatest ability to detoxify Cd and is considered least sensitive. Intestine and gill are of intermediate sensitivity. Histopathological assessment verifies this argument—scorpionfish kidneys are necrotic and lack cellularity with marked tubular degeneration; whereas, the liver, intestine, and gill show no significant pathological trends at the dosages used.

Although the exact role of ascorbic acid is not clear (Thomas, *et al.*, 1981), ascorbic acid is known to exert a protective effect on symptoms of Cd poisoning—limiting accumulation and intestinal uptake of Cd (Sahagian, *et al.*, 1967; Fox, 1975). Decreased ascorbic acid levels are observed in juvenile mullet (*Mugil cephalus*) exposed to 10 ppm Cd for up to 6 wk in seawater (33 ‰) (Thomas, *et al.*, 1982). Hepatic ascorbic acid levels decrease to 40% of control levels in the liver, 50% of control levels in the gill, and 38% of control levels in the kidney. Cadmium accumulation increases simultaneously to 1260 ppm (dry weight) in liver, to about 360 ppm in gill, and to about 80 ppm in kidney. In contrast, brain accumulates about 7.5 ppm Cd after 6 wk and shows marked fluctuation in levels of ascorbic acid. Therefore, results indicate that tissue levels of ascorbic acid can be decreased markedly during Cd accumulation. Liver and gill may have a greater requirement for ascorbic acid since these tissues accumulate large quantities of Cd. Ascorbic acid may be required for metallothionein production, and overall resistance of fish might be severely decreased during marginal ascorbic acid deficiency due to Cd poisoning.

Cadmium inhibits phase I and phase II biotransformation reactions in fish, probably because Cd alters enzymes responsible for these reactions. In phase I reactions the parent drug can be converted to a more polar metabolite by introducing or unmasking a functional group (e.g. -COOH, -OH, $-NH_2$, -SH) (Sipes and Gandolfi, 1986). These functional groups permit the compound to undergo phase II reactions. This can sometimes inactivate the parent drug or modify the activity of the metabolite. Cytochrome P-450-containing monoxygenases are involved in phase I reactions; whereas, glutathione peroxidase can play a major role in phase II reactions. In phase II metabolism, conjugation or synthesis reactions occur. Endogenous substrates (e.g. glucuronic acid, sulfuric acid, acetic acid, or amino acids) can combine with a newly established functional group to form a highly polar conjugate. Ultimately both types of biotransformation reactions promote the excretion of the chemical by the urinary and/or biliary-fecal routes. Basically, Cd ions might inhibit the enzymes of phase I and II reactions by interacting with essential protein groups which promote the catalytic function. In fact, Cd toxicity may be mediated through inhibition of key antioxidant defense enzymes such as selenium-dependent GSH-Px, GSH-Px II, and others (Reddy, *et al.*, 1981).

Forlin and colleagues (1986) find marked inhibition of microsomal cytochrome P-450 monoxygenase activities, accompanied by inhibition of conjugation reactions, in rainbow trout (*Salmo gairdneri*) after 4-wk aqueous exposures and/or intraperitoneal injections (0.5 mg Cd/kg). In vitro incubations show 50% reduction in phase I and phase II reactions at 5 μM and 0.2–0.5mM, respectively. For comparison, in mammalian studies non-selenium glutathione peroxidase (GSH-Px, important in phase II reactions) is markedly inhibited by Cd (Reddy, *et al.*, 1981). Fifty percent inhibition of transferase activity in calf liver is observed at concentrations as low as 0.01 mM (0.112 ppm) Cd; whereas, 50% inhibition of peroxidase activity is observed at about 0.001 mM (0.011 ppm) Cd. Although Cd-induced inhibition of transferase and peroxidase activities is known to be noncompetitive, exact mechanisms of metal interaction with GSH-Px are not clear. Nonetheless, the exposure of fish to Cd can result in *de novo* synthesis of Cd-BP and/or metallothionein, decreased levels of ascorbic acid, and inhibition of phase I and II biotransformation reactions.

Summary

Roughly, the toxicity of Cd in fish is equivalent to that of Pb, but twenty-five times less than that of Hg. Species-specific variation is the rule. Cadmium-induced toxicity is a function of water quality (e.g. salinity, water hardness, pH, alkalinity, and temperature) and parasitism. Apparently, mucous production, asphyxiation, inhibition of acetylcholinesterase, depression of the respiratory center, hypocalcemia, and gill injury are causal factors in death, depending on the species.

Gill concentrations of Cd are used to monitor acute exposures of fish; whereas, hepatic, renal, and gastrointestinal Cd levels are useful in evaluating chronic exposures. Renal and hepatic tissues are critical in detoxification, as evidenced by marked Cd accumulation in these organs. In fact, the kidney continues to accumulate Cd long after exposure has ceased, probably as a result of hepatic to renal translocation and of long-term hepatic storage of Cd bound to metallothionein. Cadmium may be translocated to the kidney complexed with this or other metal-binding proteins.

Equilibrium times of 1- to 4-mo are obtained for Cd, depending on species and conditions of exposure. At this time, metal uptake is equal to metal elimination; therefore, death may be a consequence of the failure of excretion mechanisms, stimulation of uptake mechanisms, or storage of a specific amount of Cd in critical organ.

Plasma changes result from Cd exposure of fish. Cadmium alters plasma levels of ions, glucose, and lactate, as well as other substances. Hypermagnesemia and/or hypokalemia can occur in flounder, rainbow trout, and perch. Hypocalcemia occurs in flounder and rainbow trout. Low or high plasma Cl, low plasma Na, and hyperphosphatemia also are found in flounder and rainbow trout. Hyperglycemia persists in rainbow trout, but not in fresh-water catfish. Blood electrolytes may be altered by stress as a consequence of increased adrenaline levels which alter water permeability at gill membranes. Salt tolerance of the species involved plays a major role in the effect which Cd has on plasma electrolyte balance.

Hematological changes following Cd exposure include anemia and erythrocyte abnormalities. Among changes reported are chromatin condensation at the nuclear periphery, increased interchromatin spaces in erythrocyte nuclei, nuclear puffs, hypochromia, cytoplasmic erythrocyte stippling, cell membrane deterioration, schistocytosis, lymphocytosis, thrombocytosis, neutropenia, and eosinophilia. Thrombocytes show nuclear pleomorphism and/or vacuolation and alteration of the plasma membrane. Therefore, Cd-induced effects on fish are pronounced—justifying the stringent water quality criteria standards which have been imposed (Appendix 1).

References

Almer, B., W. Dickson, C. Ekstrom, E. Hornstrom, and U. Miller, Effects of acidification on Swedish Lakes, Ambio, 3 (1974) 30.

Arme, C., and R.W. Owen, Infections of the three-spined stickleback, *Gasterostus aculeatus* L., with the plerocercoid larvae of *Schistocephalus solidus* (Muller, 1776), with special reference to pathological effects, Parasitol., 57 (1967) 301.

Atchison, G.J., B.R. Murphy, W.E. Bishop, A.W. McIntosh, and R.A. Mayes, Trace metal concentration of bluegill (*Lepomis macrochirus*) from two Indiana lakes, Trans. Am. Fish. Soc., 106 (1977) 637.

Beattie, J.H., and D. Pascoe, A cadmium-binding protein in rainbow trout, Tox. Lett., 4 (1979) 241.

Bendell-Young, L.I., H.H. Harvey, and J.F. Young, Accumulation of cadmium by white suckers (*Catostomus commersoni*) in relation to fish growth and lake acidification, Can. J. Fish. Aquat. Sci., 43 (1986) 806.

Bengtsson, B.-E., C.H. Carlin, A. Larsson, and O. Svanberg, Vertebral damage in minnows, *Phoxinus phoxinus* L., exposed to cadmium, Ambio, 4 (1975) 166.

Benoit, D.A., E.N. Leonard, G.M. Christensen, and J.T. Fiandt, Toxic effects of cadmium on three generations of brook trout (*Salvelinus fontinalis*), Trans. Am. Fish. Soc., 105 (1976) 550.

Beveridge, J.M.R., Sulfur distribution in fish flesh proteins, J. Fish. Res. Bd. Can., 7 (1947) 51.

Bilinski, E., and R.E.E. Jonas, Effects of cadmium and copper on the oxidation of lactate by rainbow trout (*Salmo gairdneri*) gills, J. Fish. Res. Bd. Can., 30 (1973) 1553.

Bohn, A., and R.O. McElroy, Trace metals (As, Cd, Cu, Fe, and Zn) in Artic cod, *Boreogadus saida*, and selected zooplankton from Strathcona Sound, Northern Baffin Island, J. Fish. Res. Bd. Can., 33 (1976) 2836.

Bohn, A., and B.W. Fallis, Metal concentrations (As, Cd, Cu, Pb and Zn) in shorthorn sculpins, *Myoxocephalus scorpius* (Linnaeus), and Arctic char, *Salvelinus alpinus* (Linnaeus), from the vicinity of Strathcona Sound, Northwest Territories,, Water Res., 13 (1978) 659.

Bonnell, J.A., J.H. Ross, and E. King, Renal lesions in experimetnal cadmium poisoning, Brit. J. Industr. Med., 17 (1960) 69.

Brown, D.A., S.M. Bay, J.F. Alfafara, G.P. Hershelman, and K.D. Rosenthal, Detoxification/toxification of cadmium in scorpionfish (*Scorpaena guttata*): acute exposure, Aquat. Toxicol., 5 (1984) 93.

Brown, K.I., and K.E. Nestor, Some physiological responses of turkeys selected for high and low adrenal response to cold stress, Poult. Sci., 52 (1973) 1948.

Brown, V.M., The calculation of the acute toxicity of mixtures of poisons to rainbow trout, Water Res., 2 (1968) 723.

Bryan, S.E., and H.A. Hildalgo, Nuclear ^{115}Cadmium: Uptake and disappearance correlated with cadmium-binding protein synthesis, Biochem. Biophy. Res. Comm., 68 (1976) 858.

Bussey, R.E., D.E. Kidd, and L.D. Potter, The concentrations of ten heavy metals in some selected Lake Powell game fishes, University of New Mexico, Albuquerque, Lake Powell Research Project Bulletin, (1976).

Carpenter, K.E., The lethal action of soluble metallic salts on fish, Br. J. Exptl. Biol., 4 (1927) 378.

Carroll, J.J., S.J. Ellis, and W.S. Oliver, Influences of hardness constituents on the acute toxicity of cadmium to brook trout (Salvelinus fontinalis), Bull. Environ. Contam. Toxicol., 22 (1979) 575.

Cearley, J.E., and R.L. Coleman, Cadmium toxicity and bioconcentration in largemouth bass and bluegill, Bull. Environ. Contam. Toxicol., 11 (1974) 146.

Christensen, G.M., Effects of metal cations and other chemicals upon the in vitro activity of two enzymes in the blood plasma of the white sucker, *Catostomus commersoni* (*Lacepede*), Chem.-Biol. Interactions, 4 (1971/72) 351.

Christensen, G., E. Hunt, and J. Fiandt, The effect of methylmercuric chloride, cadmium chloride, and lead nitrate on six biochemical factors of the brook trout (*Salvelinus fontinalis*), Toxicol. Appl. Pharmacol., 42 (1977) 523.

Cusimano, R.F., D.F. Brakke, and G.A. Chapman, Effects of pH on the toxicities of cadmium, copper, and zinc to steelhead trout (*Salmo gairdneri*), Can. J. Fish. Aquat. Sci., 43 (1986) 1497.

Davis, J.C., and R.A.W. Hoos, Use of sodium pentachlorophenate and dehydrobietic acid as reference toxicants for salmonid bioassays, J. Fish. Res. Bd. Can., 32 (1975) 411.

Dwyer, F.J., C.J. Schmitt, S.E. Finger, and P.M. Mehrle, Biochemical changes in longear sunfish, *Lepomis megalotis*, associated with lead, cadmium and zinc from mine tailings, J. Fish. Biol., 33 (1988) 307.

Eaton, J.G., Chronic cadmium toxicity to the bluegill (*Lepomis macrochirus* Rafinesque), Trans. Amer. Fish. Soc., 103 (1974) 729.

Eisler, R., Cadmium poisoning in *Fundulus heteroclitus* (Pisces: Cyprinodontidae) and other marine organisms, J. Fish. Res. Bd. Can., 28 (1971) 1225.

Eisler, R., Radiocadmium exchange with seawater by *Fundulus heteroclitus* (L.) (Pisces: Cyprinodontidae), J. Fish. Biol., 6 (1974) 601.

El Nabawi, A., B. Heinzow, and H. Kruse, As, Cd, Cu, Pb, Hg, and Zn in fish from the Alexandria region, Egypt, Bull. Environ. Contam. Toxicol., 39 (1987) 889.

Epstein, F.H., and A.I. Katz, Sodium- and potassium-activated adenosine triphosphatase of gills: Role in adaptation of teleosts to salt water, Science, 156 (1967) 1245.

Fleischer, M., A.F. Sarofim, D.W. Fassett, P. Hammond, H.T. Shacklette, I.C.T. Nisbet, and S. Epstein, Environmental impact of cadmium: A review by the panel on hazardous trace substances, Environ. Hlth. Persp., (1974) 253.

Flick, D.F., H.F. Kraybill, and J.M. Dimitroff, Toxic effects of cadmium: A review, Environ. Res., 4 (1971) 71.

Forlin, L., C. Haux, L. Karksson-Norrgren, P. Runn, and A. Larsson, Biotransformation enzyme activites and histopathology in rainbow trout, *Salmo gairdneri*, treated with cadmium, Aquat. Toxicol., 8 (1986) 51.

Fox, M.R.S., Protective effects of ascorbic acid against toxicity of heavy metals, Ann. N.Y. Acad. Sci., 258 (1975) 144.

Frazier, J.M., and J. Puglese, Dose dependence of cadmium kinetics in the rat liver following intravenous injection, Toxicol. Appl. Pharmacol., 43 (1978) 461.

Friberg, L., M. Piscator, G.F. Nordberg, and T. Kjellstrom, Cadmium in the Environment, 2nd ed., CRC Press, Cleveland, Ohio, (1974) 1.

Gardner, G.R., and P.P. Yevich, Toxicological effects of cadmium on *Fundulus heteroclitus* under various oxygen, pH, salinity, and temperature regimes, Amer. Zool., 9 (1969) 1096.

Gardner, G.R., and P.P. Yevich, Histological and hematological responses of an estuarine teleost to cadmium, J. Fish. Res. Bd. Can., 27 (1970) 2185.

Gerson, R.J., and Z.A. Shaikh, Differences in the uptake of cadmium and mercury by rat hepatocyte primary cultures, Biochem. Pharmacol., 33 (1984) 199.

Giesy, J.P., Jr., G.J. Leversee, and D.R. Williams, Effects of naturally occurring aquatic organic fractions on cadmium toxicity to *Simocephalus serrulatus* (Daphnidae) and *Gambusia affinis* (Poeciliidae), Water Res., 11 (1977) 1013.

Giesy, J.P., Jr., and J.C. Wiener, Frequency distributions of trace metal concentrations in five freshwater fishes, Trans. Am. Fish. Soc., 106 (1978) 393.

Giles, M.A., Electrolyte and water balance in plasma and urine of rainbow trout (*Salmo gairdneri*) during chronic exposure to cadmium, Can. J. Fish. Aquat. Sci., 41 (1984) 1678.

Gill, C.D., and D.M. Fisk, Vertebral abnormalities in sockeye, pink, and chum salmon, Trans. Am. Fish. Soc., 95 (1966) 177.

Gill, T.S., and J.C. Pant, Erythrocytic and leucocytic responses to cadmium poisoning in a freshwater fish, *Puntius conchonius*, Environ. Res., 36 (1985) 327.

Gill, T.S., and J.C. Pant, Chromatin condensation in the erythrocytes of fish following exposure to cadmium, Bull. Environ. Contam. Toxicol., 36 (1986) 199.

Gottofrey, J., I. Bjorklund, and H. Tjalve, Effect of sodium isopropylxanthate, potassium amylxanthate and sodium diethyldithiocarbamate on the uptake and distribution of cadmium in the brown trout (*Salmo trutta*), Aquat. Toxicol., 12 (1988) 171.

Grahl, K., P. Franke, and R. Hallebach, The excretion of heavy metals by fish, Symp. Biol. Hung., 29 (1985) 357.

Greichus, Y.A., A. Greichus, B.D. Ammann, and J. Hopcraft, Insecticides, polychlorinated biphenyls, and metals in African lake ecosystems. III. Lake Nakuru, Kenya., Bull. Environ. Contam. Toxicol., 19 (1978) 454.

Hall, W.S., R.L. Paulson, L.W. Hall Jr., and D.T. Burton, Acute toxicity of cadmium and sodium pentochlorophenate to Daphnids and fish, Arch. Environ. Contam. Toxicol., 37 (1986) 308.

Hamilton, S.J., P.M. Mehrle, and J.J. Jones, Evaluation of metallothionein as a biological indicator of stress from cadmium in brook trout, Trans. Am. Fish. Soc., 116 (1987) 551.

Haux, C., and A. Larsson, Long-term sublethal physiological effects on rainbow trout, *Salmo gairdneri*, during exposure to cadmium and after subsequent recovery, Aquat. Toxicol., 5 (1984) 129.

Havre, G.N., B. Underdal, and C. Christiansen, Cadmium concentrations in some fish species form a coastal area in southern Norway, Oikos, 24 (1973) 155.

Holcombe, G.W., D.A. Benoit, E.N. Leonard, and J.M. McKim, Long-term effects of lead exposure on three generations of brook trout (*Salvelinus fontinalis*), J. Fish. Res. Bd. Can., 33 (1976) 1731.

Horowitz, A., and B.J. Presley, Trace metal concentrations and partitioning in zooplankton, neuston, and benthos from the south Texas Outer Continental Shelf, Arch. Environ. Cont., 5 (1977) 241.

Huang, P.C., B. Smith, P. Bohdan, and A. Corrigan, Effect of zinc on cadmium influx and toxicity in cultured CHO cells, Biol. Trace Element Res., 2 (1980) 211.

Hubbs, C., High incidence of vertebral deformities in two natural populations of fishes inhabiting warm springs, Ecology, 40 (1959) 154.

Jackim, E., J.M. Hamlin, and S. Sonis, Effects of metal poisoning on five liver enzymes in the killifish (*Fundulus heteroclitus*), J. Fish. Res. Bd. Can., 27 (1970) 383.

Jackim, E., Influence of lead and other metals on fish ∂-aminolevulinate dehydrase activity, J. Fish. Board Can., 30 (1973) 560.

James, V.A., and T. Wigham, The effects of cadmium on prolactin cell activity and plasma cortisol levels in the rainbow trout (*Salmo gairdneri*), Aquati. Toxicol., 8 (1986) 273.

Jones, J.R.E., The relation between the electrolytic solution pressures of the metals and their toxicity to the stickleback (*Gasterosteus aculeatus* L.), J. Exp. Biol., 16 (1939) 425.

Jones, J.R.E., The metals as salts, Fish and River Pollution, Butterworth & Company, London, (1964) 66.

Kinkade, M.L., and H.E. Erdman, The influence of hardness components (Ca^{2+} Mg^{2+}) in water on the uptake and concentration of cadmium in a simulated freshwater ecosystem, Environ. Res., 10 (1975) 308.

Kito, H., T. Tazawa, Y. Ose, T. Sato, and T. Ishikawa, Protection by metallothionein against cadmium toxicity, Comp. Biochem. Physiol., 73C (1982) 135.

Klaassen, C.D., and F.N. Kotsonis, Biliary excretion of cadmium in the rat, rabbit, and dog, Toxicol. Appl. Pharmacol., 41 (1977) 101.

Larsson, A., B.-E. Bengtsson, and C. Haux, Disturbed ion balance in flounder, *Platichthys flesus* L., exposed to sublethal levels of cadium, Aquat. Toxicol, 1 (1981) 19.

Lovett, R.J., W.H. Gutenmann, I.S. Pakkala, W.D. Youngs, D.J. Lisk, G.E. Burdick, and E.J. Harris, A survey of the total cadmium content of 406 fish from 49 New York State fresh waters, J. Fish. Res. Bd. Can., 29 (1972) 1283.

Lowe-Jinde, L., and A.J. Niimi, Short-term and long-term effects of cadmium on glycogen reserves and liver size in rainbow trout (*Salmo gairdneri* Richardson), Arch. Environ. Contam. Toxicol., 13 (1984) 759.

Lucas, H.F., Jr., D.N. Edgington, and P.J. Colby, Concentrations of trace elements in Great Lakes fishes, J. Fish. Res. Bd. Can., 27 (1970) 677.

Marafante, E., Binding of mercury and zinc to cadmium-binding protein in liver and kidney of goldfish (*Carassius auratus* L.), Experientia, 15 (1976) 149.

Mathis, B.J., and N.R. Kevern, Distribution of mercury, cadmium, lead, and thallium in a eutrophic lake, Hydrobiologia, 46 (1975) 207.

Mazeaud, M.M., F. Mazeaud, and E.M. Donaldson, Primary and secondary effects of stress in fish: Some new data with a general review, Trans. Amer. Fish Soc., 106 (1977) 201.

McCann, J.A., and R.L. Jasper, Vertebral damage to bluegills exposed to acutely toxic levels of pesticides, Trans. Am. Fish. Soc., 101 (1972) 317.

McCarty, L.S., J.A.C. Henry, and A.H. Houston, Toxicity of cadmium to goldfish, *Carassius auratus*, in hard and soft water, J. Fish. Res. Bd. Can., 35 (1978) 35.

McFarlane, G.A., and W.G. Franzin, An examination of Cd, Cu, and Hg concentrations in livers of northern pike, *Esox lucius*, and white sucker, *Catostomus commersoni*, from five lake near a base metal smelter at Flin Flon, Manitoba, Can. J. Fish. Aquat, Sci., 37 (1980) 1573.

Merlini, M., and G. Pozzi, Lead and freshwater fishes: Part I—Lead accumulation and water pH, Envir. Poll., 12 (1977a) 167.

Merlini, M., and G. Pozzi, Lead and freshwater fishes: Part 2—Ionic lead accumulation, Environ. Pollut., 13 (1977b) 119.

Moore, J.W., and S. Ramamoorthy, Heavy Metals in Natural Waters, Springer-Verlag, New York, (1984) 28.

Morgan, W.S.G., and P.C. Kühn, A method to monitor the effects of toxicants upon breathing rate of largemouth bass (*Micropterus salmoides* Lacépéde), Water Res., 8 (1974) 67.

Mount, D.I., and C.E. Stephen, A method for detecting cadmium poisoning in fish, J. Wildlife Manag., 31 (1967) 168.

Murphy, B.R., G.J. Atchison, and A.W. McIntosh, Cadmium and zinc content of fish from an industrially contaminated lake, J. Fish Biol., 13 (1978) 327.

Ney, J.J., and J.H. Van Hassel, Sources of variability in accumulation of heavy metals by fishes in a roadside stream, Arch. Environ. Contam. Toxicol., 12 (1983) 701.

Noel-Lambot, F., C. Gerday, and A. Disteche, Distribution of Cd, Zn, and Cu in the liver and gills of the eel *Anguilla anguilla* with special reference to metallothioneins, Comp. Biochem. Physiol., 61C (1978) 177.

Olsson, P.-E., and C. Haux, Increased hepatic metallothionein content correlates to cadmium accumulation in environmentally exposed perch (*Perca fluviatilis*), Aquat. Toxicol., 9 (1986) 231.

Overnell, J., I.A. Davidson, and T.L. Coombs, A cadmium-binding glycoprotein from the liver of the plaice (*Pleuronectes platessa*), Biochem. Soc. Trans., 5 (1977) 267.

Pascoe, D., and P. Cram, The effect of parasitism on the toxicity of cadmium to the three-spined stickleback, *Gasterosteus aculeatus* L., J. Fish Biol., 10 (1977a) 467.

Pascoe, D., and D.L. Mattey, Studies on the toxicity of cadmium to the three-spined stickleback *Gasterosteus aculeatus* L., J. Fish Biol., 11 (1977b) 207.

Pascoe, D., S.A. Evans, and J. Woodworth, Heavy metal toxicity to fish and the influence of water hardness, Arch. Environ. Contam. Toxicol., 15 (1986) 481.

Pennycuick, L., Quantitative effects of three species of parasites on a population of three-spined sticklebacks, *Gasterosteus aculeatus*, J. Zool. Lond., 165 (1971) 143.

Pfeiler, E., and L.B. Kirschner, Studies on gill ATPase of rainbow trout (*Salmo gairdneri*), Biochim. Biophys. Acta, 282 (1972) 301.

Pickering, Q.H., and C. Henderson, The acute toxicity of some heavy metals to different species of warmwater fishes, Air Water Pollut. Int. J., 10 (1966) 453.

Pickering, Q.H., and M.H. Gast, Acute and chronic toxicity of cadmium to the fathead minnow (*Pimephales promelas*), J. Fish. Res. Bd. Can., 29 (1972) 1099.

Powers, E.B., Influence of temperature and concentration on the toxicity of salts to fishes, Ecology, 1 (1920) 95.

Puvion, E., and M. Lange, Functional significance of perichromatin granule accumulation induced by cadmium chloride in isolated rat liver cells, Exp. Cell Res., 128 (1980) 47.

Puvion-Dutilleul, F., and E. Puvion, Relationship between chromatin and preichromatin granules in cadmium-treated isolated hepatocytes, J. Ultrastuct. Res., 74 (1981) 341.

Ramamoorthy, S., and K. Blumhagen, Uptake of Zn, Cd, and Hg by fish in the presence of competing compartments, Can. J. Fish. Aquat. Sci., 41 (1984) 750.

Reddy, C.C., R.W. Scholz, and E.J. Massaro, Cadmium, methylmercury, mercury, and lead inhibition of calf liver glutathione s-transferase exhibiting selenium-independent glutathione peroxidase activity, Toxicol. Appl. Pharmacol., 61 (1981) 460.

Reichert, W.L., D.A. Fiderighi, and D.C. Malins, Uptake and metabolism of lead and cadmium in coho salmon (*Oncorhynchus kisutch*), Comp. Biochem. Physiol., 63C (1979) 229.

Roberts, K.S., A. Cryer, J. Kay, J.F. De L. G. Solbe, J.R. Wharfe, and W.R. Simpson, The effects of exposure to sub-lethal concentrations of cadmium on enzyme activities and accumulation of the metal in tissues and organs of rainbow and brown trout (*Salmo gairdneri*, Richardson and *Salmo trutta fario* L.), Comp. Biochem. Physiol., 62C (1979) 135.

Roch, M., and E.J. Maly, Relationship of cadmium-induced hypocalcemia with mortality in rainbow trout (*Salmo gairdneri*) and the influence of temperature on toxicity, J. Fish. Res. Bd. Can., 36 (1979) 1297.

Roch, M., J.A. McCarter, A.T. Matheson, M.J.R. Clark, and R.W. Olafson, Hepatic metallothionein in rainbow trout (*Salmo gairdneri*) as an indication of metal pollution in the Campbell River System, , Can. J. Fish. Aquat. Sci., 39 (1982) 1596.

Roch, M., and J.A. McCarter, Hepatic metallothionein production and resistance to heavy metals by rainbow trout (*Salmo gairdneri*)—I. Exposed to an artificial mixture of zinc, copper, and cadmium, Comp. Biochem. Physiol., 77C (1984a) 71.

Roch, M., and J.A. McCarter, Hepatic metallothionein production and resistance to heavy metals by rainbow trout (*Salmo gairdneri*)—II. Held in a series of contaminated lakes, Comp. Biochem. Physiol., 77C (1984b) 77.

Rowe, D.W., and E.J. Massaro, Cadmium uptake and time dependent alterations in tissue levels in the white catfish *Ictalurus catus* (Pisces: Ictaluridae), Bull. Environ. Contam. Toxicol., 11 (1974) 244.

Ryan, P.M., and H.H. Harvey, Growth of rock bass, *Ambloplites rupestris*, in relation to the morphoedaphic index as an indicator of environmental stress, J. Fish. Res. Bd. Can., 34 (1977) 2079.

Ryan, P.M., and H.H. Harvey, Growth responses of yellow perch *Perca flavescens* (Mitchill) to lake acidification in the La Cloche Mountain Lakes on Ontario, Environ. Biol. Fishes, 5 (1980) 97.

Sahagian, B.M., I. Harding-Barlow, and H.M. Perry Jr., Transmural movements of zinc, manganese, cadmium, and mercury by rat small intestine, J. Nutr., 93 (1967) 291.

Sangalang, G.B., and M.J. O'Halloran, Cadmium-induced testicular injury and alterations of androgen synthesis in brook trout, Nature, 240 (1972) 470.

Sangalang, G.B., and H.C. Freeman, Tissue uptake of cadmium in brook trout during chronic sublethal exposure, Arch. Environ. Contam. Toxicol., 8 (1979) 77.

Sastry, K.V., and K. Subhadra, Effect of cadmium on some aspects of carbohydrate metabolism in a freshwater catfish *Heteropneustes fossilis*, Toxicol. Lett., 14 (1982) 45.

Schmitt, C.J., F.J. Dwyer, and S.E. Finger, Bioavailability of Pb and Zn from mine tailings as indicated by erythrocyte ∂-aminolevulinic acid dehydratase (ALA-D) activity in suckers (*Pisces: Catostomidae*), Can. J. Fish. Aquat. Sci., 41 (1984) 1030.

Schofield, C.L., Acid precipitation: Effects on fish, Ambio, 5 (1976) 228.

Schreck, C.B., and H.W. Lorz, Stress response of coho salmon *(Oncorhynchus kisutch)* elicited by cadmium and copper and potential use of cortisol as an indicator of stress, J. Fish. Res. Bd. Can., 35 (1978) 1124.

Siegal, H.S., Adrenals, stress and the environment, World's Poult. Sci. J., 27 (1971) 327.

Sipes, I.G., and A.J. Gandolfi, Biotransformation of toxicants, in Klaassen, C.D., M.O. Amdur, and J. Doull (Eds.), Toxicology: The Basic Science of Poisons, 3rd ed., Macmillan Publ. Co., New York, (1986) 64.

Sjobeck, M.-L., C. Haux, A. Larsson, and G. Lithner, Biochemical and hematological studies on perch, *Perca fluviatilis*, from the cadmium-contaminated river Eman, Ecotoxicol. Environ. Safety, 8 (1984) 303.

Smith, B.P., E. Hejtmancik, and B.J. Camp, Acute effects of cadmium on *Ictalurus punctatus* (catfish), Bull. Environ. Cont. Toxicol., 15 (1976) 271.

Sorensen, E.M.B., R.E. Henry, and R. Ramirez-Mitchell, Arsenic accumulation, tissue distribution and cytotoxicity in teleosts following indirect aqueous exposures, Bull. Environ. Contam. Toxicol., 21 (1979) 162.

Spencer, S.L., Internal injuries of largemouth bass and bluegill caused by electricity, Prog. Fish-Cult., 29 (1967) 168.

Tafanelli, R., and R.C. Summerfelt, Cadmium-induced histopathological changes in goldfish, in Ribelin, W.E., and G. Migaki (Eds.), The Pathology of Fishes, Univ. Wisc. Press, Madison, (1975) 613.

Thomas, P., R.S. Carr, and J.M. Neff, Biochemical stress responses of mullet *Mugil cephalus* and polychaete worms *Neanthes virens* to pentachlorophenol, in Vernberg, V.J., A. Calabes, F.P. Thurberg, and W.B. Vernberg (Eds.), Biological Monitoring of Marine Pollutants, Academic Press, New York, (1981) 73.

Thomas, D.G., M.W. Brown, D. Shurben, J.F. del G. Solbe, A. Cryer, and J. Kay, A comparison of the sequestration of cadmium and zinc in the tissues of rainbow trout (*Salmo gairdneri*) following exposure to the metals singly or in combination, Comp. Biochem. Physiol., 82C (1985) 55.

Thomas, P., M. Bally, and J.M. Neff, Ascorbic acid status of mullet, *Mugil cephalus* Linn., exposed to cadmium, J. Fish Biol., 20 (1985) 183.

Tjalve, H., J. Gottofrey, and J. Bjorklund, Tissue disposition of Cd-109 in the brown trout (*Salmo trutta*) studied by autoradiography and impulse counting, Toxicol. Environ. Chem., 12 (1986) 31.

U.S. Environmental Protection Agency, Methods for acute toxicity tests with fish, microinvertebrates, and amphibians, U.S. Environmental Protection Agency, EPA 660 13-75-009, Corvallis, Oregon, (1975) 67.

Vallee, B.L., and D.D. Ulmer, Biochemical effects of mercury, cadmium, and lead, Annu. Rev. Biochem., 41 (1972) 91.

Varanasi, U., and D. Markey, Uptake and release of lead and cadmium in skin and mucus of coho salmon (*Oncorhynchus kisutch*), Comp. Biochem. Physiol., 60C (1978) 187.

Vinikour, W.S., R.M. Goldstein, and R.V. Anderson, Bioconcentration patterns of zinc, copper, cadmium, and lead in selected fish species from the Fox River, Illinois, Bull. Environ. Contam. Toxicol., 24 (1980) 727.

Walkey, M., and R.W. Meakins, An attempt to balance the energy budget of a host-parasite system, J. Fish Biol., 2 (1970) 361.

Watson, C.F., and W.H. Benson, Comparative activity of gill ATPase in three freshwater teleosts exposed to cadmium, Ecotox. Environ. Safety, 14 (1987) 252.

Watson, T.A., and F.W.H. Beamish, Effects of zinc on brachial ATPase activity in vivo in rainbow trout, *Salmo gairdneri*, Comp. Biochem. Physiol., 66C (1980) 77.

Webb, M., and K. Cain, Functions of metallothionein, Biochem. Pharmacol., 31 (1982) 137.

Westfall, B.A., Coagulation film anoxia in fishes, Ecology, 26 (1945) 283.

Wiener, J.G., and J.P. Giesy Jr., Concentrations of Cd, Cu, Mn, Pb, and Zn in fishes in a highly organic softwater pond, J. Fish. Res. Board Can., 36 (1979) 270.

Winge, D.R., R. Premakumar, R.D. Wiley, and K.V. Rajagopalan, Copper-chelatin: Purification and properties of a copper-binding protein from rat liver, Arch. Biochem. Biophys., 170 (1975) 253.

Wofford, H.W., and P. Thomas, Interaction of cadmium with sulhydryl-containing compounds in striped mullet (*Mugil cephalus* L.), Mar. Environ. Res., 14 (1984) 119.

Woodworth, J., A.S.A. Evans, and D. Pascoe, The production of cadmium-binding protein in three species of freshwater fish, Toxicol. Lett., 15 (1983) 289.

Woodworth, J., and D. Pascoe, Cadmium uptake and distribution in sticklebacks related to the concentration and method of exposure, Ecotoxicol. Environ. Safety, 7 (1983) 525.

CHAPTER VII
COPPER

Introduction

Copper is an essential element and part of about thirty enzymes and glycoproteins (e.g., amine oxidases, catalase, ceruloplasmin, cytochrome oxidase, dopamine beta-hydroxylase, feroxidases, peroxidase, superoxide dismutase, tyrosinase, and uricase (Prosser and Brown, 1966; Williams, 1969; Aaseth and Norseth, 1986; Goyer, 1986)). Since oxidative enzymes (e.g. catalase, peroxidase, and cytochrome oxidase) require Cu, the metal is involved in hydrogen peroxide/organic substance destruction and energy production. Copper promotes Fe absorption from the gastrointestinal system, is involved in the transport of Fe from tissues into plasma, helps maintain myelin in the nervous system, is important in formation of bone and brain tissue, is necessary for hemoglobin synthesis (Cu is a constituent of erythrocuprein in erythrocytes), and other important functions. Copper is bound to α-globulin (ceruloplasmin) in the blood and transported to the liver, kidney, heart, central nervous system, bone, and muscle for storage.

Because Cu is essential to so many functions in the vertebrate body, complicated homeostatic mechanisms are undoubtedly involved in regulation of internal levels. As long as trace and storage levels of Cu are input from external sources, internal homeostasis is maintained. When levels of external Cu exceed certain values, however, defense mechanisms to protect against excess Cu are overcome and toxicity results, as current reasoning goes. In brown bullheads, this level appears to occur at 27–53 ppb Cu in the water (Brungs, *et al.*, 1973). Therefore, the toxicology of essential elements such as Cu, Zn, and Se tends to be more complicated than that of nonessential elements such as Hg.

Toxicity

Identification of specific, toxic Cu species is a controversial issue. The list of toxic Cu species often includes $CuOH^{+}$ (Pagenkopf, *et al.*, 1974; Howarth and Sprague, 1978), $CuCO_3$ (Shaw and Brown, 1974), and $Cu_2(OH_2)^{+2}$ (Howarth and Sprague, 1978). However, without question, Cu^{+2} is considered to be toxic (Stiff, 1971a; 1971b; Pagenkopf, *et al.*, 1974; Sylva, 1976; Howarth and Sprague, 1978). Only a fraction of total Cu is present as Cu^{+2} in various aqueous systems. Stiff (1971a) indicates that only 0.009–0.250 of Cu is present as Cu^{+2} in 0.001–0.005 M HCO_3^{-}. Even lower fractions of Cu^{+2} (0.000004–0.001700) are reported in another study, from seventeen water samples from different systems (Stiff, 1971b). Speciation of Cu (II) is expected to be highly variable from system to system and from one time to another in the same system (Sylva, 1976). Generally, only soluble Cu species are thought to be available to fish (Stiff, 1971b); however, precipitates of Cu may also be important and must be tested before definitive conclusions are drawn.

Sylva (1976) models major species of Cu (II) over a 4–9 pH range, when the total concentration of Cu is 2 ppm (Fig. 1). Most rapid speciation changes occur at pH 6.5–8.0, a pH range in which Cu^{+2} free ion decreases rapidly as $CuCO_3$, $CuOH^{+}$, and $Cu_2(OH)_2^{+2}$ increase. Precipitation reactions alter speciation of soluble Cu complexes, particularly at higher pH values. Malachite [$Cu_2(OH)_2CO_3$] would begin to precipitate at pH values greater than 6.0 if malachite is considered alone. Similarly, azurite [$Cu_3(OH)_2(CO_3)_2$] and $Cu(OH)_2$ will begin to precipitate at pH values of about 6.2 and 6.8, respectively, if also considered alone. However, in this system, Cu (II) carbonate does not precipitate at any pH. In the absence of significant levels of organic complexing agents, hydrolysis and precipitation reactions dominate the chemistry of copper (II) at pH values expected in most natural water systems.

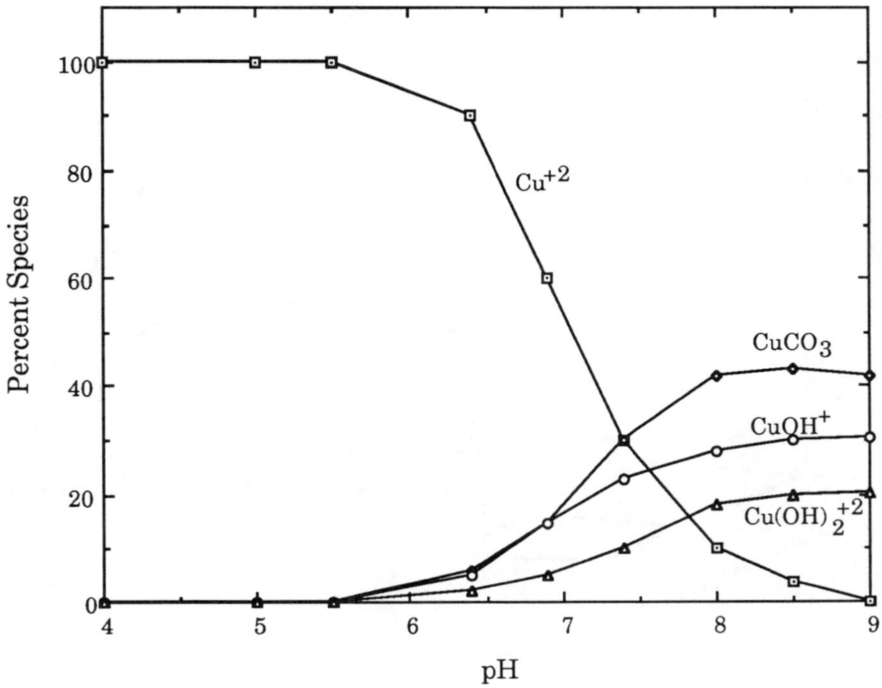

Figure 1. Speciation of 2 ppm Cu (II) as a function of pH. (Modified and redrawn from Sylva (1976).)

Howarth and Sprague (1978) expose rainbow trout (*Salmo gairdneri*) to Cu in various combinations of hardness and pH (15°C, pH 5–9, 30–360 ppm hardness). Accepting Cu^{+2}, $CuOH^{+}$, and $Cu_2(OH)_2^{+2}$ as toxic species in this modeling study provides a simple, fairly smooth, response surface with a good fit to observation (Fig. 2). Copper-induced lethality in trout spans a large range from 0.09 to 232.00 ppb in soft water and hard water, respectively. Copper is especially toxic at pH 9, possibly because CO_2 excretion at the gills lowers pH in the immediate area, ionizes hydroxides, and liberates free Cu^{+2} which can be internalized by gill epithelium. This mechanism also is proposed for Zn at high pH (Mount, 1966), but whether sufficient CO_2 for this effect is released at the gill remains to be elucidated.

One of the most debated issues is whether total hardness or alkalinity of water is the major modulating factor controlling the toxicity of Cu to fish. Stiff (1971a) measures Cu^{+2} concentration and pH in a series of three

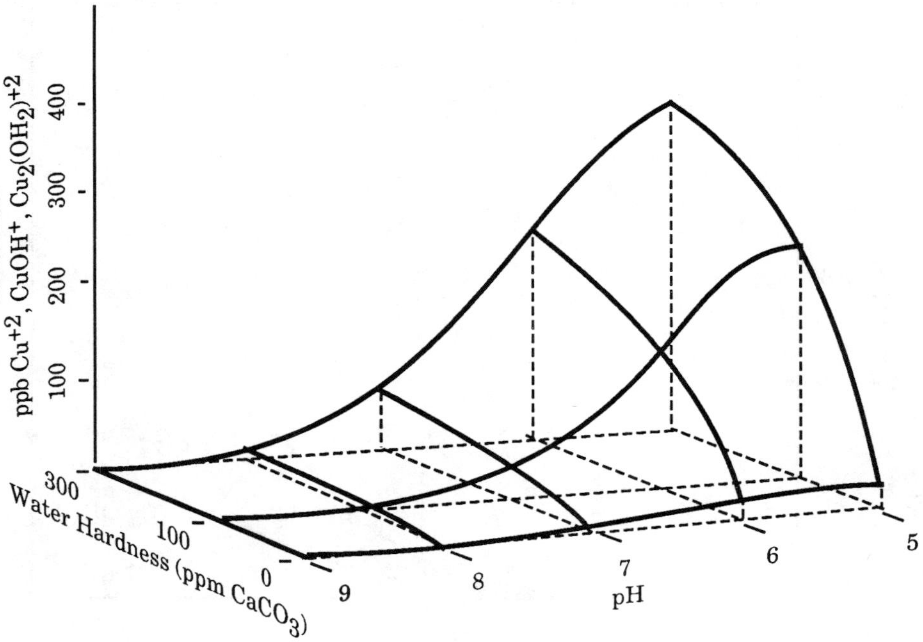

Figure 2. Relationship among pH, water hardness (ppm as $CaCO_3$), and 96-h LC50 for toxic Cu species (Cu^{+2}, $CuOH^+$, and $Cu_2(OH)_2^{+2}$). (Modified and redrawn after Howarth and Sprague (1978).)

bicarbonate test systems (0.001, 0.002, and 0.005 M) and four concentrations of total Cu (0.05, 0.10, 0.50, and 1.00 ppm) to simulate the water composition of most natural waters (those in which HCO_3^- levels are 0.001–0.005 M, alkalinities are 50–250 ppm, and pH is 6.5–8.5). If free cupric ion were the toxic form of Cu and $CuCO_3$ were relatively nontoxic, this would account for the difference in Cu toxicity in water of different hardness. Toxicity differences would then be related, not to hardness, *per se*, but to the difference in alkalinity which it accompanies.

Other workers (Pagenkopf, *et al.*, 1974; Lauren and McDonald, 1986) also consider alkalinity of more importance than hardness in protecting fish from Cu. Based on equilibrium model calculations using data from Mount (1968), Mount and Stephan (1969), Pickering and Henderson (1966), and Lloyd (1961), Pagenkopf (1974) determine that Cu^{+2} (and possibly $CuOH^+$) is (are) the chemical species which is (are) toxic to fathead minnows and rainbow trout. Copper is highly complexed with carbonate and hydroxide ions in natural waters and, as a result, free Cu^{+2} ion is usually low in concentration. Complexation of Cu^{+2} with inorganic carbon increases with alkalinity and pH; therefore, at high alkalinity and a pH of 8 or more, most Cu^{+2} is complexed. Toxicity tests using alkalinity, pH, total hardness, and total Cu concentration show that alkalinity, rather than total hardness may be the factor controlling Cu^{+2} concentration. Relationships are complex, however, increased total hardness is accompanied by increased alkalinity and increased complexing capability (with decreased Cu^{+2}).

If pH and total water hardness are known, alkalinity can be estimated by means of an equation which accounts for 96.7% of the variability (Howarth and Sprague, 1978), as follows:

$$\begin{aligned}\text{Alkalinity} = \text{antilog}\,[&-4.720 + 1.397\ \text{pH} - 1.381\ \text{Log}_{10}\ \text{Hardness} \\ &+ 0.1601\ \text{pH}\ \text{Log}_{10}\ \text{Hardness} - 0.08175\ \text{pH}^2 \\ &+ 0.1753\ (\text{Log}_{10}\ \text{Hardness})^2] - 1.0\end{aligned}$$

In contrast, other scientists believe that total hardness (level of aqueous Ca) is more important than carbonate-bicarbonate alkalinity in protecting fish from Cu-poisoning (Miller and Mackay, 1980). Fingerling rainbow trout (*Salmo gairdneri*) are used to determine effects of water hardness, alkalinity, and pH on the incipient lethal concentration which causes 50% mortality (ILC50) for Cu. At low water hardness (12 ppm), the ILC50 of Cu does not change despite a five-fold change in alkalinity (10 to 50 ppm) (Fig. 3). However, at 10 ppm alkalinity, a ten-fold increase in hardness (12–100 ppm) causes a five-fold increase in ILC50. Moreover, at a hardness of 100 ppm, increasing the alkalinity significantly reduces Cu toxicity. Apparently, at $pH<5.4$, Cu^{+2} and H^{+} toxicity are antagonistic; whereas, the ions are synergistic at pH 5.4 to neutrality. Antagonism may be caused by stimulation of mucous secretion at $pH<5.4$; so that, mucus chelates Cu. At pH 5.5, synergism or potentiation of Cu toxicity may be due to liberation of cupric ion from inorganic complexes in the water.

Therefore, although increasing alkalinity reduces Cu toxicity in fish exposed in hard water, Ca level (or total hardness) is thought to be more important than the carbonate-bicarbonate alkalinity in protecting fish from

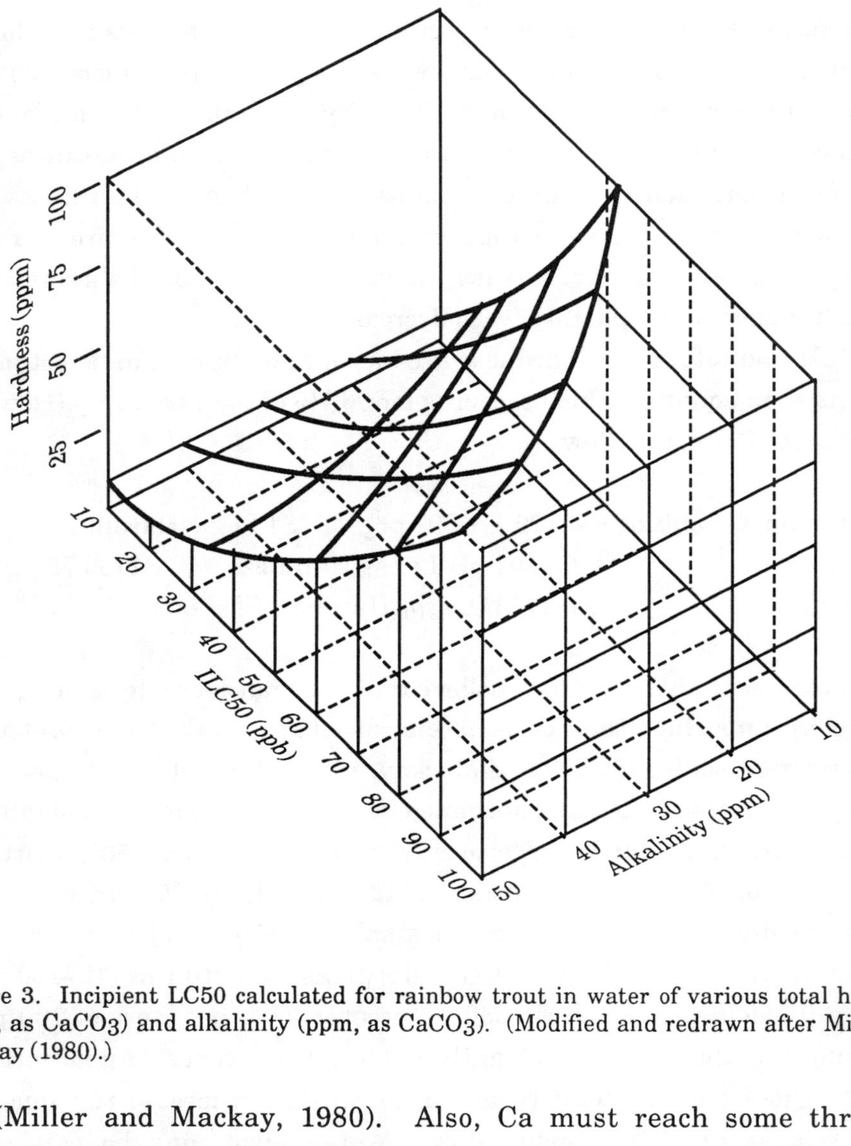

Figure 3. Incipient LC50 calculated for rainbow trout in water of various total hardness (ppm, as $CaCO_3$) and alkalinity (ppm, as $CaCO_3$). (Modified and redrawn after Miller and Mackay (1980).)

Cu (Miller and Mackay, 1980). Also, Ca must reach some threshold concentration between 12 and 28 ppm before alkalinity begins to reduce Cu toxicity. The ILC50 value for Cu is greatest when both Ca and alkalinity levels are highest (Fig. 3). Calcium and magnesium indirectly affect carbonate and bicarbonate complexation. The more carbonate and bicarbonate complexed by Ca and Mg, the less available for complexation and the greater the levels of Cu^{+2} in aqueous solutions.

A multitude of other factors control Cu toxicity. Most natural water systems contain a vast array of organic complexing agents which can substantially reduce free Cu^{+2} concentrations in aqueous systems. This list includes humic and fulvic acids derived from rotting vegetation (Zitko, *et al.*, 1973; Brown, *et al.*, 1974; Sylva, 1976), amino acids (Brown, *et al.*, 1974), spent sulfite liquor (Wilson, 1972), suspended solids (Brown, *et al.*, 1974), and nitrogenous waste products from endemic species (Sylva, 1976). Copper also can be complexed by cyanide and polypeptides and can be associated with particulate, colloidal, and soluble phases in natural water systems (Stiff, 1971b). Moreover, absorption of free Cu^{+2} by sand, gravel, and suspended clay is important in these systems (Sylva, 1976). Because many complexing agents are normally present in water, synthetic water solutions (prepared from deionized, distilled water) are best suited for laboratory testing.

As indicated, organic substances decrease Cu toxicity in fish. Exposure of rainbow trout (*Salmo gairdneri*) to variable pH, cupric ion, and nitrilotriacetic acid [NTA, $N(CH_2COOH)_3$, a Cu-chelator], suggests that Cu-induced toxicity in trout is related to total concentration of Cu^{+2} and $CuCO_3$ in the water (8–11°C, pH 6.5–7.5, and total hardness 250 ppm as $CaCO_3$) (Shaw and Brown, 1974). Use of NTA as a chelating agent for Cu results in a considerable increase in survival of trout in the presence of Cu. Using NTA at a 3:1 ratio to total Cu, prevents mortality at 0.17 ppm total free Cu^{+2} and $CuCO_3$. At totals of 0.50 ppm Cu^{+2} and $CuCO_3$; however, 50–90% trout die. Therefore, in solutions of pH 6.5–8.5, with 0.001–0.005 M HCO_3^-, and in the absence of complexing agents other than H_2CO_3, Cu toxicity to rainbow trout may best be defined in terms of concentration of Cu^{+2} and $CuCO_3$ combined, rather than the concentration of either Cu^{+2} or $CuCO_3$ alone.

Zitko and colleagues (1973) study effects of humic acid on Cu toxicity in fish. The presence of humic acid (complex organic acids obtained from humus) increases the level of Cu referred to as the incipient lethal level (ILL, toxic level beyond which 50% of the population cannot live for an indefinite period of time (Sprague, 1969). These organic acids probably bind Cu, decreasing Cu-induced toxicity. A Cu^{+2} selective electrode is used to predict ILL's of Cu at different concentrations of humic acid (3.8–4.8°C, 14 ppm total hardness). Predicted ILL's for juvenile Atlantic salmon (*Salmo salar*) are 58 and 110 ppb Cu at 5 and 10 ppm humic acid, respectively. These predicted values are lower than experimentally derived ILL's (90 and 165 ppb Cu), but agreement is considered adequate. Ratios of predicted to calculated ILL's decrease with increasing total hardness of water—probably due to decreased binding of Cu^{+2} by humic acid at higher hardness.

Wilson (1972) uses parr (2–3 yr old) Atlantic salmon (*Salmo salar*) to evaluate the effect of mining wastes on Cu toxicity (18–21°C, pH 5.0–6.7, and 8–10 ppm total hardness). The 96-h LC50 of Cu alone is 125 ppb in Exploits River water, known to contain organic complexing agents which are capable of removing free Cu^{+2} from solution. Assuming the 96-h LC50 is the incipient lethal level, the value (125 ppb) is four or five times higher than values previously published (Lloyd and Herbert, 1962; Sprague and Ramsay, 1965). Addition of 350 ppm spent sulfite liquor (SSL) prevents mortality at 180 ppb Cu. No Cu concentration permitted fish to survive at higher levels than the 96-h LC50 value for SSL. Therefore, death of fish exposed to both Cu and SSL can be assumed due primarily to the presence of Cu alone, at sublethal SSL levels. At least one fraction of SSL can chelate free Cu^{+2} and remove these ions from solution. Because the number of potential chelators occurring in river water is beyond useful definition, simple yet accurate predictive models of Cu toxicity in receiving waters may never exist. This study also emphasizes the importance of the use of synthetic solutions made from deionized, distilled water, rather than river water, well water, or water from any natural source.

Brown and co-workers (1974) expose rainbow trout (*Salmo gairdneri*) to 2–5 ppm Cu in the presence or absence of surfactant-free domestic sewage or one of three possible constituents thereof, namely glycine (an alpha-amino acid), humic substances (humic acid or hot-water extracts of peat), or suspended solids from an experimental activated-sludge plant (250 ppm total hardness). Regardless of the organic matter added, Cu-induced toxicity in trout decreases as a function of the amount of added organic. Therefore, neither total Cu concentrations, nor that of "soluble" Cu (i.e. filterable at an average membrane pore size of 0.45 μm) could be used to determine the toxicity attributable to Cu. Also, in cases in which natural surface waters are used for dilution purposes in Cu-toxicity tests, true Cu-induced toxicity cannot be determined or applied to other natural waters unless concentrations of toxic chemical species are known. The exact composition of humic substances, domestic sewage, and suspended solids is not precisely defined; nonetheless, the study shows that organic substances can bind Cu and decrease Cu toxicity in trout (Brown, *et al.*, 1974).

Salinity of testing solutions is important in establishing toxic Cu levels in fish: increased salinity exerts an ameliorative effect. For example, Birdsong and Avault (1971) expose juvenile pompano (*Trachinotus carolinus*) to Cu at 10, 20, or 30 ppt salinity (20–25°C, pH 7.8–8.2, and 5,468 ppm total hardness at about 30 ppt). The 96-h LC50 values for Cu are 1.42 ppm at 10 ppt salinity and 1.97 ppm at 30 ppt salinity ($p<0.01$) (Table 1). Therefore, higher salinities reduce Cu-induced mortalities in pompano.

Table 1. Copper toxicity in various species of fish, based on 96-h LC50 data. Water hardness, pH, and temperature are provided

Common Name *Species*	Cu 96-h LC50 (ppb)	Hardness (ppm, $CaCO_3$)	Temp. (°C)	pH	Reference
Coho salmon *Oncorhynchus kisutch*	60–74	89–88	10–12	6.8–7.5	Lorz & McPherson, 1976
Blue gourami *Trichogaster trichopterus*	90	-	26–28	-	Roales & Perlmutter, 1977
Brook trout *Salvelinus fontinalis*	100	45	10.6	7.5	McKim & Benoit, 1971
Atlantic salmon *Salmo salar*	125	8–10	18–21	6.5–6.7	Wilson, 1972
Brown bullhead *Ictalurus nebulosus*	170–190	202	5–25	7.6	Brungs *et al.*, 1973
Rainbow trout *Salmo gairdneri*	330 250–680	374 365	15 10	7.7 8	Dixon, 1981a&b Letti, 1976
Fantail darter *Etheostoma flabellare*	330–392	-	20	-	Lydy & Wissing, 1988
Fathead minnow *Pimephales promelas*	430 460–490	198 200	16–25 20–26	7.9 7.5–8.2	Mount, 1968 Pickering, 1977
Johnny darter *Etheostoma nigrum*	483–602	-	20	-	Lydy & Wissing, 1988
Pompano *Trachinotus carolinus*	1970	5468	20–25	8.2	Birdsong, 1971
Bluegills *Lepomis macrochirus*	1100 2400	45 35	13–28 25	7–8 7.7	Benoit, 1975 O'Hara, 1971

In contrast, elevated temperatures increase the toxicity of Cu (Lydy and Wissing, 1988). Fantail darters (*Etheostoma flabellare*) and johnny darters (*E. nigrum*) are exposed to Cu to determine 96-h LC50 values at 20°C. At this temperature, fantail darters are more sensitive to Cu and have LC50 values of 330–392 ppb Cu; whereas, johnny darters have values of 483–602 ppb Cu. In both species, Cu lowers the thermal tolerance; however, fantail darters are twice as sensitive to Cu as are johnny darters. Thermal tolerance is determined using the CTMax method, which uses a 1°C/min thermal elevation until the end point (loss of equilibrium) is reached. Thermal tolerance is determined in fantail darters exposed 96 h to 34–149 ppb Cu and in johnny darters exposed 96 h to 52–292 ppb. For fantail darters, CTMax is 7.6°C below control values if fish are exposed to the highest Cu level (149 ppb). Johnny darters show depressed CTMax 5.2°C below that of controls after 96-h exposures to the highest Cu level (292 ppb).

Time to loss of thermal resistance is tested (Lydy and Wissing, 1988). A 1-d exposure to 125 or 385 ppb Cu causes lower thermal tolerance ($p<0.05$) in johnny and fantail darters, respectively, and changes become more pronounced during the 4-d exposure period. Also, following 96-h exposures to Cu, darters are moved to control solutions and tested for CTMax. For low concentrations of Cu, recovery of CTMax is noted in both species. Within 1 and 2 d, johnny darters and fantail darters recover from 96-h, 152 or 82 ppb exposures, respectively. However, no recovery of CTMax values is noted even 3 d after transfer of fish to clean water in johnny or fantail darters previously exposed 4 d to 278 or 120 ppb Cu, respectively. Reversible Cu-induced hematologic changes (e.g. anemia), coupled with the increased oxygen demand associated with thermal stress might account for the short-term changes noted in the Lydy and Wissing (1988) study.

Size of fish alters 96-h LC50 levels (Howarth and Sprague, 1978; Lauren and McDonald, 1986). Howarth and Sprague (1978) expose rainbow trout (*Salmo gairdneri*) to Cu in various combinations of hardness and pH (15°C, pH 5–9, 30–360 ppm hardness). Copper is more toxic to small fish than to large fish. A 10-g trout is about three times more resistant to Cu poisoning than a 0.71 g trout. This may be due to the reduced metabolic rate per g of heavier fish. Large fish would accumulate less Cu during a given exposure and would require a more severe exposure to achieve a toxic level of Cu in the target tissue or tissues (if one or more exists).

Lauren and McDonald (1986) also observe size-related differences in Cu toxicity in fish. Rainbow trout (*Salmo gairdneri*) are exposed to 25–400 ppb Cu at variable pH, water hardness, and alkalinity at 15°C. Copper causes about twice the net loss of exchangeable Na^+ in juveniles than in adult rainbow trout due to two-fold higher Na^+ efflux rates in juveniles than in adults. Therefore, juveniles die faster than adults—presumably because 50–55% of whole body exchangeable Na^+ has been lost. This occurs at about 200 ppb Cu in juveniles—a level at which only 23% Na^+ has been lost in adults. Toxicity of Cu is not correlated with the amount of Cu that enters rainbow trout—presumably because of surface binding of Cu or because of differential uptake of toxic and nontoxic Cu species at different pH values (i.e. more toxic Cu species are taken up at pH 5.0; while nontoxic Cu carbonates are taken up from hard water at high alkalinity). Total Cu uptake could be identical, but trout taking up the more toxic Cu species would succumb more readily.

Previous exposure to Cu can influence the toxicity of Cu in fish (Dixon and Sprague, 1981a). Rainbow trout (*Salmo gairdneri*) become more tolerant to Cu if exposed 3 wk to 0.29, 0.40, or 0.59 of the incipient lethal level (ILL) of 330 ppb Cu. At the 0.09 ILL for Cu, trout are sensitized to Cu, and tolerance to Cu is reduced 20%. At 0.18 ILL, tolerance is unaffected; therefore, 0.18 ILL is considered the threshold for onset of acclimation. Acclimation at 94-194 ppb Cu (0.29–0.59 ILL) results in about two times greater tolerance to Cu, and most of the increase occurs the first 7 d. The three-dimensional response surface for tolerance of rainbow trout to Cu resembles a ski jump, with concentration- and time-dependent increases in tolerance above the threshold and sensitization below the threshold (Fig. 4). Increased tolerance is not a sustained response: when trout are moved to clean water, acclimation to Cu is lost in 7 d. Moreover, Cu acclimation for 4 wk at 0.59 ILL results in sensitization to lethal levels of Zn. Zinc ILL decreases by 56% from 6.2 to 2.7 ppm Zn. Therefore, on the basis of this study, a pulse of Cu would afford little protection against a second pulse of Cu unless the time lapse is less than a week and Cu levels exceed the 0.09 ILL, at least in rainbow trout. Also, lack of cross-acclimation between Cu and Zn argues for metal-specific physiological adaptations (e.g. *de novo* synthesis of metallothionein or isometallothionein) to increase tolerance to a specific metallic pollutant.

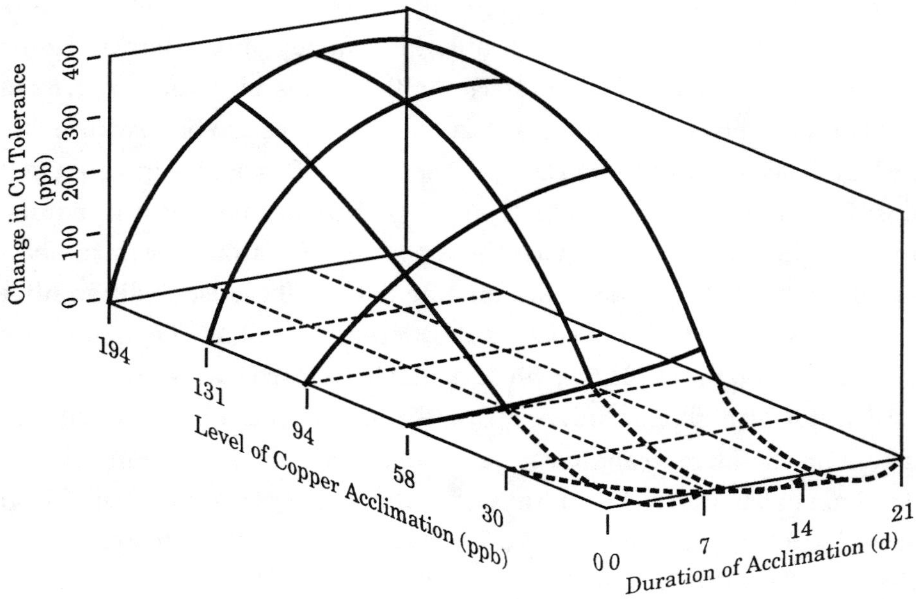

Figure 4. Relationship among change in Cu tolerance (experimental LC50 - control LC50), duration of acclimation, and level of Cu used in acclimation (0–194 ppb aqueous Cu). (Modified and redrawn, after Dixon and Sprague (1981a).)

Accumulation

Deposition of Cu in tissues and organs is variable. One-thousand-fold differences exist in livers of related species of fish collected from uncontaminated areas (Frazier, 1984). Table 2 indicates that white perch (*Morone americana*) can accumulate hepatic levels of 2795 ppm Cu, compared to the normal level of about 3 ppm Cu found in striped bass (*Morone saxatilis*). As is the case for Bedlington terriers (Aaseth and Norseth, 1986), the unexposed white perch is suggested as a model of Wilson's disease in humans because hepatic levels often exceed 1000 ppm (wet weight). In an analytical comparison, livers from rats and striped bass (*Morone saxatilis*) are shown to have about 3 ppm Cu and 25 ppm Zn. White perch have hepatic levels of 2795 ppm Cu and 55 ppm Zn. Subcellular distribution patterns are similar for Zn for all species, but 70% more Cu is found in the nuclei and

Table 2. Accumulation of copper in whole body and tissues of fish. Where possible, pH, temperature (in °C), and hardness (ppm, as $CaCO_3$) are given. Concentration values are presented (where possible) on the basis of wet weight (ww) or dry weight (dw).

Species	Exposure Time	Exposure Level	Whole Body (ppm)	Liver (ppm)	Kidney (ppm)	Gill (ppm)	Muscle (ppm)	Blood (ppm)	pH/Hardness/ °C	Reference
Brook trout	24 mo	9 ppb	-	238	16	7	17		-/44/9.6	McKim & Benoit, 1974
		3 ppb*	-	239	20	5	15		-/44/9.6	"
Bluegill	24 mo	162 ppb	-	480	44	13	-	-	-/45/13–28	Benoit, 1975
	"	72 ppb	-	57	12	6	-	-	-/45/13–28	"
	"	3 ppb*	-	7	22	3	-	-	-/45/13–28	"
Rainbow trout	28 d	55 ppb	~7 ww	110 ww	-	~1.6 ww	-	-	-	Lauren & McDonald, 1987b
Rainbow trout	21 d	30 ppb	2 ww	-	-	-	-	-	7.7/374/15	Dixon & Sprague, 1981b
	"	58 ppb	6 ww	-	-	-	-	-	7.7/374/15	"
	"	94–194 ppb	7–8 ww	-	-	-	-	-	7.7/374/15	"
Coho salmon	28–30 d	140 ppb	-	~320 dw	~11 dw	~8 dw	-	0.50	-	Buckley *et al.*, 1982
		70 ppb	-	~110 dw	~8 dw	~5 dw	-	0.50	-	"
		~0 ppb*	-	70 dw	~9 dw	~3 dw	-	0.47	-	"
Brown bullhead	20 mo	50 ppb	-	130 dw	-	25 dw	-	-	7.6/202/5–25	Brungs *et al.*, 1973
	"	~0 ppb*	-	29 dw	-	4 dw	-	-	7.6/202/5–25	"
	30 d	104 ppb	-	116 dw	23 dw	34 dw	-	-	7.6/202/5–25	"
	"	~0 ppb*	-	23 dw	8 dw	6 dw	-	-	7.6/202/5–25	"
Brown trout	38 wk	290 ppb	-	-	-	-	-	25	7.8/207/15.5	O'Neill, 1981
Mirror carp	38 wk	290 ppb	-	-	-	-	-	44	7.8/207/15.5	"
Bluefish	Envir**	3 ppb*	-	-	-	-	0.61	-	-	Cross *et al.*, 1973
Striped bass	Envir**	-	-	3 ww	-	-	-	-	-	Frazier, 1984
White perch	"	-	-	2795 ww	-	-	-	-	-	"

*Control levels for laboratory studies or normal reference levels for environmental studies. **Envir, environmentally exposed.

plasma membrane fraction of white perch than is the case for rats or striped bass. Copper-containing granules are observed at optical and ultrastructural levels. Cystolic copper in both species of *Morone* associates with a protein similar in molecular weight to metallothionein, but in the rat liver, little Cu is associated with metallothionein. Mechanisms to explain elevated hepatic Cu in white perch have not been elucidated; however, possibilities include accelerated metallothionein synthesis with enhanced storage capacity (e.g. cytosolic or lysosomal) and/or enterohepatic redistribution of Cu, as with As in sunfish (Sorensen, *et al.*, 1979). Sequestration could stabilize a considerable intracellular burden of the toxic ion (Cu^{+2}) in a less toxic form, reducing injury to vulnerable mitochondrial and plasma membranes.

The liver is a vital organ in maintaining internal Cu homeostasis, regardless of whether or not exposures are excessive. In contrast to renal uptake, hepatic accumulation is both time- and concentration-dependent. However, this is only the case in the early period of exposure (28–30 d) and at aqueous Cu levels greater than 9–10 ppb, (Brungs, *et al.*, 1973; McKim and Benoit, 1974; Benoit, 1975; Buckley, *et al.*, 1982). Time-dependent hepatic accumulation stabilizes after 28–30 d in brown bullheads and coho salmon (Table 2) (Brungs, *et al.*, 1973; Buckley, *et al.*, 1982). Initially it would seem logical that concentration-dependent Cu uptake in the trout liver does not occur at aqueous exposure levels of 9 ppb Cu or less (Table 2) (McKim and Benoit, 1974). However, in comparing levels of hepatic Cu in these brook trout with normal and low Cu levels in coho salmon (Buckley, *et al.*, 1982), rainbow trout (Lauren and McDonald, 1987b), and striped bass (Frazier, 1984), the control value of 239 ppm Cu in brook trout liver (McKim and Benoit, 1974) seems excessive (Table 2). One possible explanation for this elevated level of hepatic Cu is that trout might have been inadvertently exposed to Cu-contaminated water prior to testing. Typically, a rigorous exposure (e.g. 28–30 d to 140 ppb) is required to produce comparable hepatic Cu levels (Table 2).

Nonetheless, normally the teleost liver actively processes and stores large Cu loads; whereas, the kidney does not. With few exceptions, hepatic Cu levels are greater than renal levels following exposure to aqueous Cu (Table 2). Also, no discernable concentration- or time-dependent trends are evident in fish kidney during Cu exposure. Therefore, the kidney is not thought to play a major role in Cu detoxification *per se* (Buckley, *et al.*, 1982).

Almost the same can be said of the gill, except that slight concentration-dependent effects have been shown (Buckley, *et al.*, 1982). Blood, brain, spleen, gonad, and skeletal muscle also generally accumulate lower levels of Cu than the liver (Cross, *et al.*, 1973; McKim and Benoit, 1974; O'Neill, 1981; Buckley, *et al.*, 1982). The brain, spleen, gonad, and skeletal muscle of Cu-exposed fish do not appear to be involved in Cu accumulation, since Cu levels in exposed fish are similar to those of controls. Control bluegills average 29, 18, 5, and 1 ppm Cu in brain, spleen, gonad, and muscle, respectively (Benoit, 1975).

Elevated levels of Cu in tissues are associated with development of enhanced tolerance to subsequent Cu exposure (Dixon and Sprague, 1981b). Rainbow trout are exposed to 0–194 ppb Cu (0–0.59 of the incipient lethal level, ILL) in water for up to 21 d (15°C, pH 7.75, 374 ppm hardness). Subsequently, fish are sacrificed or exposed to an even higher level of Cu (570 ppb). Whole body levels of Cu, protein content of the partially-purified hepatoprotein, and C-14 leucine incorporation into this hepatic metal-binding protein are measured.

At levels of 0.29–0.59 ILL (94–194 ppb Cu), whole body Cu levels stabilize in 21 d at 34–37 ppb (dry weight), compared with control levels of 3.2–4.5 ppb Cu prior to exposure (Dixon and Sprague, 1981b). Fish exposed 21 d to 30 ppb Cu (0.09 ILL) accumulate about 11 ppm, dry weight (2 ppm, wet weight) whole body levels of Cu; fish exposed 21 d to 58 ppb Cu (0.18 ILL) accumulate levels about 26 ppm Cu, dry weight (6 ppm, wet weight) (Table 2). The three-dimensional response surface showing the relationship of whole body Cu, exposure time, and exposure level resembles a haystack or mound, flattened to a plateau at the top with increasing Cu concentration and exposure duration. No significant differences in whole body Cu levels are noted for fish exposed to 94, 131, or 194 ppb Cu for 14 or 21 d (corresponding to the plateau region), in which case the whole body levels are 34–37 ppm, dry weight (7–8 ppm, wet weight) Cu.

Prior Cu exposure enables fish to tolerate higher subsequent Cu exposures, and this tolerance seems to be associated with elevated levels of Cu in the body (Dixon and Sprague, 1981b). A 21-d exposure of untreated, control fish to 0 ppb Cu, followed by a 2-d exposure to 570 ppb Cu results in 7.4 ppm Cu (dry weight) whole body levels. However, if fish are pre-exposed to 94 ppb Cu for 21 d, before the 570 ppb exposure, levels reach 33.4 ppm Cu (dry weight).

To evaluate C-14 leucine incorporation into the protein and the protein content of the partially-purified, metal-binding hepatoprotein, Dixon and Sprague (1981b) expose fish to 0.43 ILL Cu (141 ppb Cu), 0.65 ILL (21 ppb Cd), or increased handling stress (overcrowded conditions, low water levels, and continuous lighting). Both Cu- and Cd-exposed fish synthesize significantly more metal-binding hepatoprotein ($p<0.01$). Leucine is incorporated into the protein from Cu-exposed fish but not into that from Cd-exposed fish. Based on lower liver and whole body weights, trout exposed 48 h to 0.43 ILL Cu are more severely affected than those exposed 48 h to 0.65 ILL Cd. Stressed fish do not show differences in protein content or C-14 incorporation, compared with control fish. This study indicates that elevated levels of Cu in tissues are associated with development of enhanced tolerance to subsequent Cu exposure.

Although the partially-purified, metal-binding hepatoprotein of Dixon and Sprague (1981b) cannot be called metallothionein, isolation methods seem to insure inclusion of metallothionein because a heat-denaturation method is used to partially purify the homogenates. However, although heat stability is assured, most metallothioneins do not contain leucine—bringing the isolation procedure into question. At present, amino acid sequencing in fish has been done on isometallothionein I, but not isometallothionein II (McCarter, *et al.*, 1982), and in the coho salmon, but not in rainbow trout. Hepatic proteins analogous to metallothionein could be responsible for increased Cu tolerance in fish.

Using coho salmon (*Oncorhynchus kisutch*), McCarter and colleagues (McCarter, *et al.*, 1982; McCarter and Roch, 1983) isolate the two, Cu-inducible isometallothioneins which bind Cu. Salmon are exposed 14 wk to 70 or 140 ppb Cu (0.25 and 0.50 of the 168-h LC50 value) (McCarter, *et al.*, 1982). The high molecular weight (HMW) fraction binds constant levels of Cu the first 8–10 wk, but levels steadily increase thereafter to 14 wk. Levels of Zn in the HMW fraction usually are higher than levels of Cu. The low molecular weight (LMW) fraction is heat-resistant, high in SH-group activity, and high in Cu. This LMW fraction contains at least two Cu isometallothioneins (12,000–13,000 molecular weight), based on amino acid composition, resistance to heat denaturation, and absorbancy at 280 nm (indicating the absence of aromatic amino acids). Levels of Cu in the LMW fraction from fish exposed to 70 ppb Cu are stable the first 6 wk, but levels increase thereafter. In contrast, coho exposed to 140 ppb bind substantial Cu in the LMW fraction the first 2–4 wk, but are stable until 14 wk. Apparently, Cu induces *de novo* synthesis of isometallothioneins in livers of coho salmon, and higher and earlier induction occurs for fish exposed to 140 ppb Cu than for those exposed to 70 ppb Cu.

In another study, McCarter and co-workers (1983) expose coho salmon (*Oncorhynchus kisutch*) to 1, 50, 100, or 150 ppb Cu in water for 10 wk (11–12°C, pH 7.6, total hardness 280 ppm as $CaCO_3$). Hepatic metallothionein (probably isometallothionein) increases three-fold in 4-wk exposures of coho, but stabilizes thereafter. Concentrations of metallothionein are related to the logarithm of the Cu level to which fish are exposed. The 168-h LC50 values for coho salmon exposed to Cu increase to maximum values at 1 wk (an overshoot), decrease until 4 wk, and remain stable until 8 wk of exposure. At these exposures, the capacity of fish to bind Cu remains unsurpassed, as evidenced by the fact that Cu-acclimated fish withstand still higher aqueous levels (i.e. up to the limits measured at 168-h LC50 values). Good correlation is observed between LC50 values and metallothionein values. In fact, hepatic levels of metallothionein/isometallothionein in coho salmon can be used to predict LC50 values of the coho salmon population exposed to Cu—especially after metallothionein/isometallothionein and LC50 values are stable (~4 wk). Induction of metallothionein/isometallothionein synthesis is considered a protective mechanism for trout and salmon exposed to Cu, markedly increasing tolerance of fish to subsequent Cu exposures.

Hepatic proteins (including metallothionein) are induced in fish exposed to Cu, namely acid-soluble thiols (AST) and glutathione (GSH) (Lauren and McDonald, 1987b). Acid soluble thiols include oxidized and reduced glutathione, cysteine, and an acid-soluble, sulfhydryl-rich, Cd-inducible, metal-binding protein of about the same molecular weight as metallothionein (Wofford and Thomas, 1984). This protein is probably comparable to the isometallothioneins isolated by McCarter and colleagues (McCarter, *et al.*, 1982; McCarter and Roch, 1983). The sum of glutathione and cysteine does not equal the total AST; moreover, the acid-soluble, sulfhydryl-rich, Cd-inducible, metal-binding protein of about 10,000 daltons is present in the livers of control fish (Wofford and Thomas, 1984; Lauren and McDonald,

1987a). Hepatic detoxification of Cu might involve metallothionein/isometallothionein induction with Cu displacement of Zn and/or other elements originally bound nonspecifically to HMW. In this way, a progressively increasing proportion of metal-binding sites would be occupied by Cu, increasing the total amount of Cu bound by the liver without reducing survival. Moreover, variability in levels of hepatic Cu indicates that marked species-specific differences exist in fish in response to the Cu challenge (e.g. in *de novo* synthesis of isometallothionein and/or displacement of nonspecifically bound elements by Cu).

Other defense mechanisms protect fish against Cu poisoning. Prolactin-stimulated mucification (Notter, *et al.*, 1976) provides a protective coat over brachial and other epidermal surfaces upon direct contact of those surfaces with Cu (Lewis and Lewis, 1971). If only the head and gills are exposed to Cu-laden water, mucus appears only in those areas. Mucification of the remainder of the body occurs if that portion is exposed to aqueous Cu; whereas, osmolarity changes occur only if the head and gills are exposed to Cu. Mucus chelates Cu (Miller and Mackay, 1982) and normally would be expected to decrease the competition between Cu^{+2} and H^{+} for the same binding sites [i.e. possibly intracellular tight junctions, especially those of the brachial epithelium (Lauren and McDonald, 1985)].

Miller and MacKay (1982) study mucification, focusing on chelating potential and pH effects. Rainbow trout (*Salmo gairdneri*) fingerlings are exposed to 85 or 170 ppb Cu to compare the effect of Cu on mucous secretion (9°C, pH 7.3, 50 ppm total hardness). About 33% more mucus is produced in 8 h after 85 ppb Cu exposure than in control fish. In contrast, 30–41% more mucus is secreted by trout exposed to acid (pH 3.6–4.0) than by trout exposed to Cu (86 or 170 ppb). Mucus is a strong Cu chelator (especially at neutral pH), but lowering the pH decreases the ability of mucus to bind cupric ion. Ammonia is secreted by trout, counteracting the problem of reduced chelation at low pH. For example, a 75–100 g trout in 2 l of water, increases the pH from 4.0 to 6.5 in 8 h and to pH 7.0 in 24 h. Increases in secretion of mucus (and ammonia) are viewed as important defense mechanisms.

Renal and extrarenal changes also protect fish from Cu poisoning (Lewis and Lewis, 1971; Shephard and Simkiss, 1978; Lauren and McDonald, 1985; 1987a; 1987b; Reid and McDonald, 1988). From these studies, it is apparent that Cu exposure inhibits brachial Na^{+}-K^{+}-ATPase, whereupon

Na^+, K^+, and Cl^- are lost from the gill (Lauren and McDonald, 1985) and probably also the renal tubules, based on the work of Notter and associates (1976). Prolactin secretion from prolactin cells (PC) of the pituitary (initially, probably preformed R_f 0.32 and R_f 0.45) may facilitate renal Na^+ reabsorption by increasing Na^+-K^+-ATPase activity of the kidney tubules and decreasing Na^+ efflux at the gill membrane (Dharmamba and Maetz, 1972; Notter, *et al.*, 1976; Bonga and Van der Meij, 1981). These are early events, probably occur within the first week of an exposure, because whole body Na^+ levels are known to return to normal in about 7 d (Lauren and McDonald, 1987b).

Once within the peripheral circulation, most Cu is transferred to hepatocytes (Brungs, *et al.*, 1973; Benoit, 1975; Buckley, *et al.*, 1982). Initially, Cu may displace Zn on HMW molecules, as is the case for Cd (Bryan and Hildalgo, 1976; Frazier and Puglese, 1978). This would permit early sequestration prior to *de novo* synthesis of isometallothioneins. Within 2 wk, Cu is bound to the HMW fraction (I), corresponding to molecules of molecular mass $>1.9 \times 10^4$ (McCarter, *et al.*, 1982). At 4 wk, Cu is bound to the LMW fraction (12,000–13,000 daltons) at four-fold times the 2-wk Cu levels in coho salmon exposed to 140 ppb (McCarter, *et al.*, 1982). Apparently, both Cu displacement of nonspecifically bound Zn and *de novo* synthesis of the isometallothioneins play important roles in storage of Cu in the liver. Likely, Zn displacement represents an early event, as is the case with the presumed hormone-mediated, Na^+-K^+-ATPase changes. As *de novo* synthesis of isometallothioneins proceeds, toxic cupric ions would be bound by nascent, metal-binding hepatoproteins and intracellular Cu burdens would increase. Sequestration of Cu as a less toxic complex would regulate Cu accumulation. Much remains to be understood about the mechanisms; however, these complex, interrelating events complement one another to protect fish from a Cu challenge. Isometallothionein synthesis, Zn-displacement, hormone-mediated, Na^+-K^+-ATPase changes, mucification, ammonia secretion, sequestration processes, and other changes might prolong fish survival.

Biological Effects

Histopathology: Numerous tissues and organs are reported to be damaged by Cu, including mechanoreceptors of the lateral line canals, chemoreceptors (olfactory organs), blood vessels, secondary lamellae of the gill, liver, kidney, and hematopoietic tissue. Gardner and LaRoche (1973) report that aqueous, 21-d Cu exposures (0.5, 1.0, and 5.0 ppm) of adult mummichogs (*Fundulus heteroclitus*) and Atlantic silversides (*Menidia menidia*) result in lesions of mechanoreceptors of lateral line canals and chemoreceptors (olfactory organs). Degeneration of mechanoreceptors is apparent from nuclear pyknosis and karyorrhexis. In addition, mummichogs show damaged epithelium in lateral line canals.

Mummichogs exposed to Cu show damage to olfactory organs—cystic formations, necrotized chemoreceptors, and hyperplastic sustentacular epithelium (Gardner and LaRoche, 1973). Even brief encounters with Cu (6 h) may disrupt the ability of a species to perceive peripheral events—such as the location of enemies, prey, or mates. Detection of underwater vibrations, olfaction, and gestation apparently are impaired by Cu. Interestingly, damage to mechanorecptors and chemoreceptors is caused by aqueous, but not intraperitoneal, exposures to Cu. Apparently, direct irritation of receptors rather than vascular system delivery of Cu, causes lesions of the lateral line and olfactory organs.

Damaged chemoreceptors could impair migratory ability as reported to occur at 5 ppb Cu (Lorz and McPherson, 1976). Giattina and colleagues (1982) also report results consistent with the idea that Cu damages chemoreceptors or alters chemoreceptor sensitivity, especially at high aqueous levels of Cu. Their study shows that rainbow trout (*Salmo gairdneri*) avoid about 4.4–6.4 ppb Cu (12°C, pH 7.3, total hardness 28 ppm as $CaCO_3$), but prefer high Cu levels (≥334 ppb). Preference or attraction of trout to markedly Cu-contaminated solutions may be due to narcotic effects, changes in chemoreceptor sensitivity, or damage to chemoreceptors.

Copper exposure of fish apparently alters secondary lamellae of the gill in a similar way as that reported for Zn (Skidmore and Tovell, 1972; Matthiessen and Brafield, 1973). Bilinski and Jonas (1973) find that 20% of rainbow trout (*Salmo gairdneri*) exposed to 64 ppb Cu for 48 h show epithelial layer detachment. A 96-h exposure of trout to 64 ppb Cu causes 33% mortality for 129 g fish at 10–11°C and 4 ppm total hardness, but only a small percentage of fish exposed to 64 ppb Cu for 48 h show epithelial layer detachment. Therefore, Cu may interfere primarily with enzymatic activity, rather than cellular structure (Bilinski and Jonas, 1973). However, clear distinctions dividing structural *versus* enzymatic changes would seem difficult, if not impossible, to make when changes are not visible optically.

Interestingly, Lydy and Wissing (1988) indicate that gill damage probably does not occur in fantail darters (*Etheostoma flabellare*) and johnny darters (*E. nigrum*) exposed to Cu and variable temperatures (1°C/min thermal elevation until loss of equilibrium is reached). Fantail darters have maximum temperatures at this end point (CTMax) 8°C below control values (at 149 ppb Cu); johnny darters have CTMax 5°C below that of controls (at 292 ppb Cu). However, Bilinski and Jones (1973) show that gill damage can occur within short time periods (24–96 h) and even lower Cu levels (64 ppb Cu). For, this reason histopathological evaluations—especially of the epithelial layer of secondary lamellae—can enhance the information derived from studies of Cu poisoning in fish.

Baker (1969) combines optical microscopy and electron microscopy to evaluate effects of Cu on winter flounder (*Pseudopleuronectes americanus*). Flounder are exposed to 0.56–3.20 ppm Cu for ~29 d at pH 8 and 9.4°C. Baker (1969) suggests that Cu-induced hemolytic anemia leads to secondary lesions in the liver and kidney due to the anoxic state of the organism. Fatty metamorphosis in the liver, necrosis in the kidney, and destruction of hematopoietic tissue are noted. In optical studies, gills show gross changes in architecture—such as decreases in numbers of mucous cells and increases in numbers of chloride cells on lamellae. It is suggested that mucous cells can become chloride cells. Ultrastructural evaluation indicates that 0.18 ppm Cu causes increases in numbers of chloride cells relative to mucous cells. Granules (possibly Cu) are noted within mitochondria and on the external membrane of epithelial cells. Vacuolation of epithelial cells, myelin-like figures, membrane-bound vesicles, and apical homogeneous layers of reduced thickness also are noted in gill lamellae.

In silversides, blood vessels of meninges are dilated and hemorrhages occur in brain and periorbital connective tissues (Gardner and LaRoche, 1973). Moreover, renal lesions are noted after aqueous exposures to 1 or 5 ppm Cu in mummichogs, but not in silversides. Hepatic lesions are not observed after aqueous exposures of even 5 ppm Cu; however, some mummichogs receiving injections of 100 ppm Cu have focal necrotic areas in the liver.

Frazier (1984) notes Cu-containing granules in livers of white perch (*Morone americana*) and hepatic levels which often exceed 1000 ppm wet weight in the unexposed condition (i.e. environmental levels of Cu are not excessive). In an analytical comparison, livers from rats and striped bass (*Morone saxatilis*) are shown to have about 3.5 ppm Cu and 25.0 ppm Zn. White perch have hepatic levels of 2795 ppm Cu and 55 ppm Zn. Subcellular distribution patterns are similar for Zn for all species, but 70% more Cu is found in the nuclei and plasma membrane fraction of white perch than in that of rats or striped bass. Copper-containing granules are observed at optical and ultrastructural levels. Cystolic copper in both species of *Morone* associates with a protein similar in molecular weight to metallothionein, but in the rat liver, little Cu is associated with metallothionein.

To summarize, Cu exposure of fish can result in degeneration of mechanoreceptors and chemoreceptors (Gardner and LaRoche, 1973) which may cause attraction of trout to areas having toxic levels of Cu (Giattina, 1982). Detachment of the epithelial layer of the secondary lamellae of the gill can result from Cu exposure (Bilinski and Jonas, 1973), but whether this alteration substantially alters survival has been questioned (Giattina, 1982; Lydy and Wissing, 1988). Numbers of chloride cells of the gill increase and numbers of mucous cells decrease in Cu-exposed fish. Renal, hepatic, and hematopoietic tissue can undergo fatty changes and/or necrosis (Baker, 1969; Gardner and LaRoche, 1973). Hepatic granules either are known to contain Cu (Frazier, 1984) or have been postulated to contain Cu (Baker, 1969). Blood vessels can undergo hemorrhage in fish exposed to Cu (Gardner and LaRoche, 1973). Obviously, a number of teleost organs and systems can be altered detrimentally because of the presence of excess Cu in the aqueous environment.

Hematology: Fish exposed to Cu show hemotological changes in numbers of blood cells, hemoglobin levels, hematocrit values, protein concentrations, plasma salt levels, antibody titers, glucose concentrations, ammonia levels, and as well as changes in concentrations of other constituents. Sometimes changes are transient; other times changes seem to be long-lived. Experimental differences in experimental design, purpose, Cu level, species, and water quality often determine the outcome of a particular study and compound the difficulty of data interpretation. Nonetheless, some published information seems especially relevant to an understanding of Cu-induced effects on teleost hematology.

One of the earliest studies published in this area shows that Cu can damage hematopoietic tissue (Baker, 1969). Destruction of hematopoietic tissue occurs in winter flounder (*Pseudopleuronectes americanus*) exposed to 0.56–3.20 ppm Cu for ~29 d at pH 8 and 9.4°C. Damage to this tissue (which surrounds renal tubules) is noted following exposure to 1000 or 3200 ppb Cu until just before fish die, but not in fish exposed in control solutions or 560 ppb Cu. Reported hepatic and renal (and perhaps hematopoietic tissue) lesions may be caused by the anoxic state of the organism—a factor which may be secondary to Cu-induced hemolytic anemia (Baker, 1969).

As is the case for all anemias, those caused by metal poisoning in fish are diagnosed based on low hemoglobin, reduced hematocrit, and/or decreased numbers of erythrocytes (Sorensen and Bauer, 1983). Anemia may be found in some species of Cu-exposed fish (O'Neill, 1981), but not in others (McKim, *et al.*, 1970). In fact, increased hemoglobin, higher hematocrits, and elevated numbers of erythrocytes are reported for 6- and/or 21-d exposures of brook trout (*Salvelinus fontinalis*) (McKim, *et al.*, 1970). A moderate potentiation of hemopoiesis by Cu in these trout may fit in the same category with the occasional study which shows that certain levels of Cu can benefit some fish species (Benoit, 1975). However, Cu-induced changes in brook trout are short-lived. Following 48-wk exposures to 2–33 ppb Cu, brook trout are not different from controls, based on erythrocyte number, hemoglobin level, or hematocrit value (McKim, *et al.*, 1970).

Evaluating McKim's study shows that brook trout exposed to aqueous Cu for 6, 21, or 337 d show transient changes in seven blood characteristics (12°C, pH 7.4, 46 ppm total hardness). About 38 or 68 ppb Cu causes increases in the number of erythrocytes, hemoglobin, plasma glutamic oxalacetic transaminase (PGOT), hematocrit (at 6-d only) and total protein (6-d only). However, plasma Cl^- and osmolarity decrease after 6- and 21-d exposures, as reported by other workers (Lewis and Lewis, 1971; Dharmamba and Maetz, 1972; Lauren and McDonald, 1985; Reid and McDonald, 1988). In brook trout, 337-d exposures to 17.4 or 32.5 ppb Cu result in decreased PGOT, possibly secondary to hepatic or cardiac damage (McKim, *et al.*, 1970). And although changes are transient, they may be used to determine toxic Cu concentrations for a population of fish.

As indicated, Cu decreases osmolarity and plasma Na^+ (or Cl^-); however, these Cu-induced ionoregulatory disturbances subsequently stabilize. The kinetics of this response can be evaluated, as can the acclimation to higher toxic levels of Cu following Cu pre-exposures. Several studies focus on Cu-induced ionoregulatory disturbances and provide useful information (Lewis and Lewis, 1971; Dharmamba and Maetz, 1972; Lauren and McDonald, 1985; Reid and McDonald, 1988).

Lewis and Lewis (1971), for example, expose channel catfish (*Ictalurus punctatus*), and golden shiners (*Notemigonus crysoleucas*) to 2.5–5.0 ppm Cu in static tests (19°C, pH 6.6–6.9, total hardness 206–236 ppm). Ionic content of whole blood is decreased below 230 mOsm in shiners which succumb to Cu poisoning. Intermediate levels of Cu (2.5 ppm) in catfish result in decreased osmolality followed by a return to normal levels. Addition of NaCl (235 mOsm) to water containing Cu prevents distress symptoms (piping at the surface) and mortality associated with exposure to these heavy metals. Blood osmolality increases over that of fish held in 235 mOsm NaCl alone. Also, confinement allowing exposure of only the head and gills or only the body, shows that mucus production coincides with the area actually exposed to Cu, but that osmolality changes are similar to those of whole fish exposures only when the head and gills of fish are exposed to Cu.

The work of Dharmamba and Maetz (1972) shows the relationship between handling stress, hypophysectomy, and declines in osmolarity and plasma levels of Na^+. Normally, hypophysectomized Mozambique mouthbrooders (*Tilapia mossambica*) survive only 6–10 d in fresh water at 26°C because of the severe negative Na^+ balance which results from increased Na^+ efflux and decreased Na^+ influx. Decline of plasma Na^+ levels and osmolarity of hypophysectomized cichlids can be prevented by prolactin injection (10 μg/g body weight). Prolactin is produced by prolactin cells which are located in the pituitary of fish. Prolactin prevents loss of Na^+ by decreasing the rate of Na^+ efflux, restoring it to the level found in the intact fish. Prolactin controls Na^+ leakage which may be a remnant of Na^+ extrusion activity characterizing the teleost gill in seawater. In fact, prolactin activity may be related to reduction of the Na^+-K^+-dependent ATPase activity of the gill which accompanies the fresh-water adaptation of this species. Prolactin treatment of seawater-adapted *Fundulus heteroclitus* (mummichogs) also is accompanied by prolonged reduction of Na^+ efflux.

Therefore, the role of prolactin during fresh-water adaptation of euryhaline fish seems critical to the "shutdown" of the seawater Na^+ extrusion pump, counteracting the electrolyte imbalance which occurs during Cu exposure.

An interesting part of this study (Dharmamba and Maetz, 1972) is that intact, normal cichlids transferred to small plexiglass chambers for flux measurements show negative Na^+ balance during the first 24 h, since Na^+ efflux exceeds Na^+ influx (Dharmamba and Maetz, 1972). However, within 48 h of confinement, fish reattain a positive Na^+ balance—primarily by decreasing Na^+ efflux. Handling, therefore, induces a shock-effect similar to that effect observed in a hypophysectomized cichlid.

Other workers (Notter, *et al.*, 1976; Bonga and Van der Meij, 1981) show that prolactin is more important to fish under conditions of low Ca^{++} than high Ca^{++}. Prolactin secretion seems to be related to water flux, irrespective of the direction of the flux. Also, prolactin cells of the pituitary release protein substances (R_f 0.32 and R_f 0.45) which facilitate Na^+ reabsorption in kidney tubules and decreased Na^+ efflux at the gill. Prolactin cells work synergistically with Stannius corpuscle cells of the kidney (important in Ca^{++} homeostasis) to maintain electrolyte balance.

Yet another study evaluates Cu-induced effects on ionoregulation, as well as provides plasma glucose and ammonia measurements following Cu-exposure of rainbow trout. Lauren and McDonald (1985) expose rainbow trout (*Salmo gairdneri*) to hard water (~2028 μeq/l Ca^{+2}) or soft water (~57 μeq/l Ca^{+2}) prior to exposure to 12.5, 25.0, 50.0, or 200.0 ppb Cu for 2, 12-h periods at pH 5.0 or 7.8, and 15°C, in an effort to define the toxic mechanism of Cu in fish and to determine the basis for the modifying effects of both water hardness and pH on this mechanism. Despite problems with Cu absorption by the acrylic material used in flux chambers, Lauren and McDonald (1985) show that exposure of rainbow trout to 12.5–200.0 ppb Cu causes the loss of Na^+, Cl^-, and K^+, regardless of pH or water hardness. As little as 12.5 ppb Cu leads to losses in the three ions, and losses are dependent on the aqueous Cu level but not on prior acclimation in hard water or soft water. In the 12.5–50.0 ppb Cu concentration range, NaCl losses result from inhibition of J_{in} (influxes of Na and Cl); whereas, at 100–200 ppb Cu, J_{out} (difference between net flux and Na and Cl influx) also is stimulated. Plasma NaCl is inversely correlated with aqueous Cu levels, while plasma glucose and ammonia increase with increasing levels of Cu in the water. At pH 5 and Cu aqueous levels of 12.5–50.0 ppb, H^+ contributes significantly to total ion loss. That is, NaCl losses are greater (30–60%) than that attributed to Cu alone. At 100 and 200 ppb Cu, however, ion losses are no greater at pH 5 than at neutral pH.

Regarding the toxic mechanism of Cu action in rainbow trout, it is possible that H^+ and Cu^{+2} compete for the same binding sites in the brachial epithelium, possibly intracellular tight junctions. The possible reason that Cu^{+2} and H^+ effects are additive at low aqueous Cu concentrations (but not at high Cu levels) seems to be that Cu^{+2} may displace H^+ from gill sites at low concentrations, but that only Cu exerts an effect at high Cu levels.

Elevation of plasma glucose and ammonia levels seems typical of cortisol- and catecholamine-mediated stress responses in fish, which result in stimulation of both protein catabolism and gluconeogenesis (Mazeaud, *et al.*, 1977). However, Schreck and Lorz (1978) show that cortisol levels (although elevated) are not strongly correlated with Cu exposure levels. Lauren and McDonald (1985) suggest that measurement of glucose and ammonia (after 24 h) might be a better indicator of stress than measurement of cortisol.

One additional study is valuable in understanding more about Cu-induced ionoregulatory disturbances because it compares Cu-induced effects with Cd-induced effects (Reid and McDonald, 1988). Reid and McDonald (1988) show that Cu exposure of trout decreases plasma levels of Na^+. Rainbow trout (*Salmo gairdneri*) adults or juveniles are exposed 24 h to pH 7.8 or pH 4.8 alone or in combination with 6.5 μM/l Cd or Cu at 14°C and under low Ca (39 μM/l) or high Ca (995 μM/l) conditions. Plasma Ca^{+2} is not a reliable indicator of disturbances in brachial Ca^{+2} flux. Copper disturbs Na^+ balance; whereas, Cd disturbs Ca^{+2} balance. Cadmium inhibits Ca influx, but has minor effects on Na flux and no effect on Ca efflux. Copper decreases plasma Na^+ levels and has a transient effect on Ca^{+2} homeostasis. Whereas, inhibition of Ca^{+2} influx by Cd is modulated by water levels of Ca^{+2} (total hardness of the water), effects of Cu on Na^+ flux are not modulated by water Ca^{+2} or by pH. Hydrogen ion stimulates Ca^{+2} efflux, but reduces Ca^{+2} influx only slightly. Each of the three ions (Cu^{+2}, H^+, and Cd^{+2}) disrupt transepithelial ion exchange in rainbow trout. However, effects of each are different with respect to site of action, rapidity of action, and persistence of the disturbance. From this study, one may postulate that Cd acts primarily by inhibiting Ca^{+2}-ATPase and Cu acts by inhibition of Na^+-K^+-ATPase;

whereas H^+ exerts a weaker, less specific action on the gills than either Cu or Cd. A companion study (Lauren and McDonald, 1987a) shows that Cu probably causes mixed inhibition of Na^+-K^+-ATPase.

Antibody titers in teleost whole blood decrease in fish exposed to Cu. Antibodies are immunoglobins possessing terminal variable regions specific for antigen binding and are derived from a prior antigenic challenge of progenitor cells. O'Neill (1981) shows that antibody levels are decreased as a result of Cu exposure in trout and carp. Brown trout (*Salmo trutta*) and mirror carp (*Cyprinus carpio*) are exposed 38 wk to aqueous solutions containing 0.29 ppm Cu at pH 7.8, total hardness 207 ppm (as $CaCO_3$), and 15.5°C. Humoral antibody response to MS2 bacteriophage (picornavirus, group 1 RNA-phage) is determined using a 50% viral neutralization assay. Trout immune response is suppressed, especially the secondary response, suggesting that immune memory cells are suppressed by Cu in both species after 38-wk exposures to 0.29 ppm Cu. Plateau titers also are significantly reduced in Cu-exposed trout. Copper-induced toxicosis in carp causes total suppression of the immune response (secondary and plateau peaks) and a significant decrease in the primary response, as well. Copper reduces viral clearance in both species. In trout, 7 d rather than ~3 d are required for blood clearance of the primary MS2 bacteriophage; whereas, carp are suppressed for more than 2 wk.

Other effects of Cu include significant elevated serum concentrations of Cu ($p<0.001$, both species), reduced hematocrits ($p<0.001$, both species), reduced serum protein levels ($p<0.001$, carp), and weight loss ($p<0.001$, both species) (O'Neill, 1981). From this study, one might conclude that no single action of Cu is responsible for delayed viral clearance, suppression of antibody titers, and other effects. Copper, like other heavy metals, interacts with immunocompetent cells in a complex way. Chronic, Cu-induced suppression of the immune response undoubtedly is capable of devastating the normal teleost response to viral antigens, such as those from bacteriophage MS2, increasing susceptibility to those antigens.

As with the O'Neill (1981) study, the Roales and Perlmutter (1977) study shows that antibody production is decreased in gouramis exposed to aqueous Cu. Blue gouramis (*Trichogaster trichopterus*) of 12–14 g body weight respond to Cu at 10% of the 96-h LC50 (9 ppb) at 26–28°C. In this study, antigens are injected 1 and 3 wk following Cu exposure. Viral neutralization tests show that 9 ppb Cu decreases antibody production appreciably. The exact mechanism of action is unknown, but could involve breakdown of antibodies, interference with complement production, alteration of protein synthesis, cytotoxic effects on the hemopoietic centers in fish (e.g. spleen), or hematologic changes. These effects are exerted at levels of Cu less than those affecting growth and reproduction.

In conclusion, hematological effects of Cu in in vivo studies varies markedly and can include one or more of the following alterations: direct injury of hematopoietic tissue surrounding kidney tubules (Baker, 1969), increased hematocrit (McKim, *et al.*, 1970), decreased hematocrit (O'Neill, 1981), increased then decreased plasma glutamic oxalacetic transaminase (McKim, *et al.*, 1970), decreased antibody titers (Roales and Perlmutter, 1977; O'Neill, 1981), increased hemoglobin (McKim, *et al.*, 1970), increased numbers of erythrocytes (McKim, *et al.*, 1970), increased total protein (McKim, *et al.*, 1970), decreased serum protein (O'Neill, 1981), decreased osmolarity (McKim, *et al.*, 1970; Lewis and Lewis, 1971; Dharmamba and Maetz, 1972), decreased plasma Na^+ (Lauren and McDonald, 1985; Reid and McDonald, 1988), increased plasma glucose (Lauren and McDonald, 1985), and increased plasma ammonia (Lauren and McDonald, 1985). Obviously, besides the hematopoietic cells of fish, a great number of other cell types are dramatically altered by Cu exposure. Effects can either be transient (Lewis and Lewis, 1971; McKim, *et al.*, 1970; Reid and McDonald, 1988) or longer-lived (Baker, 1969; O'Neill, 1981). Characterization and functionality of protein fractions R_f 0.32 and R_f 0.45 (prolactins) could provide some understanding of the reversal of Cu-induced hematological changes. To understand the mechanism of action of Cu on fish, relationships among the actions of the prolactins, Cu-induced inhibition of Na^+-K^+-ATPase, and competition between Cu^{+2} and H^+ at the gill surface would be useful.

Behavior: Fish are sensitive to low levels of Cu in the water. Giattina and co-workers (1982) show that rainbow trout (*Salmo gairdneri*) are capable of detecting as little as 1.4–2.7 ppb Cu (12°C, pH 7.3, total hardness 28 ppm as $CaCO_3$). These values are based on declines in residence times in shallow-gradient tests. In shallow and steep concentration gradient tests, trout avoid about 4.4–6.4 ppb Cu. Residence times and activity of trout in Cu-contaminated solutions are inversely related to Cu at low concentrations. Trout spend over 30% less time in 6 ppb Cu than in control solutions.

However, rainbow trout apparently are attracted to higher Cu levels in the water (334–386 ppb). Therefore, except at elevated Cu concentrations, trout have the ability to avoid Cu at levels below most chronic toxicity levels. Rainbow trout swimming upstream into Cu-contaminated effluent would avoid lower concentrations of Cu (1.4–6.4 ppb), but would appear to prefer higher levels (334–386 ppb). Preference or attraction of trout to markedly Cu-contaminated solutions may be due to narcotic effects, changes in chemoreceptor sensitivity, or damage to chemoreceptors (Gardner and LaRoche, 1973).

Once a fish has detected Cu, hyperactivity and reduced exploratory behavior are two of the major behavioral changes noted. Brook trout increase locomotor activity at 9–60 ppb Cu within the first few hours (Drummond, *et al.*, 1973); Atlantic silversides increase swimming speed by 43% when exposed to a pulse of Cu less than 100 ppb (Koltes, 1985); and sea catfish show increased hourly entries, decreased time spent per entry, and decreased time spent per pathway after 72 h at 100 or 200 ppb Cu (Steele, 1983). Each of the measures used in these three studies provides an indicator of hyperactivity (e.g. increased locomotor activity, increased swimming speed, and increased entries). Reduced exploratory behavior also is apparent, in that fish swim longer distances between turns (Steele, 1983) and make fewer directional changes (Koltes, 1985). A closer look at each of three studies shows additional changes which occur in fish exposed to Cu in the water (Drummond, *et al.*, 1973; Steele, 1983; Koltes, 1985).

Drummond and co-workers (1973) use potential difference between electrodes and a polygraph recording system during exposure of brook trout (*Salvelinus fontinalis*) to 0–115 ppb Cu (8.5°C, 45 ppm total hardness, pH 7.6). Marked variability is noted in response to Cu, in terms of locomotor activity and feeding responses. Increase in locomotor activity is a very sensitive indicator of Cu exposure provided observations are made within the first few hours of exposure. In controls, 4% of the time fish are active, compared with 22% when fish are exposed to 9–60 ppb Cu. Change in feeding behavior lasts the longest (24 h at 6 ppb and 14 d at 12 ppb); however, measurement of feeding behavior is less objective than that of locomotor activity. The gradual (2-wk) resumption of feeding at 9 ppb Cu probably explains the transient growth retardation in the first 23 wk at 9.5 ppb and lower Cu (McKim and Benoit, 1971).

Steele (1983) exposes hardhead sea catfish (*Arius felis*) 72 h to 0–0.2 ppm Cu to determine Cu-induced effects on locomotor and orientation behavior (21–24°C, 30 ‰ salinity). Behavior is monitored electrically while fish swim in a cylindrical tank incompletely partitioned by hollow dividers, oriented radially from the periphery into sixteen compartments. Control fish and those exposed to 0.005–0.050 ppm Cu are hypoactive after 72-h exposures, based on decreases in hourly entries (an entry and exit of a compartment), increased time spent per entry, and increased time spent per pathway (time spent traveling between the compartments). In contrast, sea catfish exposed 72 h to 0.1 or 0.2 ppm Cu are hyperactive, based on increased hourly entries, decreased time spent per entry, and decreased time spent per pathway. Orientation angles (those angles between the compartment of origin and that of destination) increase for all fish except for controls and 0.005 ppm-Cu-exposed individuals. That is, fish swim longer distances between turns if they are exposed to Cu, or frequency of turning is decreased by Cu exposure. These changes in activity and preferred turning angles might be due to avoidance behavior, gill damage (an unlikely possibility for a marine species at these Cu levels—see Table 1), and/or neurotoxic changes (Gardner and LaRoche, 1973). Likely, such Cu-induced alterations represent changes in exploratory behavior.

Koltes (1985) uses a pulse of water containing sublethal Cu levels (<100 ppb) to test behavior of a school of Atlantic silversides (*Menidia menidia*). Fish respond immediately to the presence of Cu by increasing swimming speed (by 43%), decreasing nearest-neighbor distance (by 11%), decreasing rate of change of direction (by 34%), and swimming more in parallel orientation. That is, schools become more uniform or structured while swimming more rapidly around the 1400-l tank. Behavior when exposed to Cu resembles that of spring and fall migrations—times in which Atlantic silversides form well-organized, active schools in order to migrate. Increased school organization and activity as a result of Cu exposure may be interpreted as an avoidance response.

Copper affects migratory ability in coho salmon (*Oncorhynchus kisutch*) (Lorz and McPherson, 1976). Yearling (10–18 mo old) salmon are exposed to a maximum of 80 ppb Cu to evaluate Cu-induced effects on migrating ability (10–12°C, pH 6.8–7.5, and total hardness 89–99 ppm). Fish are exposed to Cu, released into a tributary, and trapped downstream. One of the most striking results of this study is that Cu-exposure (even at 5 ppb—a level which does not alter ATPase activity or seawater survival) has a marked impact on migrating ability. Exposure to 30 ppb for as short as 72 h results in 52% migration (a 41% decrease, compared to 93% migration in controls). A level of 5 ppb for 165 d results in 70% migration; whereas, 165-d exposures to 30 ppb allowed about 30% migration. Therefore, 5–20 ppb Cu provides an application factor range of 0.09–0.35 of the 96-h LC50 value for Cu in coho salmon smolts (Table 3).

Table 3. Copper application factors in fish. Water hardness (in ppm $CaCO_3$), temperature, and pH are provided.

Species	Application Factor	Hardness (ppm, $CaCO_3$)	Temp. (°C)	pH	Reference
Bluegill *Lepomis macrochirus*	0.02–0.04	45	13–28	7–8	Benoit, 1975
Brook trout *Salvelinus fontinalis*	0.10–0.17	45	10–12	7.5	McKim & Benoit, 1971
Fathead minnow	0.13–0.22	31	19–25	6.9–7.2	Mount and Stephen, 1969
Pimephales promelas	0.03–0.08	200	16–25	7.9	Mount, 1968
Coho salmon *Oncorhynchus kisutch*	0.09–0.35	89–99	10–12	6.8–7.5	Lorz & McPherson, 1976

To summarize, Cu-induced behavioral alterations, combined with a sensitive chemoreceptor detection system, are likely to reduce Cu exposure of feral fish provided fish can escape from the contaminated area. Impairment of mobility and/or lack of adequate uncontaminated areas could render these behavioral and physiological adaptations useless. If fish are unimpaired and have clean ecosystems to inhabit, low level detection of Cu and subsequent avoidance of low concentrations could reduce deleterious Cu-induced effects on individual survival and reproduction. Marked impairment of migratory ability in coho salmon after exposure to 5 ppb Cu, attests the potential for Cu-induced behavioral impairment at levels previously considered to be no-effect levels.

Fish are capable of an immediate behavioral response to Cu. Immediately after Cu release, schools of fish swim faster and in a more organized fashion in what appears to be avoidance of the contaminant. This may be likened to migratory behavior in oceanic species (Koltes, 1985) or to the seaward migration of coho salmon (Lorz and McPherson, 1976). What can be considered an avoidance response is possible for silverside schools at 100 ppb Cu (Koltes, 1985); for individual rainbow trout at 9–60 ppb Cu (Drummond, *et al.*, 1973); and for individual sea catfish at 100 or 200 ppb Cu (Steele, 1983). However, these represent measurable responses using behavioral test chambers, video monitoring, photocell systems, and/or electrode systems to make measurements. Likely, many fish begin to respond immediately upon detection of Cu, as results from silverside schools suggest (Koltes, 1985). In such a case, avoidance might occur at considerably lower Cu levels (possibly 1–3 ppb Cu in the water) for many individuals of some species. However, a lower level of response, as well as individual variation in response (the nemesis of the behaviorist), would reduce the chances of monitoring the change at lower Cu levels. Regardless of whether methodology allows the scientist to show avoidance in the laboratory, those fish which rapidly detect and avoid Cu have a greater chance of survival and of genetically transferring these adaptive mechanisms to subsequent generations.

Physiology: Introduction of Cu into an aqueous environment can result in increased coughing frequencies, higher ventilation rates, higher opercular/buccal amplitudes, and elevated oxygen utilization in fish (Jones, 1947; O'Hara, 1971; Spoor, *et al.*, 1971; Drummond, *et al.*, 1973; Sellers, *et al.*, 1975; Klaverkamp, 1982). Coughing rate increases in brook trout (*Salvelinus fontinalis*) exposed to 9–115 ppb Cu in the water (8.5°C, 45 ppm total hardness, pH 7.6) (Drummond, *et al.*, 1973). Marked variability is noted in responses to Cu, but cough frequency generally is a function of Cu concentration. Control brook trout cough an average of 1.7 times per min, compared with 2.1 times per min at 9–15 ppb Cu; but, this change is not measurable until 4-h. In contrast, at 55–115 ppb Cu, fish begin coughing within a few minutes at a rate of 6.3–7.4 coughs/min. Five-fold increases in locomotor activity occur within the first few hours, but opercular measurements are recorded as breaks in the ventilation rhythm and can be differentiated from other movements recorded by the polygraph. Klaverkamp (1982) notes increased cough frequency for fish at a level of 0.21 ppm Cu. Zinc introduction into aqueous solutions also increases coughing rates in fish (Sellers, *et al.*, 1975; Hughes and Adeney, 1977), probably secondary to irritation of buccal and gill membranes.

Ventilation rate increases in fish exposed to Cu (Spoor, *et al.*, 1971; Sellers, *et al.*, 1975), Zn (Sellers, *et al.*, 1975; Hughes and Adeney, 1977), or Cd (Morgan and Kühn, 1974). To measure ventilation rates, stainless steel electrodes can be attached to aquarium walls to record changes in potential difference due to respiratory (and other) movements of fish (Spoor, *et al.*, 1971). Fish are given freedom of movement within the respiratory chamber. Moving water around the electrodes (from adducting opercula or other motion) disturbs the electrochemical equilibrium at the electrode surface and changes the potential from a few μV to several hundred μV, according to the amount and velocity of water movement. One electrode is always more affected than the other, causing development of a potential difference between them (recorded, in this case, as a wave). Several hundred oscillations can be recorded per min.

Normally, bluegills (*Lepomis macrochirus*) respire at 20–30 opercular beats/min following acclimation to the respiratory chamber (Spoor, *et al.*, 1971). If exposed to 0.5 ppm Cu, opercular movement increases to about 100 beats/min within 4 d. If death does not occur at this time, the opercular frequency decreases within the next 4 d to normal values. In contrast to these findings, Klaverkamp (1982) notes absence of a Cu-induced change in ventilation rate, as well as in heart rate and buccal pressure amplitude.

In the Sellers *et al.* (1975) study, rainbow trout (*Salmo gairdneri*) are cannulated in order to monitor opercular and buccal pressure changes (ventilation/buccal amplitude, ventilation rate, and coughing frequency) during exposure to Cu. Intermediate concentrations of Cu (0.07 ppm) elicit greater responses (in ventilation amplitude and ventilation rate) than higher concentrations. Higher Cu levels cause more variation in ventilation pressure changes than lower Cu levels. Copper causes marginal (almost nil) changes in coughing rate. However, increases in ventilation frequency, ventilation amplitude, and buccal amplitude result from Cu exposure. Marked individual variability in ventilatory activity is characteristic of this species. The general irritation of buccal and gill membranes possibly results in aberrant movement. Amplitude may be low and ventilation rate may be high for one subject (or for one concentration) but not for a second subject (or concentration). Unfortunately, no information is provided of total respiratory activity for individual fish using ventilation rate and amplitude date (e.g.

volume of water breathed per unit time). Perhaps the individual variation would be reduced in such comparisons.

Arterial oxygen tension and pH also are monitored (Sellers, *et al.*, 1975). In 21-h exposures to 0.29 ppm Cu (the LC50 at 51–68 ppm hardness) no change in oxygen tension occurs; however, 86-h exposures reduce oxygen tension as low as 70–80 mm Hg (from normal levels of ~120 mm Hg). As is the case for Zn, decreased oxygen tension can result from a block in the respiratory pathway, possibly because of an increased water to blood diffusion distance (Skidmore and Tovell 1972). Copper (0.29 ppm for 21 h) causes a transient increase in pH of arterial blood (Sellers, *et al.*, 1975).

O'Hara's study (1971) parallels that of others (Spoor, *et al.*, 1971; Sellers, *et al.*, 1975), except that oxygen consumption also is measured. Bluegills (*Lepomis macrochirus*) are held in a flowing water respirometer chamber for dissolved oxygen measurements on water passing into and out of the chamber (25°C, pH 7.7, 35 ppm total hardness). In bluegills, only opercular movements increase after Cu addition, suggesting that respiratory demand (rather than locomotor activity) is increased.

Copper-induced effects (at 0.1–5.0 ppm Cu) on oxygen consumption, are transient for some individuals (O'Hara, 1971). Regardless of whether fish subsequently succumb or survive, within 3–6 h after addition of 0.1–5.0 ppm Cu, they require maximum levels of oxygen. Within ~4 h, oxygen consumption maximizes and returns to normal for fish held in 0.5 or 1.0 ppm Cu, and fish survive. During the first 18 h, oxygen consumption maximizes and returns to normal for fish held in 2 or 3 ppm Cu, and fish survive. However, after 14–15 h, fish held in 4 or 5 ppm Cu die, despite a similar pattern of oxygen consumption as fish held at 2 or 3 ppm Cu.

After the first 24 h, the percentage increase in oxygen utilization is eight times higher in fish held in 3.0 ppm Cu than those in 0.5 ppm Cu, and twelve times higher in fish held at 5 ppm Cu than those held at 0.5 ppm Cu (O'Hara, 1971). Oxygen consumption is correlated strongly with Cu levels in the water up to the 96-h LC50 of 2.4 ppm Cu. Higher Cu concentrations produce greater responses, but these tend to be erratic.

Aqueous solutions of 0.1 ppm Cu produce no change in oxygen consumption in bluegill sunfish; therefore, the maximum acceptable toxicant concentration (MATC or "no effect" level) is 0.1 ppm Cu in this study. Multiplication of the bluegill 96-h LC50 (2.4 ppm) by 0.13 and 0.22

application factors derived using the fathead minnow (Mount and Stephen, 1969), allows computation of approximate MATC values for the bluegill (0.31–0.53 ppm). These MATC are only slightly higher than the measured 0.1 ppm Cu level (O'Hara, 1971). Both Mount (1969) and O'Hara (1971) use comparable soft water solutions.

In summary, bluegills and sticklebacks show increased oxygen consumption as a result of Cu exposure (Jones, 1947; O'Hara, 1971). The requirement for increased oxygen undoubtedly is met (partially or totally) from increased opercular rate, increased opercular/buccal amplitude, and/or increased coughing rate (O'Hara, 1971; Spoor, *et al.*, 1971; Drummond, *et al.*, 1973; Sellers, *et al.*, 1975; Klaverkamp, 1982). Some species (e.g. bluegills) show increased ventilation rate, but not locomotor activity, after introduction of Cu (O'Hara, 1971). Other species, such as sticklebacks (Jones, 1947) and brook trout (Drummond, *et al.*, 1973), show higher locomotor activity (see the section on the effects of Cu on behavior). Therefore, increased ventilation rate after Cu addition cannot be attributed exclusively to increased oxygen demand in all species, because of the compounding effects of locomotor activity in those species. Therefore, a number of studies show that aqueous Cu-exposures increase cough frequency (Drummond, *et al.*, 1973; Klaverkamp, 1982), ventilation rate (O'Hara, 1971; Spoor, *et al.*, 1971; Sellers, *et al.*, 1975), and oxygen consumption (Jones, 1947; O'Hara, 1971) in fish. Other studies, however, show no change in ventilation rate (or volume) in fish exposed to Cu (Klaverkamp, 1982).

<u>Reproduction</u>: Copper affects fish in most life stages, but the reproductive and embryonic stages can be considered "bottlenecks" in the life cycle, since reproductive adults and embryos are susceptible to considerably lower levels of Cu than are other life stages. A wealth of information is published to attest to the hazard of Cu to embryos, sac-fry, fry, and reproductive adults. At the embryonic stage, Cu can alter hatching time (McKim and Benoit, 1971; Gardner and LaRoche, 1973; Scudder, *et al.*, 1988) without changing developmental rate (Scudder, *et al.*, 1988).

Hatchability (% total hatch) decreases in the presence of increasing levels of Cu in some studies (McKim and Benoit, 1971; Gardner and LaRoche, 1973; McKim and Benoit, 1974; Scudder, *et al.*, 1988), but not in others (Pickering, *et al.*, 1977). Copper-induced decreases in hatchability are concentration-dependent: 0.1–0.3 ppm Cu kills king salmon eggs, but 0–0.08

ppm Cu does not (Hazel and Meith, 1970). Mummichog and silverside fry do not hatch in 3.0 or 10.0 ppm Cu, but they do hatch at 0.5, 1.0, or 5.0 ppm Cu, albeit with lower hatchability as Cu level increases (Gardner and LaRoche, 1973). Moreover, declines in hatchability are observed for fathead minnow eggs at 621 ppb Cu, but not at 338 ppb or lower concentrations (Scudder, *et al.*, 1988).

Embryonic deformities are reported in some (Scudder, *et al.*, 1988), but not all (Benoit, 1975), studies. Yolk reabsorption is delayed 4 wk in brook trout alevins hatched and exposed in 32.5 ppm Cu (McKim and Benoit, 1971), but not in fathead minnow larvae exposed to 0.621 ppm Cu (Scudder, *et al.*, 1988). Embryos and sac-fry are more sensitive to Cu than are adults; and larval sensitivity to Cu increases as exposure time increases (Hazel and Meith, 1970; Gardner and LaRoche, 1973). Fertilized fish eggs (i.e. embryos) are generally more resistant to Cu than are sac-fry. Growth of king salmon sac-fry is reduced by Cu exposure: 12% and 21% decreases in size occur at 0.02 or 0.04 ppm Cu, respectively (Hazel and Meith, 1970). Therefore, embryonic and larval stages of fish are at risk when aqueous levels of Cu are elevated.

This is doubly true for reproductive adults—especially females. Copper exposure retards sexual development and growth (Mount and Stephen, 1969), egg production per female can be decreased (McKim and Benoit, 1971; Pickering, *et al.*, 1977), spawning can be blocked partially or entirely (Mount, 1968; Mount and Stephen, 1969; Pickering, *et al.*, 1977), and migration can be impeded (Lorz and McPherson, 1976). Again, effects are concentration-dependent. A Cu level of 18.4 ppb, for example, prevents spawning (eliminating egg production) in fathead minnows, but a level within the 4.4–10.6 ppb Cu concentration range does not produce detrimental effects on these parameters (Mount and Stephen, 1969). Obviously, elevated levels of Cu in the water impact reproduction in a multitude of ways. Additional information details interrelationships among changes at the embryonic and reproductive-adult stages, beginning with early life stages and progressing to the reproductive stage in adults.

Amazingly, no Cu is passed from the parent to the eggs of brook trout at aqueous levels ≤9.4 ppb Cu at 9.6°C and 44 ppm total hardness, as $CaCO_3$. However, 3-wk exposures of control embryos to 9.4 ppb Cu results in a five-fold increase in Cu concentration from control levels of 6 ppb to 28 ppb Cu

(McKim and Benoit, 1974). Most Cu is thought to be bound to the outer shell membrane, never distributing to the embryo; however, more recent research indicates that the chorion is permeable to Cu at specific developmental stages (Scudder, *et al.*, 1988).

Copper exposure can either delay or accelerate hatching time (McKim and Benoit, 1971; Gardner and LaRoche, 1973; Scudder, *et al.*, 1988). Gardner and LaRoche (1973) observe a 4-d delay in hatching of silversides and mummichogs in 0.5, 1.0, or 5.0 ppm Cu, but not for those in control and 0.25 ppm Cu. In contrast, McKim and Benoit (1971) find that premature hatching occurs in brook trout: mean incubation times decrease 17% with exposure to 32.5 ppb Cu, but no changes are noted at 1.9–17.4 ppb Cu. A study by Scudder and associates (1988) also shows a Cu-induced, 1-d, premature hatching.

These findings (Scudder, *et al.*, 1988) are particularly interesting. Embryos of fathead minnows are exposed to 0.6–621.0 ppb Cu (21°C, pH 8.17). Those embryos which are exposed from 5–10 h post-fertilization through 2 d post-hatch, hatch faster as Cu levels in water increase, but do not show a change in embryonic development rate. Therefore, fish hatch in a more premature state than normal. In 338 and 621 ppb Cu, fish hatch in about 6 d, rather than 7 d. Accelerated hatching may be due to the quiescence of embryos exposed to Cu. Instead of tumbling, shivering, and twitching, embryos exposed in Cu remain almost motionless for long periods. In contrast, normal embryonic activity is believed to distribute enzymatic or other chemical substances from the pharynx (or anterior surface) to all parts of the perivitelline space. In Cu-exposed embryos, decreases in activity might allow concentration of hatching substances in the head region allowing more rapid chorionic degeneration and accelerated hatching.

During development, days 6–9 seem to be critical for the fathead minnow, because chorion permeability appears to increase at that time as a result of embryonic changes in the outer layer of the chorion (Scudder, *et al.*, 1988). Abnormalities affect 92% of surviving fathead minnow embryos, compared with 0% in control embryos and 20% in embryos exposed to 338 ppb Cu (Scudder, *et al.*, 1988). The most common abnormalities are kyposis (abnormal dorsal curvature of the spine), general eye deformities [orientation and placement defects, lens defects, and abnormally small eyes (microphthalmia)], deformed maxilla and mandibles, and slightly misshapen fins (usually the pectorals). Abnormalities are greater during days 6–9 of development, than during days 1–5. This change in abnormalities with developmental time is probably secondary to increased permeability of the chorion, since days 6–9 are characterized by emergence of indentations within surface elevations of the outer layer of the chorion. Increased permeability of the chorion to Cu probably results in increased Cu-induced effects during this period of development.

In 204–621 ppb Cu exposures, misshapen pectoral fins and premature hatching probably contribute to the long periods of time fathead minnow larvae spend resting and the reduced time they spend actively searching for food (Scudder, *et al.*, 1988). In 28-d post-hatch tests, the length, weight, and condition decreases significantly in larvae exposed to 61–338 ppb Cu. No larvae survived 621 ppb Cu longer than 15 d. No difference is noted in yolk utilization in fathead minnows, contrasting with brook trout which do show decreased yolk resorption after Cu exposure (McKim and Benoit, 1971). Therefore, this study (Scudder, *et al.*, 1988) shows that increases in numbers of abnormal embryos, decreased hatching time, decreased survival, and decreased growth are sensitive measures of Cu-induced toxicity in embryos and larvae of fathead minnows.

No conclusive explanation exists for the inverse relationship between hatchability and Cu levels in some studies (McKim and Benoit, 1971; Gardner and LaRoche, 1973; McKim and Benoit, 1974; Scudder, *et al.*, 1988), but not in others (Pickering, *et al.*, 1977). Absence of such a relationship might occur because no-effect or MATC levels of Cu are used. For example, 0.1–0.3 ppm Cu levels kill king salmon embryos; whereas, 0–0.08 ppm Cu does not alter percent hatch (Hazel and Meith, 1970). At 3.0 or 10.0 ppm Cu, no mummichog or silverside fry hatch, but fry hatch (albeit delayed) from embryos at 0.5, 1.0, or 5.0 ppm Cu (Gardner and LaRoche, 1973). In addition, if embryos from control fish are used, hatchability is affected only by 621 ppb Cu (Scudder, *et al.*, 1988) and not by 338 ppb or less Cu. Whereas, hatching is abolished at 18.4 ppb Cu after 19-mo exposures of adults (Mount and Stephen, 1969) because fish do not spawn at these Cu levels. Under latter exposure conditions, hatchability is normal only at aqueous Cu levels of 10.6 ppb or less because spawning is normal at those levels. Obviously, if fish do not spawn, no eggs are available for determining percent hatch.

Hatchablility is a function of Cu-exposure level, species, spawning, and probably a number of other factors yet to be determined.

In summarizing Cu-induced effects on embryos and fry, it is clear that embryos are generally less sensitive to Cu than are fry. However, at certain developmental stages, embryonic changes in permeability undoubtedly allow increased Cu entry, such that abnormalities can occur in up to 93% of surviving embryos. Because Cu-exposed embryos hatch earlier in most species considered, and because developmental rates remain constant, generally fish hatch in a more premature state. Since these premature larvae spend much time resting on the bottom of aquaria, food intake and growth rate are reduced. Larval survival decreases as Cu level increases. Certain embryonic and larval states of several fish species are extremely sensitive to Cu levels in the water; however, reproductive fish in spawning condition seem even more sensitive.

Indeed, spawning is a critical, Cu-sensitive phase of the teleost life cycle. For example, in fathead minnows, spawning can be completely blocked after ~20-mo exposure to 18.4 ppb Cu, but not to 4.4–10.6 ppb Cu (Mount and Stephen, 1969). Mount and Stephen (1969) expose fathead minnows (*Pimephales promelas*) to Cu at nominal concentrations in the 0–20 ppb range (19–25°C, pH 6.9–7.2, 31.4 ppm total hardness). Fish are brought into breeding condition by increasing the photoperiod to 16 h. The highest measured Cu concentration (18.4 ppb) kills half the parental fish, retards growth and sexual development of survivors, and prevents spawning. Survival, growth, and reproduction are similar at lower levels (4.4–10.6 ppb Cu).

At 4.4–10.6 ppb Cu, no detectable effect is observed on spawning behavior, on the ability of male to care for eggs, or on the initiation and distribution of egg production. The 96-h LC50 is about 84 ppb; therefore, the application factor lies between 0.22 (18.4/84) and 0.13 (10.6/84) (Mount and Stephen, 1969). This compares well with Mount's (1968) data for hard water (200 ppm total hardness) which shows a 430 ppb 96-h LC50 value and 0.03 (15/430) to 0.08 (33/430) application factor limits (Table 3). The percentage of nontoxic Cu probably varies in the Mount and Stephen (1969) study and the Mount (1968) study because of the six-fold difference in total water hardness between these two studies.

Some studies show that shorter exposures (3- or 6-mo) of fish to Cu produce similar effects on spawning as no pre-exposure to Cu (Pickering, *et al.*, 1977). Pickering and colleagues (1977) use three exposure systems (0-, 3-, or 6-mo) to test the possible effect of length of pre-exposure time on reproduction of fathead minnows (*Pimephales promelas*) exposed to 4–100 ppb Cu (20–26°C, pH 7.5–8.2, and total hardness 200 ppm as $CaCO_3$). The 96-h LC50 values are 490 ppb for 6-wk old fry and 460 ppb for 6-mo old juveniles.

In all three exposure systems, all fish survive and only 6-mo exposures to 98 ppb Cu cause decreases in length and weight of females ($p<0.05$). Hatchability (calculated as percentage of larvae hatched after 7 d) is normal (87–97%) in all three systems. Length of time adult fish are exposed to Cu has no influence on hatchability; however, Cu exerts striking effects on spawning and egg production.

Copper decreases the numbers of spawnings per female (by 95%) and the number of eggs produced per female (by 98%) regardless of of exposure time (i.e. 0-, 3-, or 6-mo) (Pickering, *et al.*, 1977). Therefore, fecundity is inversely correlated with aqueous Cu levels ($p<0.01$), but not with time of prior exposure to Cu in this study. As previously mentioned, the spawning season represents a critical time of the sublethal Cu exposure—an abbreviated period which seems not to depend on the length of prior Cu exposure (at least in fathead minnows) and which must be simulated in the laboratory in acute studies to obtain no-effect (MATC) concentrations. In essence, spawning represents the primary bottleneck in Cu exposure over a complete life cycle of fathead minnows.

Egg production is reported to decrease in females exposed to Cu in some studies (McKim and Benoit, 1971; Pickering, *et al.*, 1977), but not in others (McKim and Benoit, 1974). The difference in effect may be concentration-dependent. The levels of Cu are low (2.7–9.4 ppb) in the McKim and Benoit study (1974), but high in both the Pickering *et al.* study (1977) (i.e. 4.2–98.0 ppb) and the McKim and Benoit study (1971) (i.e. 1.9–32.5 ppb). Results of the Pickering *et al.* study (1977) show good correlation between Cu level and number of eggs per female. The number of eggs laid per female varies inversely with the aqueous level of Cu. At exposure levels of 37–98 ppb Cu, numbers of eggs per female decrease as Cu level increases; whereas, 24 ppb Cu may not have a significant effect on this parameter. The median effective concentration, in terms of eggs per female (EC50), is estimated to be 40.4 ppb. The MATC is estimated at 32 ppb (0.07 of the 96-h LC50); 34 ppb is considered unsafe (Pickering, *et al.*, 1977). This agrees with Mount's (1968) data which shows that about 33 ppb Cu blocks spawning in fathead minnows.

Therefore, neither survival nor growth provide as sensitive an indicator of Cu-induced effects as does reproduction. The MATC value of 32 ppb is estimated to cause a 30% reduction in egg production in fathead minnows (Pickering, *et al.*, 1977). If other factors combined to further reduce egg production to 79% (e.g. the presence of parasites, low oxygen levels, or other synergistic metals), a population of fathead minnows would become extinct.

Summary

Generally, fresh-water species are more sensitive to metals than are marine species; however, this is not true in cases of Cu-poisoning in fish (Table 1). Reasons for equal sensitivity of markedly different species has not been elucidated. Nor have explanations been given for the pronounced species-specific differences in Cu accumulation levels (Table 2), which seem to involve capability for isometallothionein synthesis and hepatic storage of Cu.

Increased salinity and increased hardness protect fish from Cu poisoning; however, pH, alkalinity, organic level, temperature, fish size, and other factors also play important roles in Cu poisoning of fish. Generally, the greater the hardness, alkalinity, salinity, organic level, pH, and fish size, the more resistant fish are to Cu. Temperature produces complex effects, but elevated thermal levels increase fish sensitivity to Cu. Moreover, death of fish is not related to a specific Cu level accumulated by any one of several tissues normally examined, as may be the case for Zn and Cd. This may be due to differential accumulation of different species of Cu (i.e. toxic and nontoxic Cu species). Alternatively, in whole body uptake studies, nonspecific epidermal binding of Cu might mask the specific binding occurring in a target organ prior to death (Table 2).

The large number of Cu-containing enzymes and glycoproteins in fish probably accounts for the diversity of biological effects—effects spanning practically every system evaluated in the teleost. Hematology is altered, as is respiratory and cardiac physiology. Copper-induced histological alterations are found in the gill, kidney, hematopoietic tissue, mechanoreceptors, chemoreceptors, and other tissues. Reproductive effects are noted at low levels of Cu and include blockage of spawning, reduced egg production per female, abnormalities in newly-hatched fry, reduced survival of young, and other effects. From most studies, growth and survival are affected at higher Cu levels than are reproductive changes. Copper poisoning of surface waters is considered hazardous to teleost life in those ecosystems; levels of ≥4 ppb can severely alter migratory and other behaviors. In view of the chronic bioaccumulation of Cu and the many effects observed in Cu-exposed fish, revision of existing water quality standards (Appendix 1) for Cu seems prudent.

References

Aaseth, J., and T. Norseth, Copper, in Friberg, L., G.F. Nordberg, and V. Vouk (Eds.), Handbook on the Toxicology of Metals, 2nd ed., Elsevier Science Publishers, Amsterdam, (1986) 233.

Baker, J.T.P., Histological and electron microscopical observation on copper poisoning in the winter flounder (*Pseudopleuronectes americanus*), J. Fish. Res. Can., 26 (1969) 2785.

Benoit, D.A., Chronic effects of copper on survival, growth, and reproduction of the bluegill (*Lepomis macrochirus*), Trans. Am. Fish. Soc., 104 (1975) 353.

Bilinski, E., and R.E.E. Jonas, Effects of cadmium and copper on the oxidation of lactate by rainbow trout (*Salmo gairdneri*) gills, J. Fish. Res. Bd. Can., 30 (1973) 1553.

Birdsong, C.L., and J.A. Avault Jr., Toxicity of certain chemicals to juvenile pompano, Prog. Fish Cult., 33 (1971) 76.

Bonga, S.E.W., and J.C.A. Van der Meij, Effect of ambient osmolarity and calcium on prolactin cell activity and osmotic water permeability of the gills in the teleost *Sarotherodon mossambicus*, Gen. Comp. Endocrinol., 43 (1981) 432.

Brown, V.M., T.L. Shaw, and D.G. Shurben, Aspects of water quality and the toxicity of copper to rainbow trout, Water Res., 8 (1974) 797.

Brungs, W.A., E.N. Leonard, and J.M. McKim, Acute and long-term accumulation of copper by the brown bullhead, *Ictalurus nebulosus*, J. Fish. Res. Bd. Can., 30 (1973) 583.

Bryan, S.E., and H.A. Hildalgo, Nuclear ^{115}Cadmium: uptake and disappearance correlated with cadmium-binding protein synthesis, Biochem. Biophy. Res. Comm., 68 (1976) 858.

Buckley, J.T., M. Roch, J.M. McCarter, C.A. Rendell, and A.T. Matheson, Chronic exposure of coho salmon to sublethal concentrations of copper—I. Effect on growth, on accumulation and distribution of copper, and on copper tolerance, Comp. Biochem. Physiol., 72C (1982) 15.

Cross, F.A., L.H. Hardy, N.Y. Jones, and R.T. Barber, Relation between total body weight and concentrations of manganese, iron, copper, zinc, and mercury in white muscle of bluefish (*Pomatomus saltatrix*) and a bathyl-demersal fish *Antimora rostrata*, J. Fish. Res. Bd. Can., 30 (1973) 1287.

Dharmamba, M., and J. Maetz, Effects of hypophysectomy and prolactin on the sodium balance of *Tilapia mossambica* in fresh water, Gen. Comp. Endocrinol., 19 (1972) 175.

Dixon, D.G., and J.B. Sprague, Acclimation to copper by rainbow trout (*Salmo gairdneri*)—A modifying factor in toxicity, Can. J. Fish. Aq. Sci., 38 (1981a) 880.

Dixon, D.G., and J.B. Sprague, Copper bioaccumulation and hepatoprotein synthesis during acclimation to copper by juvenile rainbow trout, Aq. Toxicol., 1 (1981b) 69.

Drummond, R.A., W.A. Spoor, and B.F. Olson, Some short-term indicators of sublethal effects of copper on brook trout, *Salvelinus fontinalis*, J. Fish. Res. Bd. Can., 30 (1973) 698.

Frazier, J.M., and J. Puglese, Dose dependence of cadmium kinetics in the rat liver following intravenous injection, Toxicol. Appl. Pharmacol., 43 (1978) 461.

Frazier, J.M., A fish model of abnormal copper metabolism, Mar. Environ. Res., 14 (1984) 478.

Gardner, G.R., and G. LaRoche, Copper induced lesions in estuarian teleosts, J. Fish. Res. Bd. Can., 30 (1973) 363.

Giattina, J.D., R.R. Garton, and D.G. Stevens, Avoidance of copper and nickel by rainbow trout as monitored by a computer-based data acquisition system, Trans. Am. Fish. Soc., 111 (1982) 491.

Goyer, R.A., Toxic effects of metals, in Klaassen, C.D., M.O. Amdur, and J. Doull (Eds.), Casarett and Doull's Toxicology: The Basic Science of Poisons, MacMillian Publ. Co., New York, New York, (1986) 582.

Hazel, C.R., and S.J. Meith, Bioassay of king salmon eggs and sac fry in copper solutions, Calif. Fish and Game, 56 (1970) 121.

Howarth, R.S., and J.B. Sprague, Copper lethality to rainbow trout in waters of various hardness and pH, Water Res., 12 (1978) 455.

Hughes, G.M., and R.J. Adeney, The effects of Zn on the cardiac and ventilatory rhythms of rainbow trout (Salmo gairdneri, Richardson) and their responses to environmental hypoxia, Water Res., 11 (1977) 1069.

Jones, J.R.E., The oxygen consumption of *Gasterosteus aculeatus* L. in toxic solutions, J. Exp. Biol., 23 (1947) 298.

Klaverkamp, J.F., The physiological pharmacology and toxicology of fish cardiovascular systems, in Weber, L.J. (Ed.), Aquatic Toxicology: Volume 1, Raven Press, New York, New York, (1982) 1.

Koltes, K.H., Effects of sublethal copper concentrations on the structure and activity of Atlantic silverside schools, Trans. Am. Fish. Soc., 114 (1985) 413.

Lauren, D.J., and D.G. McDonald, Effects of copper on brachial ionoregulation in the rainbow trout, *Salmo gairdneri* Richardson, J. Comp. Physiol., B 155 (1985) 635.

Lauren, D.J., and D.G. McDonald, Influence of water hardness, pH, and alkalinity on the mechanisms of copper toxicity in juvenile rainbow trout, *Salmo gairdneri*, Can. J. Fish. Aq. Sci., 43 (1986) 1488.

Lauren, D.J., and D.G. McDonald, Acclimation to copper by rainbow trout, *Salmo gairdneri*: Physiology, Can. J. Fish. Aq. Sci., 44 (1987a) 99.

Lauren, D.J., and D.G. McDonald, Acclimation to copper by rainbow trout, *Salmo gairdneri*: Biochemistry, Can. J. Fish. Aq. Sci., 44 (1987b) 105.

Lewis, S.D., and W.M. Lewis, The effect of zinc and copper on the osmolality of blood serum of the channel catfish, *Ictalurus punctatus* Rafinesque, and golden shiner, *Notemigonus crysoleucas* Mitchell, Trans. Am. Fish. Soc., 100 (1971) 639.

Lloyd, R., The toxicity of mixtures of zinc and copper sulfates to rainbow trout (*Salmo gairdneri* Richardson), Ann. Appl. Biol., 49 (1961) 535.

Lloyd, R., and W.M. Herbert, The effect of the environment on the toxicity of poisons to fish, J. Inst. Pub. Health Eng., July (1962) 132.

Lorz, H.W., and B.P. McPherson, Effects of copper or zinc in fresh water on the adaptation to sea water and ATPase activity and the effects of copper on migratory disposition of coho salmon (*Oncorhynchus kisutch*), J. Fish. Res. Bd. Can., 33 (1976) 2023.

Lydy, M.J., and T.E. Wissing, Effect of sublethal concentrations of copper on the critical thermal maxima (CTMax) of the fantail (*Etheostoma flabellare*) and johnny (*E. nigrum*) darters, Aq. Toxicol., 12 (1988) 311.

Matthiessen, P., and A.E. Brafield, The effect of dissolved zinc on the gills of the stickleback *Gasterosteus aculeatus* (L.), J. Fish Biol., 5 (1973) 607.

Mazeaud, M.M., F. Mazeaud, and E.M. Donaldson, Primary and secondary effects of stress in fish: Some new data with a general review, Trans. Amer. Fish Soc., 106 (1977) 201.

McCarter, J.A., A.T. Matheson, M. Roch, R.W. Olafson, and J.T. Buckley, Chronic exposure of coho salmon to sublethal concentrations of copper--II. Distribution of copper between high- and low-molecular-weight proteins in liver cytosol and the possible role of metallothionein in detoxification, Comp. Biochem. Physiol., 72C (1982) 21.

McCarter, J.A., and M. Roch, Hepatic metallothionein and resistance to copper in juvenile coho salmon, Comp. Biochem. Physiol., 74C (1983) 133.

McKim, J.M., G.M. Christensen, and E.P. Hunt, Changes in the blood of brook trout (*Salvelinus fontinalis*) after short-term and long-term exposure to copper, J. Fish. Res. Bd. Can., 27 (1970) 1883.

McKim, J.M., and D.A. Benoit, Effects of long-term exposures to copper on survival, growth, and reproduction of brook trout (*Salvelinus fontinalis*), J. Fish. Res. Bd. Can., 28 (1971) 655.

McKim, J.M., and D.A. Benoit, Duration of toxicity tests for establishing "no effect" concentrations for copper with brook trout (*Salvelinus fontinalis*), J. Fish. Res. Bd. Can., 31 (1974) 449.

Miller, T.G., and W.C. Mackay, The effects of hardness, alkalinity, and pH of test water on the toxicity of copper to rainbow trout (*Salmo gairdneri*), Water Res., 14 (1980) 129.

Miller, T.G., and W.C. Mackay, Relationship of secreted mucus to copper and acid toxicity in rainbow trout, Bull. Environ. Contam. Toxicol., 28 (1982) 68.

Morgan, W.S.G., and P.C. Kühn, A method to monitor the effects of toxicants upon breathing rate of largemouth bass (*Micropterus salmoides* Lacépède), Water Res., 8 (1974) 67.

Mount, D.I., The effect of total hardness and pH on acute toxicity of zinc to fish, Air Water Pollut. Int. J., 10 (1966) 49.

Mount, D.I., Chronic toxicity of copper to fathead minnows (*Pimephales promelas*, Rafinesque), Water Res., 2 (1968) 215.

Mount, D.I., and C.E. Stephen, Chronic toxicity of copper to the fathead minnow (*Pimephales promelas*) in soft water, J. Fish. Res. Bd. Can., 26 (1969) 2449.

Notter, M.F.D., J.E. Mudge, W.H. Neff, and A. Anthony, Cytophotometric analysis of RNA changes in prolactin and Stannius corpuscle cells of acid-stressed brook trout, Gen. Comp. Endocrinol., 30 (1976) 273.

O'Hara, J., Alterations in oxygen consumption by bluegills exposed to sublethal treatment with copper, Water Res., 5 (1971) 321.

O'Neill, J.G., The humoral immune response of *Salmo trutta* L. and *Cyprinus carpio* L. exposed to heavy metals, J. Fish Biol., 19 (1981) 297.

Pagenkopf, G.K., R.C. Russo, and R.V. Thurston, Effect of complexation on toxicity of copper to fishes, J. Fish. Res. Bd. Can., 31 (1974) 462.

Pickering, Q.H., and C. Henderson, The acute toxicity of some heavy metals to different species of warmwater fishes, Air Water Pollut. Int. J., 10 (1966) 453.

Pickering, Q., W. Brungs, and M. Gast, Effect of exposure time and copper concentration on reproduction of fathead minnow (*Pimephales promelas*), Water Res., 11 (1977) 1079.

Prosser, C.L., and F.A. Brown Jr., Comparative Animal Physiology, 2nd ed., Philadelphia, (1966) 688.

Reid, S.D., and D.G. McDonald, Effects of cadmium, copper, and low pH on ion fluxes in the rainbow trout, *Salmo gairdneri*, Can. J. Fish. Aq. Sci., 45 (1988) 244.

Roales, R.R., and A. Perlmutter, The effects of sub-lethal doses of methylmercury and copper, applied singly and jointly, on the immune response of the blue gourami (*Trichogaster trichopterus*) to viral and bacterial antigens, Arch. Environ. Contam. Toxicol., 5 (1977) 325.

Schreck, C.B., and H.W. Lorz, Stress response of coho salmon *(Oncorhynchus kisutch)* elicited by cadmium and copper and potential use of cortisol as an indicator of stress, J. Fish. Res. Bd. Can., 35 (1978) 1124.

Scudder, B.C., J.L. Carter, and H.V. Leland, Effects of copper on development of the fathead minnows, *Pimephales promelas* Rafinesque, Aq. Toxicol., 12 (1988) 107.

Sellers, C.M., A.G. Heath, and M.L. Bass, The effect of sublethal concentrations of copper and zinc on ventilatory activity. Blood oxygen and pH in rainbow trout (*Salmo gairdneri*), Wat. Res., 9 (1975) 401.

Shaw, T.L., and V.M. Brown, The toxicity of some forms of copper to rainbow trout, Water Res., 8 (1974) 377.

Shephard, K., and K. Simkiss, The effects of heavy metal ions on Ca^{++}-ATPase extracted from fish gills, Comp. Biochem. Physiol., 61B (1978) 69.

Skidmore, J.F., and P.W.A. Tovell, Toxic effects of zinc sulfate on the gills of rainbow trout, Water Res., 6 (1972) 217.

Sorensen, E.M.B., R.E. Henry, and R. Ramirez-Mitchell, Arsenic accumulation, tissue distribution and cytotoxicity in teleosts following indirect aqueous exposures, Bull. Environ. Contam. Toxicol., 21 (1979) 162.

Sorensen, E.M.B., and T.L. Bauer, Hematological dyscrasia in teleosts chronically exposed to selenium-laden effluent, Arch. Environ. Contam. Toxicol., 12 (1983) 135.

Spoor, W.A., T.W. Neiheisel, and R.A. Drummond, An electrode chamber for recording respiratory and other movements of free-swimming animals, Trans. Am. Fish. Soc., 100 (1971) 22.

Sprague, J.B., and B.A. Ramsay, Lethal levels of mixed copper-zinc solutions for juvenile salmon, J. Fish. Res. Bd. Can., 22 (1965) 425.

Sprague, J.B., Measurement of pollutant toxicity to fish. I. Bioassay methods for acute toxicity, Water Res., 3 (1969) 793.

Steele, C.W., Effects of exposure to sublethal copper on the locomotor behavior of the sea catfish, *Arius felis*, Aq. Toxicol., 4 (1983) 83.

Stiff, M.J., Copper/bicarbonate equilibria in solutions of bicarbonate ion at concentrations similar to those found in natural water, Water Res., 5 (1971a) 171.

Stiff, M.J., The chemical states of copper in polluted fresh water and a scheme of analysis to differentiate them, Water Res., 5 (1971b) 585.

Sylva, R.N., The environmental chemistry of copper (II) in aquatic systems, Water Res., 10 (1976) 789.

Williams, S.R., Nutrition and Diet Therapy, C. V. Mosby Company, St. Louis, (1969) 686.

Wilson, R.C.H., Prediction of copper toxicity in receiving waters, J. Fish. Res. Bd. Can., 29 (1972) 1500.

Wofford, H.W., and P. Thomas, Interaction of cadmium with sulhydryl-containing compounds in striped mullet (*Mugil cephalus* L.), Mar. Environ. Res., 14 (1984) 119.

Zitko, P., W.V. Carson, and W.G. Carson, Prediction of incipient lethal levels of copper to juvenile Atlantic salmon in the presence of humic acid by cubric electrode, Bull. Environ. Contam. Toxicol., 10 (1973) 265.

CHAPTER VIII
MERCURY

Introduction

Lewis Carroll's Mad Hatter is the fictitious victim of the use of mercuric nitrate in the felt-hat industry. This character depicts the vague, neurological and behavioral symptoms of human mercurial poisoning. Unfortunately, such gaiety surrounds the Hatter that the "specter of death" shrouding mercurials passes unrecognized. Kings, emperors, and nations alike have fallen before the sinister specter. The deaths of Napoleon, Ivan the Terrible, and Charles II of England could have resulted from Hg-poisoning (Goldwater, 1971). And many nations have lost additional human life, as well as wildlife, to Hg-poisoning in the quest for a bigger market share.

Anthropomorphic redistribution of mercurials has ravaged a substantial portion of our planet. Japan, Sweden, Canada, the Great Lakes, Iraq, Pakistan, Ghana, and Guatemala form a partial listing of contamination sites and provide a historical basis from which to review some of the teleost studies. From 1953–1960, one-hundred-eleven people near Minamata, Japan died or suffered neurological injury from eating fish and shellfish from Hg-contaminated areas in the bay (Abelson, 1970). Of the one-hundred-eleven people, nineteen were congenitally defective babies of mothers who had eaten these foods. Inorganic Hg was used as a catalyst at a chemical plant, which released both the inorganic and methylated mercurials directly into Minamata Bay.

Consumption of Hg-contaminated fish and shellfish caused Minamata disease in humans and other vertebrates, including fish (Matida, *et al.*, 1972). In this disease, muscles become weakened, vision is lost, cerebral functions are impaired, paralysis occurs, and coma and death can result. Mercurials also pass the human placenta, are associated with embryonic and fetal nervous tissue, causing chromosomal disorders, neurological damage, and teratogenicity. In fact, fetal accumulation of Hg can provide some maternal protection against the detrimental effects of Hg.

In Sweden and Canada, mercurials (e.g. PMA, phenyl mercuric acetate) were used as slimicides to keep machinery from fouling and to prolong the life of wood fibers in the paper and pulp mill industries (MacLeod and Pessah, 1973). This practice has been banned in Sweden since 1966, but residual contamination continues to pose a problem, as illustrated by studies ten or more years later. Other major sources of Hg are the chloralkali industry and the electric generating industry (Goyer, 1986). In fact, fossil fuels may contain as much as 1 ppm Hg and release about 5000 tons of Hg annually when burned.

In Iraq, Pakistan, Ghana, and Guatemala, thousands of cases of Hg poisoning have resulted from human consumption of seed treated with seed dressings (e.g. CH_3Hg^+), despite warnings that the grain was poisonous (Abelson, 1970; Goldwater, 1971; Klaassen, 1985). Utah (Smith, *et al.*, 1974), Wisconsin (Kleinert and Degurse, 1972), the Elbe River in Germany (Köhler, 1989), and Maine (Hamilton and Haines, 1989) are other areas of mercurial poisoning. Teleosts of virtually every ecosystem on the Earth have elevated levels of Hg; even those in the deeper reaches of oceans are not shielded from this deadly specter (Barber, *et al.*, 1972; Williams and Weirs, 1973).

Environmental Exposures

The focus for this section is mercurial methylation and the identification of areas of severe contamination world-wide.

Methylation: Organic, rather than inorganic, mercurials are the most hazardous to fish; therefore, consideration of environmental methylation of mercurials is a necessary part of any review of Hg-poisoning in fish. Several published studies evaluate methylation of Hg^{+2} in the natural environment or components thereof (Wood, *et al.*, 1968; Jensen and Jernelöv, 1969; Olson and Cooper, 1974). Olson and Cooper (1974) study methylation of Hg^{+2} in an intertidal zone of central San Francisco Bay. Wire mesh cylinders are embedded in the sediment, and sediment from inside each cylinder is removed at a deep zone (6 cm) and at a shallow zone (surface). To prepare sediment concentrations of 10 and 100 ppm Hg, $HgCl_2$ is added. On day 28, methylation is greatest in the deep zone—13.6 ppb is formed *in situ*, compared with 1.4 ppb formed at shallow depths (at 10 ppm Hg). At 100 ppm Hg, levels of 0.0 ppb and 7.0 ppb Hg are reported on day 28 for shallow and deep sediments, respectively, compared with 28.3 and 37.6 ppb on day 18. The reason for reduced methylation of Hg^{+2} in the shallow zone is unknown, but greater losses of CH_3Hg^+ or higher rates of degradation of CH_3Hg^+ may exist at the surface.

Wood *et al.* (1968) evaluate the transfer of methyl groups to Hg^{+2} using extracts of a methanogenic bacterium isolated from a symbiotic mixed culture obtained from canal mud at Delft, Holland. Methyl groups can be transferred from methylcobalamine (CH_3-Co-5,6-dimethyl benzimidazolylcobamide) to Hg^{+2} to form CH_3Hg^+ in biological systems by both enzymatic and nonenzymatic reactions. Anaerobic conditions enhance methyl transfer, as does the presence of large numbers of bacteria which are capable of synthesizing alkylcobalamines, and the presence of increased levels of nutrients (sewage pollution). Factor III (the prosthetic group, R-Co-5OH benzimidazolylcobamide), hydrogen as an electron source, and ATP are required for enzymatic, methyl transfers.

Jensen and Jernelöv (1969) treat bottom sediments from Lake Langsjön near Stockholm with inorganic Hg^{+2} in two experiments. Both CH_3Hg^+ and $(CH_3)_2Hg$ are produced in sediments (and rotten fish). Dead fish containing 125 μg CH_3Hg^+ form $(CH_3)_2Hg$ from CH_3Hg^+. The formation of volatile $(CH_3)_2Hg$ (BP: 94°C) could contribute to redistribution of Hg from aqueous industrial sources. Addition of 10 or 100 ppm Hg^{+2} to bottom sediments, results in formation of CH_3Hg^+ in progressively increasing levels until about 7 d, when levels stabilize at about 50 and 100 ppb, respectively. If 100 ppb of Hg^{+2} is added to bottom sediments and incubated 7 d, approximately 110 ppb CH_3Hg^+ is formed, but higher and lower levels of added Hg^{+2} result in production of less CH_3Hg^+. Both elevated background CH_3Hg^+ and higher biological activity of sediments probably result in greater amounts of CH_3Hg^+ after 10 d in sediments in laboratory studies, than for sediments from Lake Langsjön.

These three studies show that Hg^{+2} can be methylated to $(CH_3)_2Hg$ and CH_3Hg^+ and that a number of factors control the amount of methylation (Wood, *et al.*, 1968; Jensen and Jernelöv, 1969; Olson and Cooper, 1974). Therefore, the longer divalent inorganic mercurials remain in the environment, the more problematic Hg contamination becomes because of the lipophilicity of the product—organic alkylmercury salt. Increased lipophilicity of this organic causes marked increases in accumulation of Hg and Hg-induced toxicity, as explained in the respective following sections.

Contamination Sites: Published results warn of upcoming problems with Hg-poisoning in humans. In years to come, fish flesh could be banned from many of the world markets because of the ubiquity, lengthy half-life, and the insidiousness of the biological effects (e.g. teratogenicity) of Hg. In fact, even now the consumption of fish flesh is like a game of Russian roulette: the risk of Hg-poisoning increases with every bite of fish from some areas. The "safely level" for Hg in sea foods (0.5 ppm) is exceeded in many places in the world (Table 1). The World Health Organization provisional maximum intake for Hg in humans is 0.3 mg/person/wk, of which not more than 0.2 mg should be CH_3Hg^+ (MacKay, *et al.*, 1975). Since the majority (often 100%) of Hg in muscle is CH_3Hg^+ (Johnels, *et al.*, 1967; Bache, *et al.*,

1971; Fimreite and Reynolds, 1973; Olson, *et al.*, 1973b; ; Westöö, 1973; Olson, *et al.*, 1975; Pentreath, 1976a), little safety margin exists.

Table 1. Representative skeletal muscle levels of Hg. Numbers in parenthesis signify wet weight (w) or no designation of the wet or dry weight basis of analysis (-). In several cases, numerical values are taken from figures, and are therefore, approximate.

Common Name *Species*	**[Hg]-skeletal muscle (ppb)**	**Location**	**Reference**
Black marlin *Makaira indica*	500–16,500 (w)	NE Australia	Makay *et al.*, 1975
Northern pike *Esox lucius*	60–6,100 (w)	Sweden, Finland, and Norway	Johnels *et al.*, 1967
Salmon - *Salmo salar* Seatrout - *Salmo ocla*	63–321 (-)	Sweden	Westöö, 1973
Flounder - *Platichthys flesus*	950 (w)	Germany	Köhler, 1989
Dogfish *Squalius acanthias*	50–1,960 (w)	British Columbia	Forrester *et al.*, 1972
Northern pike - *Esox lucius* Burbot - *Lota lota* Walleye - *Stizostedion vitreum* -	240–27,800 (-)	NW Ontario	Fimreite & Reynolds, 1973
Rockbass *Ambloplites rupestris*	100–2,700 (w)	Lake Ontario	Boulton & Hetling, 1972
Coho salmon *Oncorhynchus kisutch*	480 (-)	Lake Erie	Rottschafer *et al.*, 1971
White bass - *Morone chrysops*	600 (-)		
Morids - *Antimora rostrata*	220–800 (w)	Cape Hatteras (2500m)	Rottscafer *et al.*, 1971

The reason that consumption of fish flesh resembles a game of chance is that variability in Hg accumulation is pronounced for fish muscle—sometimes over two orders of magnitude (Wobeser, *et al.*, 1970; Rottschafer, *et al.*, 1971; Boulton and Hetling, 1972). Regardless of the stringency of the monitoring methodology or quality control, every fillet cannot be analyzed individually. So, one fillet could contain 0.4 ppm Hg and the next, 40.0 ppm. Such variation could be caused by phylogenetically distinct metabolic rates (Barber, *et al.*, 1972); size-related differences (Johnels, *et al.*, 1967; Bache, *et al.*, 1971; Hammond, 1971; Westöö, 1973; MacKay, *et al.*, 1975); sexual differences (Boulton and Hetling, 1972; Forrester, *et al.*, 1972); the presence of pesticides, organic wastes, or other elements (Fimreite and Reynolds, 1973; Smith, *et al.*, 1974; MacKay, *et al.*, 1975; Hamilton and Haines, 1989; Köhler, 1989); and/or depth (Williams and Weirs, 1973). Alkalinity and pH of water can determine Hg uptake levels in fish and can increase variability. Surface waters having an alkalinity less than 50 ppm and a pH of more than 7 contain fish having elevated Hg in muscle tissues (Kleinert and Degurse, 1972; Fimreite and Reynolds, 1973).

In evaluations of world-wide mercurial contaminates, Japan, Scandinavia, and the Great Lakes-Canada region are considered "hot spots." In these areas, Hg is released from chemical plants (Friberg and Vostal, 1972), pulp and paper mills (Johnels, *et al.*, 1967; Kleinert and Degurse, 1972; Fimreite and Reynolds, 1973), electric generating stations (Johnels, *et al.*, 1967), various industries (Köhler, 1989), and/or agricultural operations (Johnels, *et al.*, 1967). Environmental Hg-poisoning of fish is considered for each area separately. Table 1 provides Hg levels in muscle of various fish species in areas in Australia, Scandinavian countries, the Canada-Great Lakes region, as well as Cape Hatteras.

Japan: Minamata Bay is the site of the release of mercurial wastes from the mercurial catalysts mentioned earlier. Every form of life was altered, from the screaming cats which plunged into the surf, to crows which dropped dead in midair. Both fish-eaters and fish were affected. In fact, fish dying after a CH_3Hg^+ spill in Minamata Bay, Japan, accumulated about 20 ppm (McKim, *et al.*, 1976). In comparison, rainbow trout are known to accumulate 13 ppm (wet weight) after about 189 d exposures to 25 ppm Hg from the diet and to develop the symptoms of Minamata disease (Matida, *et al.*, 1972). Symptoms include a rolling swim, inability to stop in front of obstacles, and visual disturbances (Matida, *et al.*, 1972). Humans and other vertebrates show similar symptomatology, which is indicative of Hg-induced damage to the occipital cortex and the cerebellum.

Scandinavian Countries: One of the main sources of Hg in Scandinavian countries is mercurial fungicides used in mechanical pulp and paper mills to improve storage properties of wood fibers and to prevent slime formation in machinery of paper mills. Other sources of pollution include seed-dressings in agriculture, coal and oil combustion, electrical installations using Hg in rectifiers, and aerial fallout (Johnels, *et al.*, 1967; Ackefors, 1971). Mercury levels in northern pike (*Esox lucius*) are used as an indicator of environmental Hg pollution because of stationary habitats, long life span, and wide distribution. Mercury levels in a single pike are 1000, 850, 780, and 640 ppb (wet weight) in the heart muscle, axial muscle, liver, and kidney, respectively; however, Hg can concentrate to levels over 6000 ppb in axial muscle (Table 1) (Johnels, *et al.*, 1967). About 60–90% of Hg is organic. Concentration factors up to 3000 (water to 1-kg pike) are reminiscent of radioisotopic enrichment. At tissue levels above 100 ppb and below 1000 ppb, a correlation exists between Hg content of axial muscle and body weight.

Another Scandinavian study shows that Hg levels in muscle increase roughly with age (Westöö, 1973). Salmon (*Salmo salar*) and sea trout (*Salmo ocla*) show no age dependence in the proportion of CH_3Hg^+ to total Hg in skeletal muscle. However, total levels of Hg in muscle generally increase with age in both species (0.063–0.321 ppm) in 1–7 yr old salmon and 1–2 year old sea trout. About 93% of total Hg is CH_3Hg^+. In comparison, visceral Hg levels are 0.061–0.469 ppm, with 30–60% of Hg in visceral in the form of CH_3Hg^+. Viscera make up 10–25% of fish weight and cannot depress the CH_3Hg^+ percentage of any whole fish samples below 67% (from the average of 93%).

More than twice the percentage of Hg is accumulated from dietary sources by an upper level predator (e.g. pike) in Sweden than by a bottom feeder (Jernelöv and Lann, 1971). Pike (*Esox lucius*) and bottom feeders consumed by pike are collected from two Swedish lakes and the Delangeran River for evaluation of Hg levels. Despite cessation of discharge of Hg several years prior to sampling, about 60% of Hg continues to be accumulated through the food chain in pike. For bottom feeders, less than 25% of accumulated Hg comes from the diet. Nonetheless, dietary accumulation of Hg produces a basic level of Hg in predators—a level of the same magnitude as that in prey. To this basic level of Hg is added that from aqueous Hg uptake. Therefore, in short limnetic food chains in River Delangeran, 6.00, 3.00, and 0.07 ppm Hg are accumulated in northern pike, bottom feeders consumed by pike, and benthic fauna consumed by bottom feeders, respectively. Bottom feeders accumulate about 2.9 ppm from the water and 0.07–0.11 ppm through the food chain. Pike accumulate 1.5–3.0 ppm from water, and 3.0–4.5 ppm from the food chain.

Great Lakes and Canada: Emissions from chloride plants and slimicides used in Canada and the Great Lakes leave residual Hg-contamination (Rottschafer, *et al.*, 1971; Boulton and Hetling, 1972; Forrester, *et al.*, 1972; Fimreite and Reynolds, 1973). Six fish species are collected July–August, 1970, from northwestern Ontario (Wabigoon Lake, as well as Tetu, Rowdy, and Sumach lakes) (Fimreite and Reynolds, 1973). The main source of Hg in this study is reported to be Dryden Chemicals, a chloride plant releasing aerial emissions for 8 yr prior to collection (1962–1970). Slimicide use at the Dryden pulp mill and discharge of organic wastes from that mill also contribute to Hg pollution. Mercury levels in skeletal muscle of fish decrease with distance from Dryden Chemicals from about 16 ppm for fish in Clay Lake to about 2 ppm Hg for fish in Tetu Lake (322 km downstream from Dryden). Elevated Hg levels in skeletal muscle (0.55 to 27.8 ppm, with 85–100% methylated) are found in fish inhabiting waters ranging from 17–50 ppm alkalinity (as $CaCO_3$) and having a pH slightly higher than 7.0. Maximum levels are found in northern pike (*Esox lucius*), burbot (*Lota lota*), and walleye (*Stizostedion vitreum*)—27.8, 24.8, and 19.6 ppm, respectively, at 80–97 km from Dryden (Table 1). "Uncontaminated" lakes contain fish having more than 0.2 ppm Hg and less than 1.00 ppm Hg in muscle. Most importantly, no game fish survive in the Wabigoon River; and fish-eating mammals such as the otter (*Lutra canadensis*) and mink (*Mustela vison*) have "totally disappeared" (Fimreite and Reynolds, 1973).

Forrester (1972) evaluates Hg content in dogfish (*Squalius acanthias*) from waters off British Columbia. More than 0.5 ppm Hg in dogfish muscle was discovered in shipments prepared for export for human consumption. Four sampling stations in the Strait of Georgia are sampled May to June of 1971. Both intrauterine embryos and adult fish are sampled. Embryos accumulate from 0.01 to 0.05 ppm Hg (wet weight), with average levels of 0.022 ppm. Maternal rather than embryonic Hg accumulation in dogfish (Forrester, *et al.*, 1972) differs from that of humans, in which the fetus protects the parent from excess Hg by accumulating more Hg than the parent (Friberg and Vostal, 1972). Also, in adult dogfish over 65 cm, males show higher Hg concentrations in muscle than females; Hg levels reach 0.5 ppm in 72 cm males and 77 cm females (Forrester, *et al.*, 1972). Maximum observed concentrations are about 2.0 ppm. Although the reason for the sex-related difference in Hg levels is unknown, males probably grow more slowly—resulting in a less pronounced growth dilution effect on Hg uptake in males than in females.

Rottschafer *et al.* (1971) observe that the liver, kidney, and skeletal muscle of coho salmon (*Oncorhynchus kisutch*) from Lake Erie accumulate 0.87, 0.36, and 0.48 ppm Hg, respectively. Whole body levels from various species of fish from Lake Erie, Lake St. Clair, Lake Huron, Bonham State Park Lake, and Sandy Bottom Lake vary from 0.05 to 1.56 ppm Hg. Largemouth bass, whitefish, and yellow perch concentrate the lowest Hg levels, while walleye pike accumulate the highest.

Bache *et al.* (1971) measures total Hg and CH_3Hg^+ in lake trout (*Salvelinus namaycush*) from Cayuga Lake in Ithaca, New York, in 1970. Fish are of known age since they are tagged and stocked annually as fingerlings. Total Hg and CH_3Hg^+ increase with fish age for 1–12 yr old lake trout—possibly a reflection of exposure time. Levels of Hg in whole fish range from from 0.19–0.66 ppm (wet weight) in trout which have not been eviscerated. Percent of total Hg as CH_3Hg^+ ranges from 30.8–101.9%. Also the proportion of Hg as CH_3Hg^+ appears to be lower in younger fish. However, neither total Hg or CH_3Hg^+ are correlated with sex of fish.

Boulton and Hetling (1972) use data from thirty-three hundred fish from New York State from collections made from April to November 1970, at about one hundred fifty locations. Lakes Erie, Ontario, and Champlain, as well as the St. Lawrence River, are included as collection sites. Eight species have average Hg levels in skeletal muscle greater than 0.5 ppm—lake trout (0.73 ppm, wet weight), walleye (0.69 ppm), white perch (0.63 ppm), channel catfish (0.62 ppm), rock bass (0.61 ppm), northern pike (0.59 ppm), smallmouth bass (0.55 ppm), and goldfish (0.51 ppm). Of species evaluated, rainbow trout, lake whitefish, and crappie accumulated the lowest levels of Hg—0.14, 0.17, and 0.20 ppm, respectively. Females from seven of eight species accumulate about 20% higher muscle levels of Hg than do males. Only white perch males accumulate 8.7% more Hg in muscle than females. Eleven muscle samples of fish from Onondaga Lake contain an average of 3.21 ppm Hg, with maximum levels of 8.20 ppm. In Lake Ontario muscle Hg levels vary from a high of 0.88 ppm in white perch to a low of 0.23 ppm in black crappie (Boulton and Hetling, 1972). From these studies involving Canada and the Great Lakes, it is obvious that some of the most severe contamination problems arise from the release of mercurial slimicides and the emissions from chloride plants.

Additional Areas: Other areas containing endemic fish species contaminated with mercurials include Australia (MacKay, *et al.*, 1975), Utah (Smith, *et al.*, 1974), Wisconsin (Kleinert and Degurse, 1972), the Elbe River in Germany (Köhler, 1989), and Maine (Hamilton and Haines, 1989). MacKay (1975) studies black marlin (*Makaira indica*) from apparently unpolluted waters off northeastern Australia. Liver and skeletal muscle average 10.4 and 7.3 ppm Hg (wet weight), respectively (Table 1). The liver to muscle ratio is 1.4 for Hg, indicating that fish continue to accumulate Hg from this environment. Less than 50 g of marlin muscle can be consumed by humans per wk "safely," based on the commonly accepted maximum for Hg in sea foods (0.5 ppm). Concentrations of Cd, Cu, Pb, and As are relatively low in muscle. However, Se levels are elevated in liver and muscle—averaging 5.4 and 2.2 ppm, respectively. Only Hg and Se concentrations in muscle are correlated to fish size, suggesting a causal relationship between these elements. Selenium may protect black marlin from the damaging effects of accumulated Hg. The source of elevated metal/metalloid levels in northeastern Australia is not known.

Even deep sea fish are not shielded from Hg-poisoning (Hammond, 1971; Barber, *et al.*, 1972; Miller, *et al.*, 1972; Williams and Weirs, 1973), although studies indicate that during the past 81–112 yr, no substantial increase has occurred in Hg levels in tissues from fish collected in deeper areas of the ocean. Mercury in the deeper parts of oceans (below 100 m) is apparently derived primarily from natural sources; whereas, Hg in rivers, streams, lakes, and the upper 100 m of the ocean is believed to originate from mercurial wastes. Earlier reports show a 0.03–2.00 ppb range for Hg in oceanic waters; however, a value of 0.1 ppb may be a better approximation. Therefore, Japan, Sweden, the Great Lakes, and Canada are areas in which mercurial contamination is or has been the most severe; yet, many other areas of the Earth show marked elevations in environmental Hg.

Toxicity

General Information: Table 2 compares the toxicity of various species and life stages (or sizes) of fish. In terms of 96-h LC50 values, the toxicity ranges two orders of magnitude from 11–1800 ppb Hg in the water. Apparently, marked species-specific differences in these values are phylogetically determined. Differences in uptake can be striking despite the similarity in dietary items and aqueous exposure levels for several species. For example, uptake (hence toxicity) of Hg in deep sea halosaurs, for example, have muscle levels one order of magnitude lower than the range for macrourids and morids (Barber, *et al.*, 1972). Phylogenetic distinctions in metabolic rate determine Hg uptake and, therefore, Hg toxicity.

Chemical form alters Hg-induced toxicity. At 10°C, in fingerling rainbow trout, CH_3Hg^+ is about seven times more toxic than Hg^{+2} if fish are exposed under similar conditions (MacLeod and Pessah, 1973; Wobeser, 1975). This toxicity difference is related to the difference in the accumulation of CH_3Hg^+ compared with Hg^{+2}. Since, uptake of CH_3Hg^+ is greater than that of Hg^{+2} (Olson, *et al.*, 1973b; Wobeser, 1975; Lock, *et al.*, 1981), Hg uptake modulates Hg-induced toxicity.

Size also alters Hg-induced toxicity (Gill and Pant, 1981). Smaller fish (2 cm) are more susceptible to Hg toxicity than are larger fish (5 cm) (Table 2). These smaller fish are ten-times more susceptible to Hg than larger fish at lower water hardness (4.5 ppm) and three-times more susceptible at higher water hardness (398 ppm). Since smaller fish show higher metabolic rates than larger fish, smaller fish accumulate Hg more rapidly and show increased Hg-induced toxicity (Reinert, *et al.*, 1974).

Temperature alters Hg-induced toxicity (*via* uptake) (Rehwoldt, *et al.*, 1972; MacLeod and Pessah, 1973; Reinert, *et al.*, 1974; Rodgers and Beamish, 1981). Metabolism increases at higher temperatures, increasing the level of Hg accumulated (Reinert, *et al.*, 1974) and increasing susceptibility of fish to Hg (MacLeod and Pessah, 1973). As temperature is increased from 5°C to 20°C, active metabolism also increases almost two-fold and toxicity increases two-fold (MacLeod and Pessah, 1973) (Table 2). At 5–10°C, metabolic rate increases with Hg level in the water in a constant fashion. However, at 20°C, the active metabolism is considerably reduced if Hg levels in the water are increased from 0.1 to 0.2 ppm Hg^{+2}. Apparently, higher temperatures (i.e. 20°C for fingerling rainbow trout) potentiate the depressant action of Hg^{+2} on active metabolism (MacLeod and Pessah, 1973). Therefore, Hg-induced toxicity is related to differences in species, chemical form, specimen size, and temperature, as well as other factors.

Table 2. Mercury toxicity in various fish species based on 96-h LC50 data. Corresponding water hardness, pH, and temperature are provided.

Genus, species, and life-stage or size	Chemical Form	Hg 96-h LC50	Hardness (ppm $CaCO_3$)	Temp. (°C)	pH	Reference
Rainbow trout *Salmo gairdneri*	CH_3Hg^+					Wobeser, 1975
fry		24	101	10	8.6	
fingerlings		42	101	10	8.6	
Rainbow trout	Hg^{+2}	400	90	5	7.5–7.8	MacLeod &
Salmo gairdneri		280	90	10	7.5–7.8	Pessah,1973
fingerlings		220	90	20	7.5–7.8	
Freshwater barb *Barbus conchonius*	Hg^{+2}					
2 cm fish		11	4.5	-	-	Gill & Pant,
2 cm fish		68	398	-	-	1981
5 cm fish		101	4.5	-	-	
5 cm fish		182	398	-	-	
Freshwater barb *Barbus conchonius*	Hg^{+2}	181	398	14	7.3	Gill & Pant, 1985
Striped bass *Morone saxatilis*	Hg^{+2}	90	55	28	8.0	Rehwoldt *et al.*, 1972
Banded killifish *Fundulus diaphanus*	Hg^{+2}	110	55	28	8.0	Rehwoldt *et al.*, 1972
American eel *Anguilla rostrata*	Hg^{+2}	140	55	28	8.0	Rehwoldt *et al.*, 1972
Carp *Cyprinus carpio*	Hg^{+2}	180	55	28	8.0	Rehwoldt *et al.*, 1972
White perch *Morone americana*	Hg^{+2}	220	55	28	8.0	Rehwoldt *et al.*, 1972
Pumpkinseed *Lepomis gibbosus*	Hg^{+2}	300	55	28	8.0	Rehwoldt *et al.*, 1972
Channa punctatus (freshwater fish)	Hg^{+2}	1800	160	20	7.4	Sastry & Sharma, 1980a

Reproductive Toxicity: Mercury is extremely embryotoxic and teratogenic. A three-generation study involving brook trout exposed to CH_3Hg^+ in water results in branchiostegal-opercular deformities in all second-generation embryos (McKim, *et al.*, 1976). Exposures of this same generation to 2.93 ppb aqueous Hg results in no development beyond the neural keel and death of all embryos. Aqueous levels of inorganic Hg at 20–30 ppb Hg^{+2} reduce hatchability to 0% in ricefish embryos, and 15 ppb Hg^{+2}

decreases hatchability to 21% (Heisinger and Green, 1975). Moreover, killifish blastula succumb to 1.0 ppm Hg^{+2} before gastrulation (Weis and Weis, 1977). Mercury exposure causes anterior-ventral spinal curvatures of newly-hatched ricefish and killifish fry (Heisinger and Green, 1975; Weis and Weis, 1977). Also, killifish embryos show reduced axis formation, incomplete induction of the forebrain, partial convergence, synophthalmia (partial fusion of the optic cups with separate lenses formed), cyclopia (formation of a single orbit), and anencephaly (Weis and Weis, 1977). Therefore, embryotoxicity and teratogenicity are reported for at least three studies involving Hg poisoning of fish eggs (Heisinger and Green, 1975; McKim, *et al.*, 1976; Weis and Weis, 1977). Detailed evaluation of these studies provides additional features of Hg-induced reproductive toxicity.

McKim *et al.* (1976) expose three generations of brook trout (*Salvelinus fontinalis*) to CH_3Hg^+ at aqueous levels up to 2.97 ppb Hg (9°C, pH 6.9–7.6, and 45 ppm hardness, as $CaCO_3$). Yearling trout are exposed through spawning (the first-generation), offspring are exposed through spawning (the second-generation), and offspring from the second-generation are exposed to the early juvenile stage (the third-generation). The 144-wk study shows that the most sensitive stage is the embryo-alevin-juvenile stage. Few effects are observed at 0.03, 0.09, and 0.29 ppb Hg; only the ratios of total weight of young to weight of females are reduced 32–36% in second- and third-generation trout. Other Hg-induced effects are noted. A 39-wk exposure of yearling trout to 2.93 ppb Hg produces trunk deformities, muscle spasms, and 88% mortality in first generation trout. No spawning occurs in first-generation fish exposed to 2.90–2.97 ppb Hg or in second-generation fish exposed to 0.91–2.97 ppb Hg. Hatchability is decreased in second-generation trout exposed to 0.93 ppb Hg. In second-generation trout, 0.91 ppb Hg causes brancheostegal-opercular deformities, as indicated, and kills all but one female in 108-wk exposures. Therefore, the maximum acceptable toxicant concentration (MATC) for brook trout is said to be 0.29 and 0.93, and the application factor (MATC/96-h LC50) is 0.004–0.013 (McKim, *et al.*, 1976).

The Heisinger and Green (1975) study shows that Hg is accumulated by embryos to high levels, is embryotoxic, and causes severe defects in embryos. Japanese ricefish (*Oryzias latipes*) embryos are exposed to 10–30 ppb Hg^{+2} for 16 d at 26°C and pH 6.9. The concentration factors for embryos at 16 d exposures to 10, 15, 20, and 30 ppb Hg^{+2} in the water are 1,600,

1,900, 2,700, and 1,900, respectively, or 16, 29, 54, and 56 ppm accumulated Hg. Zero and twenty-one percent hatchability are observed in 20–30 ppb and 15 ppb, respectively, but embryos in 10 ppb show normal hatching (47–58%). Fry from embryos exposed to 15 ppb Hg^{+2} lack the ability to move caudal fins but can move pectoral fins rapidly. Tails are curved in an anterior and ventral direction, such that fish are shaped like the letter "C"and remain on their sides on the bottom of petri dishes. The 79% of embryos which do not hatch in 15 ppb Hg^{+2} (as well as those unhatched in 20 or 30 ppb) show hemorrhaging, blood vessel deterioration, and loss of circulating blood cells. Control embryos and those hatching in 10 or 15 ppb solutions do not show these abnormalities.

Severe Hg-induced embryotoxicity and teratogenicity also are reported for blastula of killifish (*Fundulus heteroclitus*) exposed to 0.01–10.0 ppm Hg^{+2} at 20°C Weis and Weis (1977). Exposure of blastula to 1.0 ppm Hg^{+2} kills all embryos before gastrulation. Axis formation is reduced in about 30% of embryos if blastula are exposed at stage 12 to 0.01–0.03 ppm Hg^{+2}, and 60% of embryos if blastula are exposed to 0.1 ppm. In comparison, no reduction in axis formation occurs if stage 14 (late blastula) embryos are exposed to 0.01–0.03 ppm Hg.

Defects in the development of the forebrain also are more severe if blastula are exposed to 0.01–0.10 ppm Hg^{+2} at stage 12, rather than stage 14. Most defects reflect altered induction. With incomplete induction of the forebrain, the lateral optic vesicles are brought closer together, producing partial convergence, synophthalmia, cyclopia, and anencephaly. If exposed at stage 14, negligible malformations occur at 0.01 and 0.03 ppm Hg. Exposure to 0.1 ppm Hg^{+2} produces 22.5% defects in the forebrain development in the stage 14 blastula and 24% defects in the stage 12 blastula. The stage 12 blastula may be inherently more sensitive; moreover, exposure to Hg^{+2} at an earlier life stage would increase the total amount of accumulated Hg during embryogeny by increasing exposure time. Greater embryotoxicity and more severe defects result under these conditions. Therefore, Hg causes undeniably severe reproductive toxicity at extremely low levels, compared with other elements.

Accumulation and Retention

Mercurial enrichment in fish tissues is similar to that of radioisotopes (Johnels, *et al.*, 1967), with up to 27,800 ppb reported in skeletal muscle (Fimreite and Reynolds, 1973). However, liver generally accumulates more Hg than muscle in situations in which Hg influx is occurring (Wobeser, *et al.*, 1970; Jernelöv and Lann, 1971; MacKay, *et al.*, 1975; Olson, *et al.*, 1978). In such a case, the ratio of the Hg level in the liver to that in muscle is greater than or equal to 1. In contrast, depuration is characterized by a liver to muscle ratio of less than or equal to 0.5. Regrettably, one cannot discern whether fish accumulate or release Hg when the liver to muscle ratio is greater than 0.5, but less than 1.0 ppm. Even more regrettable is the fact that only 0.02 g of fish muscle exceeds the weekly 0.5 ppm "safety level" for muscle containing 27.8 ppm Hg. One 20-mg (0.0007 ounce) fillet per meal hardly satisfies daily human protein requirements, and much less so for the weekly allotment.

As indicated, inter- and intra-species specific variation in Hg uptake can be altered by phylogenetic differences in metabolism (Barber, *et al.*, 1972); size-related differences (Johnels, *et al.*, 1967; Bache, *et al.*, 1971; Hammond, 1971; Westöö, 1973; MacKay, *et al.*, 1975); sexual differences (Boulton and Hetling, 1972; Forrester, *et al.*, 1972); the presence of pesticides, organic wastes, or other elements (Fimreite and Reynolds, 1973; Smith, *et al.*, 1974; MacKay, *et al.*, 1975; Hamilton and Haines, 1989; Köhler, 1989;); depth (Williams and Weirs, 1973); alkalinity; and pH (Kleinert and Degurse, 1972; Fimreite and Reynolds, 1973). Some additional factors include mucous binding of Hg (McKone, *et al.*, 1971; Burrows and Krenkel, 1973; Olson and Fromm, 1973a; Varanasi, *et al.*, 1975; Lock, *et al.*, 1981), demethylation (Imura, *et al.*, 1972; Pan, *et al.*, 1973; Olson, *et al.*, 1975; Olson, *et al.*, 1978), temperature (MacLeod and Pessah, 1973; Reinert, *et al.*, 1974; Rodgers and Beamish, 1981), exposure level and time (McKone, *et al.*, 1971; Reinert, *et al.*, 1974; Heisinger and Green, 1975; Olson, *et al.*, 1975), source of Hg (McKone, *et al.*, 1971; Weisbart, 1973; Heisinger and Green, 1975; Pentreath, 1976b; Phillips and Buhler, 1978), chemical form of Hg (MacLeod and Pessah, 1973; Pentreath, 1976b; Wobeser, 1975; Lock, *et al.*, 1981), and life stage of the test species (Heisinger and Green, 1975; Wobeser, 1975; Pentreath, 1976a). Other factors also play a role in the accumulation (and retention) of Hg in fish.

Mucous Production: Mucous secretion in teleosts is stimulated by the presence of Hg in water (McKone, *et al.*, 1971; Burrows and Krenkel, 1973; Olson and Fromm, 1973a; Varanasi, *et al.*, 1975; Lock, *et al.*, 1981). Presumably, higher aqueous levels of Hg increase mucous production and increase efficiency of chelation (or absorption) of Hg. Up to 79% of total Hg accumulated by goldfish is present in superficial mucus (McKone, *et al.*, 1971), although lower percentages (16–45%) are reported in bluegills and plaice (Burrows and Krenkel, 1973; Pentreath, 1976a). Mercury cannot be dialyzed from the mucus-Hg complex (Varanasi, *et al.*, 1975), whatever the nature of the interaction required to form the complex.

The divalent inorganic mercurial (i.e. Hg^{+2}) is more readily trapped by mucus than is the alkylmercury salt (i.e. CH_3Hg^+) (Lock, *et al.*, 1981). In fact, mucus concentrates four times as much of the inorganic as the organic mercurial. The reason for the greater potential for mucous ion resin trapping of Hg^{+2} is not known, but may involve the decreased lipophilicity of the inorganic mercurial, compared with that of the organic form.

One of the most interesting findings regarding Hg-induced mucous ion resin trapping is that interaction between Hg and mucus increases the fluidization of mucus (Varanasi, *et al.*, 1975). Varanasi *et al.* (1975) find that epidermal mucus from rainbow trout (*Salmo gairdneri*) accumulates about 1.0, 3.7, 8.0, or 7.8 ppm Hg after fish are exposed to 0.10, 0.25, 0.50, or 1.00 ppm aqueous Hg^{+2}, respectively (10°C, pH 5.8). Epidermal mucus is produced at 7.3 mg/g fish and contains 9.3 mg protein/g mucus. Both mucous production and protein levels in mucus are fairly constant. A 24-h depuration using living fish results in about 97% loss of Hg from epidermal mucus, but dialysis experiments of mucus show no significant loss of Hg from mucus *per se*. Therefore, the substantial loss of Hg from mucus following the 24-h depuration of trout is probably a consequence of sloughing of the Hg-mucus complex and synthesis of additional mucus.

Mercury exposure of trout increases the fluidization of mucus (Varanasi, *et al.*, 1975). Rotational correlation times (τ) for each electron spin resonance (ESR) spectra normally are 2×10^{-9} s, which is indicative of a partially immobilized probe (N,N-dimethyl-N-dodecyl-N-tempoylammonium bromide) in a relatively rigid mucous matrix. After exposure of fish to 0.1–1.0 ppm inorganic Hg in the water; however, τ is about $0.30–0.85 \times 10^{-9}$ s, suggesting that Hg increases mucous fluidity. This fluidization is probably due to direct interaction between mucus and Hg; perhaps Hg interacts with polar sites of mucous glycoproteins. Rheological properties (water flow patterns) would change as a result and could increase sloughing of the Hg-mucus complex from the body surface, thereby stimulating increases in the synthesis and/or release of mucus *de novo*.

Another study shows less rapid loss of Hg from fish as a consequence of the sloughing of mucus (Burrows and Krenkel, 1973). Bluegills (*Lepomis macrochirus*) are exposed at 24°C to 0.2–0.3 ppb CH_3Hg^+. About 16–45% of CH_3Hg^+ is found in the mucous coat, to account (at least in part) for the inverse relationship between CH_3Hg^+ accumulation and fish weight. Rapid loss of about 40% Hg from fish is observed, with a biological half time of 38.5 d, followed by slow loss of Hg with a half time of 130 d.

Initial rapid loss of Hg is due primarily to sloughing of the mucous layer (Burrows and Krenkel, 1973), as shown by Varanasi *et al.* (1975) who, however, obtain almost 100% loss of Hg from epidermal mucus of rainbow trout within 1 d of the initiation of depuration. Varanasi and colleagues (1975) obtain more complete sloughing of Hg because they use an inorganic divalent mercurial—effectively trapped by mucus, as shown by Olson and Fromm (1973a). Burrows and Krenkel use the alkylmercury salt which probably is neither completely trapped (i.e. covalently bound) by mucus nor capable of forming stable deposits. In addition, Varanasi *et al.* (1975) use higher Hg concentrations in the water (0.1–1.0 ppm) compared with Burrows and Krenkel (0.2–0.3 ppb). At a higher aqueous level of Hg^{+2}, greater efficiency of Hg^{+2}-mucus interaction is thought to occur.

Demethylation: An interesting result of the Burrows and Krenkel (1973) study is that the percentage of Hg as CH_3Hg^+ falls rapidly within 3 wk in the liver and kidney, but not in the whole body of bluegills (*Lepomis macrochirus*). Tissues are collected from whole fish (*sans* liver, kidney, and roe) which had been exposed in vivo 96 h to 0.2–0.3 ppb CH_3Hg^+, moved to clean water, and analyzed every 1–2 wk thereafter for 94 d using whole-body counting techniques. Levels of CH_3Hg^+ account for about 73% of total Hg residues throughout. Therefore, percent of total Hg as CH_3Hg^+ in the whole fish (and in the liver, kidney, and roe) is independent of time. However, the proportion of CH_3Hg^+ in the liver and kidneys falls rapidly in the first few weeks after the initial exposure, ultimately stabilizing at about 10%. This result suggests that demethylation is occurring in these and other organs in the bluegill.

However, demethylation may not occur in the liver of rainbow trout. In one study (Olson, *et al.*, 1978), rainbow trout (*Salmo gairdneri*) are exposed at 12°C to a single 24-h pulse of CH_3Hg^+ containing both the ^{14}C label and ^{203}Hg label to determine whether demethylation of CH_3Hg^+ occurs. One day after exposure, gills have accumulated ten times as much CH_3Hg^+ as almost any other tissue (of those evaluated). The accumulated CH_3Hg^+ is lost most rapidly from the gill and least rapidly from the liver. Liver to muscle ratios of Hg are 10.6, 4.4, 3.7, and 2.3 at 24 h, 2 wk, 4 wk, and 6 wk, respectively.

Evaluation of the stability of the CH_3-Hg^+ bond shows that Hg probably is demethylated before being bound to metallothionein in the rainbow trout gill, but not before being bound to metallothionein in the liver (Olson, *et al.*, 1978). In the gill, demethylation of CH_3Hg^+ is supported by the presence of a prominent ^{203}Hg-containing metallothionein peak, with very little radioactivity due to ^{14}C, 6 wk after the 24-h pulse. The fact that the ratio of ^{203}Hg to ^{14}C in gill metallothionein is about ten seems clear proof that Hg is demethylated before it is bound to metallothionein in the gill. However, only 20% of total Hg in the soluble fraction is bound to metallothionein in this gill fraction.

In contrast, because the hepatic ratio of ^{203}Hg and ^{14}C in metallothionein is constant and consistent with the ratio in the whole liver, CH_3Hg^+ is probably bound to metallothionein intact in the liver. Nonetheless, demethylation could occur with both products remaining within the liver. The greatest amounts of Hg-metallothionein are found in the liver cytosol, indicating that the liver has great potential for autodetoxification of Hg by metallothionein synthesis and binding. Ratios of ^{203}Hg to ^{14}C activity in various tissues demonstrate redistribution of demethylation products, suggesting sites of demethylation (and perhaps excretion). Of tissues evaluated, the liver, kidney, gill, gastrointestional tract, and spleen are the most probable sites for demethylation and/or excretion. In general, the high molecular weight fraction contains CH_3Hg^+ (^{203}Hg and ^{14}C); whereas, the metallothionein fraction is mostly inorganic (^{203}Hg) in the gill and kidney and CH_3Hg^+ (^{203}Hg and ^{14}C) in the liver. Reasons for this difference in organ binding of metallothionein to methylated or demethylated mercurials have not been elucidated.

Saturation of metallothionein with Hg and the "Spill Over" of Hg to the enzyme-containing pool (high molecular weight proteins) may mark the point at which pathological and other effects occur (Brown and Parsens, 1978). Chum salmon (*Oncorhynchus keta*) are exposed for 72 d to 1 or 5 ppb Hg in water (Brown and Parsons,1978). At 1 ppb Hg, a limit appears to be reached in the amount of Hg which can bind metallothionein to form Hg-thionein. After 72 d at 5 ppb Hg, livers of fish show a prominent increase of Hg in high molecular weight pool. With increasing Hg exposure, decreases in total hepatic Cu and Zn reflect decreases in Cu and Zn in the enzyme-containing pool, of Cu in metallothionein, and of Zn in the low molecular weight cytoplasmic pool.

Fish exposed 72 d to 1 ppb Hg weigh 12.5 g, but those exposed 72 d to 5 ppb Hg weigh only 6.0 g (Brown and Parsens, 1978). This dramatic decrease in fish growth may be related to the Spillover Theory (Winge, *et al.*, 1975; Brown and Parsens, 1978). Essentially, Hg challenge can result in saturation of metallothionein; so that, Hg spills over to the enzyme pool and initiates pathological effects. Decreases in growth of fish may be due to Hg-induced deficiency effects on two essential metals, because levels of Cu and Zn decrease as Hg levels in the enzyme-containing pool increase. Therefore, effects on growth could be related to decreases in Cu and Zn—rather than to direct, Hg-induced, inhibitory toxic effects on enzymes.

Based on the Spillover Theory, the level of Hg exposure and the rate of Hg uptake ultimately determine toxic effects. For example, an extended exposure to low levels of Hg may result in high Hg tissue levels, but low toxicity since Hg is bound in a less toxic form to metallothionein. In contrast, an acute Hg exposure within a short period can result in the same or lower Hg tissue levels but higher toxicity since most Hg probably would spill over into the high molecular weight pool, with subsequent toxic effects (e.g. those on growth). Therefore, tissue levels of Hg *per se* may be of less importance than the distribution of Hg to low or high molecular weight pools in the cell. This phenomenon apparently does not hold for other metals/metalloids which lack a specific intracellular chelation system and have different chemical, physical, and biological properties than Hg. General application of the Spillover Theory to all metal and metalloids is a controversial issue.

At least three studies suggest that demethylating ability is species-specific (Imura, *et al.*, 1972; Olson and Fromm, 1973a; Pan, *et al.*, 1973). Using rainbow trout (*Salmo gairdneri*), Olson and Fromm (1973a) provide results which contrast the brachial demethylation of CH_3Hg^+ prior to metallothionein binding (Olson, *et al.*, 1978). No measurable Hg deposits are detected in gills of rainbow trout exposed 23 h to 0.05 ppm CH_3Hg^+; however, deposits are present for trout exposed 23 h to 0.25 ppm inorganic Hg^{+2} (13°C). Fish exposed to Hg^{+2} apparently trap Hg^{+2} in interlamellar mucus. In fact, goldfish mucus can retain 79% of total Hg^{+2} in mucus if living fish are held 3 h at 1 ppm Hg^{+2} (McKone, *et al.*, 1971). The gill cartilage also appears to concentrate Hg after Hg^{+2}, but not after CH_3Hg^+, exposure.

Controversial results, such as these could be the result of species-specific differences in the processing of Hg. In fact, rainbow trout may not be a good species to use for demethylation studies involving Hg, since the so-called "Hg-methylation factor" (a vitamin B_{12} derivative) is not present in the liver of this species (Imura, *et al.*, 1972). Imura (1972) compares amounts of CH_3Hg^+ formed when hepatic (but not brachial) homogenates are incubated with or without 2 mM Hg^{+2} at 37 °C for 5 h in the dark. Heat stability of the homogenate is evaluated using a 15 min, 120°C autoclaving step at 2 atm prior to addition of 2 mM Hg^{+2}. Livers of yellowfin tuna (*Thunnus albacares*) and albacore (*Thunnus alaunga*) methylate Hg^{+2}; whereas, those from rat, rabbit, pig, rainbow trout, rockfish, and mackerel do not.

To verify the *in situ* methylation of Hg^{+2} (rather than the release of previously accumulated CH_3Hg^+), thin-layer chromatographs are made of extracts of albacore liver homogenates incubated with $^{203}Hg^{+2}$. Both CH_3Hg^+ and $(CH_3)_2Hg$ are labeled, indicating that methylation does occur *in situ* in liver homogenates from albacore. However, chemical transmethylation does not occur (i.e. methyl groups are not transferred from CH_3Hg^+ to Hg^{+2}). Although the absence of Hg-methylation factor in livers of rainbow trout (Imura, *et al.*, 1972) does not preclude brachial demethylation in the same species, several studies suggest that demethylation of CH_3Hg^+ does not occur in the rainbow trout gill (Burrows and Krenkel, 1973; Olson and Fromm, 1973a).

Temperature: Thermal elevations increase metabolic rates, resulting in increased Hg uptake (MacLeod and Pessah, 1973; Reinert, *et al.*, 1974; Rodgers and Beamish, 1981). MacLeod and Pessah (1973) expose rainbow trout (*Salmo gairdneri*) fingerlings (9–28 g) to aqueous phenyl mercuric acetate (PMA, a mercurial slimicide) or inorganic Hg^{+2} at 5, 10, or 20°C (pH 7.5–7.8 and hardness 90 ppm, as $CaCO_3$). Active metabolic rates are 380, 440, and 730 mg oxygen/kg fish/h at 5, 10, and 20°C. At a level of 0.1 ppm Hg^{+2}, concentration factors in fish are 4, 10, and 22 for the respective temperatures. For comparison, at 10°C and a PMA level of 0.005 ppm (as total Hg), concentration factors of 80 to 100 result. Therefore, concentration factors for Hg (as with other elements) are related to temperature.

Active metabolic rate (oxygen consumption of fish held under conditions of maximum energy expenditure) is directly related to temperature but inversely related to Hg^{+2} level in the water (MacLeod and Pessah, 1973). Higher temperatures potentiate the "depressant action" of Hg on active metabolism. For example, at 20°C, a level of 0.1 ppm Hg^{+2} in the water reduces active metabolism from 730 to 658 mg oxygen/kg/h (10% reduction); whereas, at 10°C the same Hg^{+2} level produces a depression of 5% (Fig. 1). Therefore, active metabolism at the highest temperature (20°C) is reduced two-fold (from 5% to 10%) by doubling the Hg^{+2} exposure level. Reduced Hg uptake would result under conditions in which Hg depresses active metabolism.

The study of Reinert *et al.* (1974) produces results which may verify this postulate: at the highest temperature (15°C) and the longest exposure time (12 wk), Hg uptake seems to be decreased (Fig. 2). Yearling (31 g) rainbow trout (*Salmo gairdneri*) are exposed 12 wk to 234–263 ppt CH_3Hg^+ in water at 5, 10, or 15°C. Whole body Hg accumulation varies with water temperature (pH 7.4–7.9, hardness 242–292 ppm). Whole body concentrations at the three temperatures measure 1.2, 1.7, and 2.0 ppm (wet weight), respectively. After 12-wk exposures, whole body concentration factors are 4532, 6622, and 8049 at 5, 10, and 15°C. Significant differences exist in the Hg accumulated with each 5°C thermal increase. Increases in metabolic rate are presumed to cause increased Hg uptake, as established or suggested by other studies (Murphy and Murphy, 1971; MacLeod and Pessah, 1973). Interlaboratory differences in fish size, temperature, and other factors could explain the absence of a pronounced effect in the Reinert *et al.* (1974) study. Perhaps at even longer exposure times and slightly higher

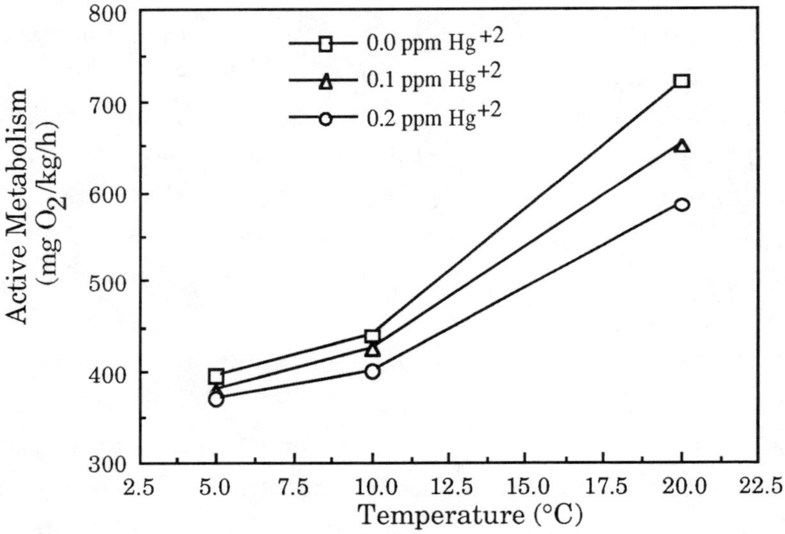

Figure 1. Temperature effects on active metabolism (oxygen use at maximum energy expenditure) for fingerling rainbow trout (*Salmo gairdneri*) exposed to 5, 10, or 20°C at three levels of Hg^{+2} (trace or 0.0, 0.1, or 0.2 ppm). Higher temperatures potentiate the depressant action of Hg^{+2} on active metabolism. (Adapted from MacLeod and Pessah (1973).)

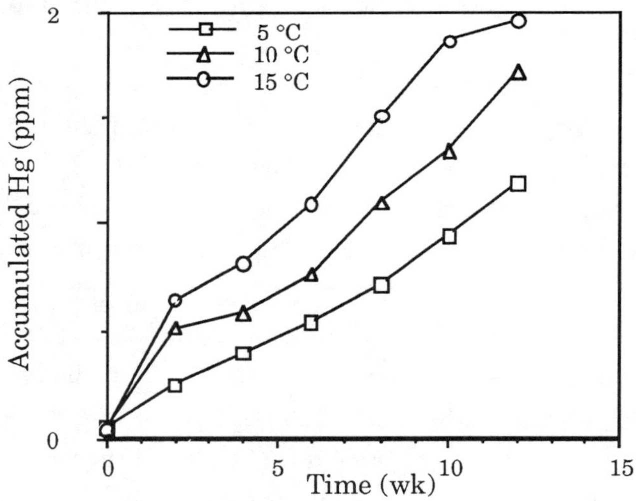

Figure 2. Whole body levels of Hg (in ppm, wet weight) accumulated from water containing 234–263 ppt CH_3Hg^+ at three temperatures over a 12-wk period. Mean and one standard deviation values are presented for yearling rainbow trout (*Salmo gairdneri*) at pH 7.4–7.9 and a hardness of 242–292 ppm. (Numerical data were obtained from Reinert *et al.* (1974) for this original figure.)

temperatures, the Hg uptake value would show a substantial decrease. Nonetheless, the similarity between plots of Hg uptake *versus* temperature (Fig. 2) and of active metabolism *versus* exposure level (Fig. 1) is striking.

The study of Rodgers and Beamish (1981) contrasts with that of MacLeod and Pessah (1973). No potentiation of the depressant action of Hg is noted at higher temperatures used in their study, which is designed to evaluate effects of metabolic rate and exposure level on CH_3Hg^+ uptake. Rainbow trout (*Salmo gairdneri*) of 4–10 g are forced to swim at 20, 30, or 40 cm/s (10 or 20°C, pH 8.0–8.2, hardness 375–390 ppm) (Rodgers and Beamish, 1981). The presence of CH_3Hg^+ (injected into inflow water at the time of measurement) does not alter oxygen consumption at the Hg exposure level and time (about 8 ppb Hg and 130 min, respectively). However, as expected, rate of oxygen consumption increases with relative swimming speed at both 10 and 20°C. Rate of CH_3Hg^+ uptake also is positively related to rate of oxygen consumption and CH_3Hg^+ exposure level. At oxygen utilization levels of 33%, fish remove 8.3% of CH_3Hg^+ from the water. Therefore, temperature affects CH_3Hg^+ and Hg^{+2} uptake by directly altering metabolic rate of rainbow trout in this and other studies. Use of an organic rather than inorganic mercurial and a four-fold higher water hardness could have produced the differences noted between the MacLeod and Pessah (1973) study and that of Rodgers and Beamish (1981).

Exposure Level and Time: The length of time fish are exposed to Hg and the concentration of Hg in the water determine Hg uptake (McKone, *et al.*, 1971; Reinert, *et al.*, 1974; Heisinger and Green, 1975; Olson, *et al.*, 1975). Reinert *et al.* (1974) expose yearling rainbow trout (*Salmo gairdneri*) to 234–263 ppt CH_3Hg^+ at 5, 10, and 15 °C (Fig. 2) over a 12-wk period and observe a progressive increase in the amount of Hg accumulated over the entire exposure. After 12-wk exposures at 5 or 15°C, whole body levels of about 1200 or 2000 ppb Hg are accumulated, respectively (Fig. 2). Therefore, Hg uptake varies as a function of exposure time.

Despite the fact that exposure time and exposure concentration are increased at the same time in the McKone *et al.* (1971) study, the importance of exposure level is illustrated. Aqueous levels of 40–50 ppm Hg are concentrated by goldfish held 100 h at 0.25 ppm; however, 81-h exposures to 0.001, 0.01, or 0.1 ppm Hg result in accumulation of 0.25, 0.53, or 3.7 ppm Hg. Therefore, use of longer exposures and more concentrated solutions results in higher Hg accumulation. Moreover, concentration factors range from 34 to 250, decreasing as Hg level in the water increases.

Embryos (i.e. developing eggs) accumulate Hg in a concentration-dependent fashion, as is the case for adult fish. Heisinger and Green (1975) expose embryos of the Japanese ricefish (*Oryzias latipes*) to 10–30 ppb Hg^{+2} at 26°C and pH 6.9. The concentration factors for eggs at 16 d of exposure to 10, 15, 20, and 30 ppb Hg in the water are 1600, 1900, 2700, and 1900 corresponding to 16, 29, 54, and 56 ppm accumulated Hg.

Olson *et al.* (1975) expose fathead minnows (*Pimephales promelas*) 48 wk to 0.018 or 0.247 ppb CH_3Hg^+ in water (25°C, pH 7.5, 45 ppm hardness, as $CaCO_3$). Whole body accumulation at 0.018 or 0.247 ppb CH_3Hg^+ is 1.47 and 10.90 ppm Hg (wet weight), respectively; whereas, control fish concentrate less than 0.21 ppm Hg residues from water containing <0.01 ppb Hg. Concentration factors of 82,000 and 44,000 are obtained for final whole body Hg residues at 0.018 or 0.247 ppb Hg in the water. Therefore, exposure level and time produce marked effects on Hg accumulation (McKone, *et al.*, 1971; Reinert, *et al.*, 1974; Heisinger and Green, 1975; Olson, *et al.*, 1975).

Source: Whether Hg is derived from the diet, water, sediments, or injection can alter both Hg concentration in specific tissues and general tissue distribution patterns (McKone, *et al.*, 1971; Weisbart, 1973; Heisinger and Green, 1975; Pentreath, 1976b; Phillips and Buhler, 1978). Phillips and Buhler (1978) expose rainbow trout (*Salmo gairdneri*) for 24 d to CH_3Hg^+ in the water (0.07–1.38 ppb), diet (8.0–380.5 ppb/d), or both (14.8–15.5°C, pH 7.3–7.6). Accumulated dietary CH_3Hg^+ does not affect whole body uptake of CH_3Hg^+ from the water. In fact, accumulation from both sources is additive, with highest CH_3Hg^+ accumulation rates (0.412 ppm/d) at the highest levels of CH_3Hg^+ in both water and diet. About 10% of CH_3Hg^+ passing over trout gills is assimilated; whereas, about 68% of CH_3Hg^+ in the diet is assimilated. No differences in growth rate are noted for any treatments; so, growth dilution effects can be eliminated. Therefore, under conditions of this study, trout are about seven times more efficient at extracting Hg from dietary sources of CH_3Hg^+ than from aqueous sources of CH_3Hg^+.

Dietary CH_3Hg^+ is preferentially accumulated (compared to Hg^{+2}) in a two-step food chain. Pentreath (1976b) force-feeds plaice (*Pleuronectes platessa*) with a marine annelid *Nereis diversicolor* which had accumulated Hg^{+2} or CH_3Hg^+ for 7 d. Five days post-intubation, plaice fed *Nereis* exposed to Hg^{+2} retain from 3–5% of the initial Hg^{+2} dose. Whereas, plaice fed *Nereis* exposed to CH_3Hg^+ retain 80–85% of the initial dose. About 25% of Hg from the annelids fed Hg^{+2} is contained in the fish gut, compared to 35% Hg from Hg^{+2} fed *Nereis*. About 7% of Hg from annelids fed CH_3Hg^+ is accumulated by fish liver and kidney combined, compared with about 1% Hg from fish fed CH_3Hg^+-fed *Nereis*. When fish are fed CH_3Hg^+ in starch or gelatin pellets, retention in the gut is about 25%; and about 11% in the liver and kidney combined. Biological half-life is 33 d and 136 d for plaice fed *Nereis* labeled with Hg^{+2} and CH_3Hg^+, respectively. Corresponding retention is 11 and 89% for the respective chemical forms. At about 40 d, Hg has redistributed in fish fed *Nereis* labeled with CH_3Hg^+ and about 81% of Hg is retained in skeletal muscle at this time. The longer biological half-life and preferential absorption of CH_3Hg^+ (compared with Hg^{+2}) would result in higher total Hg concentrations and increased percentages of CH_3Hg^+ at higher tropic levels.

Studies involving Hg accumulation from sediments are a rarity. Guppies (*Poecilia reticulata*) are exposed 60 d to Hg in sediments at 50 ppm Hg as the chloride or sulfide at 21–23°C (Gillespie and Scott, 1971). Up to 1 ppm (wet weight) is accumulated in the whole body after 60 d exposure to 0.24 ppm "control levels" of Hg in the sediments. About 2 or 3 ppm Hg (wet weight) is accumulated from sediments containing 50 ppm Hg as the sulfide or as the chloride, respectively. Presumably, Hg is released slowly into the water by microbial activity in the sediments. This work dispels the assumption that mercuric sulfides form inert precipitates in surface waters following release from chlorine plants. The sulfide form of Hg is accumulated only slightly less rapidly by guppies than the chloride form.

Compared with aqueous, dietary, or sedimentary administration of Hg, the use of intraperitoneal injections of Hg produces results which are more difficult to interpret and to compare with environmental exposures. Weisbart (1973) administers intraperitoneal injections of 10 μCi ^{203}Hg as $Hg(NO_3)_2$ to goldfish (*Carassius auratus*) averaging 10 g (5.83–13.63 g). About 0.154 ppm Hg^{+2} is given to each individual. Uptake increases during the first 24 h; however, after this time, Hg is lost from the body at a constant rate. Biological half-life is calculated as 568 h. Only the liver, head kidney, muscle, and brain show increased Hg^{+2} accumulation 672 h postinjection (at 61.08, 39.36, 6.00, and 1.52 mean % injected dose/g wet weight, respectively). The other tissues (gall bladder, gonad, spleen, eye, tail kidney, intestine, gill, heart, skin, and swim bladder) either show no change or lost of Hg at rates faster than the body as a whole. Use of the same injected dose despite marked differences in fish weight contributes to variability (i.e. 10 μCi injections to fish ranging from 5.83–13.63 g), as does the superficial Hg contamination of abdominopelvic organs due to injection into the abdominoperitoneal cavity.

A single intraperitoneal injection of 15 ppm CH^3Hg^+ in the channel catfish (*Ictalurus punctatus*) results in increased deposition of Hg in the liver with time (at 23°C) (Kendall, 1977). Levels of 59, 81, 124, and 157 ppm Hg are accumulated by the liver 24, 48, 72, and 96 h post-injection. Assumption is made that 85–93% of this Hg is methylated and that CH_3Hg^+ causes histopathological damage directly.

The primary benefit of using intraperitoneal injections in fish is the efficiency of administration, since only one injection is required to deliver a substantial dose of a mercurial to induce histopathological changes, such as those reported by Kendall (1977). However, use of intraperitoneal injections of Hg can cause superficial contamination of abdominopelvic organs and increased variability in "uptake," which represents a combination of both uptake and absorption. Both absorption and adsorption of Hg occurs and can vary markedly with exact placement of the injection. Consequently, results of this administration mode are more difficult to interpret than those of dietary or aqueous administration modes.

Chemical Form: The chemical form of Hg plays a role in Hg accumulation and tissue distribution (MacLeod and Pessah, 1973; Wobeser, 1975; Pentreath, 1976b; Lock, *et al.*, 1981). MacLeod and Pessah (1973) expose rainbow trout fingerlings (*Salmo gairdneri*) to aqueous phenyl mercuric acetate (PMA) or inorganic Hg^{+2} at 5, 10, or 20°C (pH 7.5–7.8 and hardness 90 ppm, as $CaCO_3$). At 10°C, 0.005 ppm PMA gives concentration factors of 80–100; whereas, 0.1 ppm Hg^{+2} gives a concentration factor of 10 at the same temperature. At 10°C, PMA is about thirty times more toxic than Hg^{+2}—a factor undoubtedly due to the marked difference in the accumulation of Hg when it is present as PMA.

When Hg is administered to fish in the water, CH_3Hg^+ exposure results in higher levels of Hg in fish than if Hg^{+2} exposures are used. Wobeser (1975) exposes rainbow trout (*Salmo gairdneri*) fingerlings to 0.010–0.135 ppm CH_3Hg^+ or 0.05–1.00 ppm Hg^{+2} (10°C, pH 8.55, total hardness

101 ppm, as $CaCO_3$). Trout have concentration factors of 36–480 following 24 to 96-h exposures to 0.010–0.135 ppm CH_3Hg^+; whereas, concentration factors for Hg^{+2} in fingerlings are 1.6–3.0 after 24-h exposures to 0.56–1.00 ppm Hg^{+2} in the water. A major reason for lower Hg uptake when Hg^{+2} is administered is that fingerlings exposed to Hg^{+2} produce large amounts of mucus, which traps Hg^{+2} at the brachial surface.

When dietary administration of Hg is used, CH_3Hg^+ also accumulates to higher levels than if Hg^{+2} is used. Pentreath (1976b) force-feeds plaice (*Pleuronectes platessa*) with a marine annelid *Nereis diversicolor* which had accumulated Hg^{+2} or CH_3Hg^+ for 7 d. Five days post-intubation, plaice fed *Nereis* exposed to Hg^{+2} retain from 3–5% of the initial Hg^{+2} dose. Whereas, plaice fed *Nereis* exposed to CH_3Hg^+ retain 80–85% of the initial dose. As indicated in the prior section, biological half-life is 33 d and 136 d for plaice fed *Nereis* labeled with Hg^{+2} and CH_3Hg^+, respectively. Retention is 11 and 89% for the respective chemical forms. About 81% of Hg is retained in skeletal muscle at 40 d. Preferential absorption of CH_3Hg^+ and the longer biological half-life would result in higher total Hg concentrations and increased percentages of CH_3Hg^+ at higher tropic levels.

Methylmercury is roughly ten to twenty times more potent than Hg^{+2} in collapsing the structural integrity of the gill epithelial membranes and destroying the concentration gradient at the brachial surface. Lock and co-workers (1981) expose rainbow trout (*Salmo gairdneri*) for 4 h to 1 wk to 10–2000 ppb inorganic Hg^{+2} or to 1–200 ppb CH_3Hg^+ at 12°C. Plasma levels of Na^+ and Cl^- are decreased at sublethal levels of the two mercurials. Markedly enhanced water uptake by the gills is measured at sublethal Hg levels. Apparently, the more rapid collapse of brachial functional integrity is due to the accelerated accumulation of the lipophilic CH_3Hg^+, which subsequently produces higher Hg levels than does Hg^{+2}. Less Hg^{+2} than CH_3Hg^+ enters gills; therefore, Hg-induced changes in gill permeability, probably account for reported osmoregulatory disturbances in rainbow trout. Therefore, the chemical form of Hg determines the concentration of Hg within teleosts, with more rapid accumulation occurring when organic mercurials are used (MacLeod and Pessah, 1973; Wobeser, 1975; Pentreath, 1976b; Lock, *et al.*, 1981).

Life Stage: The life stage of the organism being evaluated is important in determining ultimate levels of Hg in fish tissues (Heisinger and

Green, 1975; Wobeser, 1975; Pentreath, 1976a). Pentreath (1976a) exposes plaice (*Pleuronectes platessa*) at 10 °C to sea water containing 0.08–0.14 ppb Hg^{+2}, in addition to 0.02 ppb Hg normally present in the sea water. Embryos and adults show marked differences in concentration factors. For example, eggs and larvae have concentration factors of 60 and 800; whereas, erythrocytes and gill of 40–45 g plaice accumulate total Hg at concentration factors of 1500–2000 after 60-d exposures. For these same fish, kidney, liver, skeletal muscle, and bone show concentration factors of 8250, 6000, 3500, and 1000.

Compared with the Pentreath (1976a) study, Heisinger and Green (1975) publish markedly elevated concentration factors for Hg in embryos. Japanese ricefish (*Oryzias latipes*) embryos are exposed to 10–30 ppb Hg^{+2} for 16 d at 26°C and pH 6.9. The concentration factors at 16 d of exposure to 10, 15, 20, and 30 ppb Hg in the water are 1600, 1900, 2700, and 1900, respectively, corresponding to 16, 29, 54, and 56 ppm accumulated Hg. These values are elevated compared with the concentration factor of 60 from the Pentreath (1976a) study.

Fingerling rainbow trout (*Salmo gairdneri*) also have much lower concentration factors than the embryos in Heisinger and Green's (1975) study. Concentration factors for Hg^{+2} in fingerlings are 1.6–3.0 after 24–h exposures to 0.56–1.00 ppm Hg^{+2} and 36–480 following 24–96 h exposures to 0.010–0.135 ppm CH_3Hg^+ in the water (Wobeser, 1975). Rainbow trout accumulate both Hg^{+2} and CH_3Hg^+ rapidly from water; however, the Hg^{+2} levels accumulated are lower for rainbow trout than for plaice. The reason for the marked difference in concentration factors between embryos (Heisinger and Green, 1975) and fingerlings (Wobeser, 1975) is not known, although species-specific factors and experimental conditions probably play major roles.

Biological Effects

Histopathology: The central theme for histopathological studies involving Hg is that this metal induces necrotic changes in virtually every organ and tissue evaluated histopathologically. Mercury tenaciously binds sulfhydryl groups on the the plasma membrane and other cellular constituents (Olson, *et al.*, 1973b; Bouquegneau, 1977), ultimately destroying the permeability characteristics of the cells involved (Trump, *et al.*, 1975;

Lock, *et al.*, 1981). Mercury causes histopathological effects in the gill, kidney, exocrine pancreas, bile ducts, liver, and erythrocytes. Studies involving single organs or tissues are predominate in the literature. Olson and Fromm (1973a) and Wobeser (1975) focus on the gill; Köhler (1989) studies the liver; Trump *et al* (1975) and Kendall (1975) evaluate the kidney; and Gill and Pant (1985) monitor changes in blood cells. However, Kendall (1977) evaluates Hg-induced changes in several organs simultaneously—including the liver, exocrine pancreas, and bile duct cells.

The acute toxic action of both Hg^{+2} and CH_3Hg^+ is exerted on secondary lamellae of gill epithelium and is either related to changes in epithelial cell permeability (Lock, *et al.*, 1981) or inhibition (possibly decreased synthesis) of Na^+/K^+-ATPase (Bouquegneau, 1977). Inorganic Hg^{+2} causes severe necrosis of epithelial cells (necrosis and karyorrhexis, without separation of epithelial cells from pillar cells) in rainbow trout (*Salmo gairdneri*) (Wobeser, 1975). Changes produced by CH_3Hg^+ include epithelial hyperplasia with ballooning degeneration and desquamation of the epithelium. Wobeser exposes rainbow trout fry and/or fingerlings to 0.010–0.135 ppm CH_3Hg^+ or 0.05–1.00 ppm Hg^{+2} (10°C, pH 8.55, total hardness 101 ppm, as $CaCO_3$). Fingerlings produce large amounts of mucus when exposed to Hg^{+2}, but not to CH_3Hg^+.

Inorganic and organic forms of Hg pass the gill differently (Olson and Fromm, 1973a; Lock, *et al.*, 1981). Inorganic Hg^{+2} is trapped in mucus; whereas, CH_3Hg^+ seems to be able to cross the mucus and epithelium at a more rapid rate because of increased lipophilicity. Olson and Fromm (1973a) expose rainbow trout (*Salmo gairdneri*) for 23 h to 0.25 ppm inorganic Hg^{+2} or 0.05 ppm CH_3Hg^+ (13°C). A second addition of each chemical form of Hg is added at 23 h; fish are removed 1 h later and sacrificed. Gills are freeze-dried and carbon-coated for X-ray microprobe analysis (XRMA). No sites of Hg deposition are detected following exposure of fish to CH_3Hg^+—probably because the methylated mercurial is lost during high vacuum tissue preparation procedures which require 5 millitorr pressures. However, fish exposed to Hg^{+2} apparently trap Hg^{+2} between lamellae in the abundant supply of mucus. The gill cartilage also appears to concentrate Hg^{+2}.

Accurate ion distribution patterns are obtained from freeze-drying and XRMA procedures; therefore, binding of Hg^{+2} by interlamellar mucus and cartilage may be sufficient to account for the reduced toxicity of the inorganic mercurials. The 24-h LC50 values for Hg^{+2} and CH_3Hg^+ are 0.903 ppm and 0.125 ppm as Hg, respectively (Wobeser, 1975). Therefore, in this study, production of excess mucus and interaction between this mucus and Hg seems to confer a seven-fold greater protection against Hg^{+2}-induced toxicity than is the case for CH_3Hg^+-induced toxicity.

As with the previously mentioned Wobeser study (1975), the Rucker and Amend report (1969) shows Hg-induced damage to the gill epithelium. Hypertrophy and hyperplasia of the gill epithelium are noted. However, no histopathological abnormalities are found in erythrocytes, liver, or kidney in rainbow trout (*Salmo gairdneri*). Trout are exposed to 2 ppm Timsan (6.25% ethyl Hg phosphate) for 1 h, daily for 9 d. Levels of 22.8, 16.7, 17.3, and 4.0 ppm Hg are accumulated in the blood, liver, kidney, and muscle, respectively. The absence of histopathological alterations in these tissues and organs seems remarkable, considering the elevated level of Hg in each.

One of the most interesting findings of the Wobeser study (1975) is that CH_3Hg^+ causes a 435% increase in mitosis in the gill epithelium at aqueous levels greater than 0.032 ppm. Because repair of damaged epithelium does not continue following transfer of fish to uncontaminated water, the teleost response to epithelial damage probably is not the primary cause of increased mitotic activity in the gill. Rather, CH_3Hg^+ probably disturbs mitosis directly. This is supported by the fact at the mitotic index returns to normal within 48 h of removal of fish from CH_3Hg^+-contaminated water (following a 24-h exposure in 0.08 ppm CH_3Hg^+). Chromosomal disorders have long been associated with Hg-poisoning (Abelson, 1970; Goldwater, 1971) and may be responsible for the teratogenicity observed in fish (Heisinger and Green, 1975; McKim, *et al.*, 1976; Weis and Weis, 1977).

Kendall (1977) shows that exocrine pancreas cells and bile duct cells are more sensitive to Hg than are parenchymal hepatocytes. A single intraperitoneal injection of 15 ppm CH_3Hg^+ is administered to channel catfish (*Ictalurus punctatus*) at 23°C. Levels of 59, 81, 124, and 157 ppm Hg are accumulated by the liver 24, 48, 72, and 96 h post-injection. Abdominal distention, a reddish-yellow serous exudate, and inflamed, distended mesenteric vessels are noted. Although parenchymal hepatocytes do eventually undergo necrotic changes at 96 h, the exocrine pancreas cells and bile duct cells are thought to be more sensitive than liver cells. At 24 h, exocrine pancreas cells shrink and show a circumferential clear space around the portal vein and lining exocrine pancreas cells. At 48 h, the space enlarges and pancreas cells appear disorganized. At 72 h, only periodic acid-Schiff (PAS) positive cell remnants are visible—to show that glycogen, mucins, neutral polysaccharides, and/or glycoproteins are present. At 96 h, bile duct epithelium desquamates and the hepatic capsular surface displays a layer of suppurative, inflammatory exudate consisting of PAS positive cells and fibrin. Unfortunately, tissue is examined only at the optical level—a level at which necrotic changes sometimes are not obvious until 8 h after cells have become necrotic (Trump, *et al.*, 1975). Possibly, the parenchymal hepatocytes in Kendall's (1977) study become necrotic much sooner than observed at the optical level in this study.

Köhler (1989) focuses on the reversibility of changes in parenchymal hepatocytes from flounder (*Platichthys flesus*) removed from extremely polluted surface waters. Fish are collected from Elbe estuary downstream from Hamburg, an area said to be the most contaminated European drainage since it receives water from the highly industrialized countries of Czechoslovakia, the German Democratic Republic, and the Federal Republic of Germany. Maximum Hg levels in skeletal muscle are 950 ppb (wet weight). Other contaminants include up to 3.1 ppm polychlorinated biphenyls (PCBs) and up to 1.9 ppm hexachlorobenzene (HCB).

Morphological changes in livers from fish at collection are due to the mixture of Hg, PCBs, and HCB. Changes noted are parenchymal hepatocyte shrinkage, fatty vacuolation, karyomegaly, necrosis, pyknosis, formation of increased lipofuscin granules, rough endoplasmic reticulum disorganization, and proliferation of smooth endoplasmic reticulum. Needle-like, 37–41 nm diameter, paracristal structures (so-called "macrotubules") are observed. Unfortunately, histocytochemistry is not performed to positively identify "macrotubules" reported in this study. Neither polarization nor X-ray microanalysis are used in an effort to locate Hg or other metal deposits.

After 40 d of depuration at 15°C and pH 7.8–7.9, about 40% of flounders show liver regeneration and 30% of flounders show initial signs of regeneration, which might result in complete regeneration if depuration were extended. However, the remaining 30% of flounders show proliferation of fibrocytes and collagen fibers instead of regeneration of parenchymal hepatocytes. After the 40-d depuration, hepatic Hg levels double in non-regenerating livers (30% of flounders) and in livers showing some initial regeneration (30%). The livers which do regenerate completely have about 33% higher Hg levels on day 40 of depuration than fish on day 0 ($p<0.01$). Internal mobilization of Hg—combined with little or no elimination—could cause CH_3Hg^+ elimination from the liver into the bile and rapid intestinal reabsorption (enterohepatic redistribution), followed by accumulation of Hg in the muscle (Pentreath, 1976a). Despite the lack of measurements of metallothionein, other metal binding proteins, biotransformation reactions, or detoxification mechanisms; this study shows that depuration can elevate hepatic Hg levels, possibly altering liver regeneration. Similar studies involving Hg alone could yield significant understanding of Hg mobilization within fish and the resultant impact on regeneration during depuration.

Trump and colleagues (1975) describe effects of mercurials on the teleost kidney in vitro and in vivo using CH_3Hg^+, Hg^{+2}, or PCMBS (parachloromercuribenzenesulfonate). Isolated kidney tubules of winter flounder (*Pseudopleuronectes americanus*) and living hogchokers (*Trinectes maculatus*) are exposed to 10^{-5} to 10^{-3} M Hg. Within 6 h, a sequential series of renal changes result (stages 1–5) and illustrate the progressive damage associated with tubular necrosis. The endoplasmic reticulum and nuclear envelop dilate. Cell volume increases (hypertrophy). Mitochondrial inner compartments condense, producing a considerable increase in affinity to metal stains used in electron microscopy (e.g. Os, Pb, and U); such condensation is followed by a progressive high amplitude swelling in mitochondria. Since necrotic tissue has a higher Ca binding affinity than does normal tissue, calcified deposits appear in mitochondria. Flocculant mitochondrial deposits also appear, karolysis occurs, and the tubular cell is classified as necrotic (having reached the point of irreversible damage).

Most severe in vitro changes result from exposure of the tubular cells to 10^{-3} M Hg; all compounds produce stage 5 necrotic changes within 30 min under conditions of this study (Trump, *et al.*, 1975). The phenomenon is characteristic of agents which damage cells by direct interaction with the plasma membrane. Severity of damage is reduced as exposure level and exposure time are reduced. Generally, CH_3Hg^+ and Hg^{+2} are of equal potency; whereas, PCMBS tends to be less potent than CH_3Hg^+ and Hg^{+2}. When Hg levels are low to moderate, mitochondrial calcification occurs. When Hg concentrations are high, Hg probably enters the cell, is bound to mitochondrial transport sites, and inhibits Ca uptake.

In vivo exposures of hogchokers (given intraperitoneal injections of 16 mg/kg Hg^{+2}) produce selective necrosis which involves the second segment, and probably the first as well. However, other regions of the hogchoker nephron are not involved. As with the rat, the presence of a localized lesion such as this suggests the selective concentration of mercurials in certain tubular segments or the Hg-induced concentration of some other factor. Binding of Hg to sulfhydryl groups on surface proteins of the plasma membrane is stated to be "the important action" of Hg—in producing permeability changes, rather than some specific biochemical lesion (Trump, *et al.*, 1975).

Kendall (1975) also studies the impact of mercurials on the histopathological changes occurring in the kidney, conducting in vivo and in vitro studies of CH_3Hg^+ on channel catfish (*Ictalurus punctatus*) (at 23°C). Live fish are injected intraperitoneally at 15 ppm CH_3Hg^+. Fresh, frozen, 10 µm sections (with or without a 60-min exposure to 15 ppm CH_3Hg^+) are evaluated for succinate dehydrogenase (SDH) staining. In vitro SDH staining procedures show that kidney tubules have much reduced SDH activity following a 60-min exposure to 15 ppm Hg. Activity of SDH is assumed to be inhibited.

The in vivo, intraperitoneal injection of 15 ppm CH_3Hg^+ shows that Hg accumulates in the kidney to a level of 51 ppm in 24 h, but that levels are reduced to 14 ppm by 96 h (Kendall, 1975). The fall in Hg level in the kidney at this time corresponds to the striking loss of cellularity, desquamation, and necrosis. Declines in renal Hg levels are attributed to excretion of renal cellular constituents. Unfortunately biochemical testing (e.g. protein determination) to confirm or reject this assumption is not presented.

Gill and Pant (1985) use *Barbus (Puntius) conchonius* (a fresh-water barb from India) to show the striking changes which Hg induces in erythrocytes. Fish are exposed for up to 56 d to 1.00, 0.33, and 0.20 of the 96-h LC50 for the species. Erythrocytes of fish exposed 1–56 d to 181 ppb Hg (the 96-h LC50) are severely vacuolated and show a beaded pattern of vacuoles, as well as severe nuclear degeneration, microcytosis, and collapsed plasma membranes. After chronic 7- or 21-d exposures to 60 ppb (0.33 of the 96-h LC50), poikilocytosis, hypochromia, and fragmentation are evident in erythrocytes. Plasma membrane damage may result from Hg binding to sulfhydryl groups. Most Hg-induced changes are characteristic of degenerative changes, such as those occurring in dead or dying fish.

To summarize, the bulk of histopathological data for Hg-poisoning in fish focuses on necrotic changes in the gill (Olson and Fromm, 1973a; Wobeser, 1975), liver (Kendall, 1977; Köhler, 1989), kidney (Kendall, 1975; Trump, *et al.*, 1975), exocrine pancreas (Kendall, 1977), bile duct cells (Kendall, 1977), and erythrocytes (Gill and Pant, 1985). Mucus trapping of Hg^{+2} apparently is much greater than that for CH_3Hg^+ (Olson and Fromm, 1973a; Olson, *et al.*, 1973b; Lock, *et al.*, 1981) because of greater lipophilicity of CH_3Hg^+. Moreover, both permeability changes in the gill epithelium (Trump, *et al.*, 1975; Lock, *et al.*, 1981) and inhibition (or decreased synthesis) of Na^+/K^+-ATPase (Bouquegneau, 1977) are suggested as possible mechanisms for reduced extrusion of salts from the sea water-adapted teleost gill (Bouquegneau, 1977), collapse of the integrity of the gill epithelial membranes (Lock, *et al.*, 1981), and/or disappearance of the concentration gradient following Hg exposure (Lock, *et al.*, 1981).

Hematology: Acute, Hg exposures cause the most dramatic changes in circulating neutrophils and thrombocytes. For example, Gill and Pant (1985) expose *Barbus (Puntius) conchonius* up to 56 d to 1.00, 0.33, and 0.20 of the 96-h LC50 for inorganic Hg^{+2}. Acute exposures to 181 ppb Hg^{+2} (the 96-h LC50) for 24–48 h cause thrombocytosis (42% increase, $p<0.05$) and neutropenia (62% decrease, $p<0.01$). The average percentage of neutrophils also decreases in chronic exposures (7–56 d at 60 ppb Hg), but increases at 56 d exposures to 60 ppb. Basophilia, thrombocytosis, and lymphocytosis occur. Variation noted in percentages of circulating cells could be the result of transient stress in all exposure groups, as indicated by transient basophilia in control fish (a 300% increase in 21 d, followed by a return to normal in 14 d). Erythrocyte changes are striking in barbs exposed to Hg^{+2}. After acute exposures, erythrocytes show microcytosis and degeneration. After chronic exposures, poikilocytosis, hypochromia, and fragmentation are evident (Gill and Pant, 1985).

Transient hyperglycemia, transient reductions in hemoglobin levels, and transient suppression of erythrocyte counts occur in the same species of barb in a similar study by Gill and Pant (1981). In acute tests involving 450–850 ppb Hg^{+2} for 24 h, hyperglycemia ($p<0.05$) is noted and both liver and brain glycogen are depleted ($p<0.05$). Responses are more pronounced at higher levels of Hg and at greater exposure times. In chronic tests involving 36.3–60.5 ppb Hg^{+2} for up to 8 wk, pronounced hyperglycemia subsides after 4 wk, but glucose levels remain two to three times higher than controls at 8 wk. Levels of hepatic glycogen in fish exposed in the same chronic tests are 30–50% lower than those of controls 1 wk after exposure; however, levels return to those of controls in 4 wk. Initial (1–3 wk) decreases are noted in erythrocyte counts and hemoglobin levels, but values are elevated after 8-wk exposures. Changes in levels of blood glucose and hemoglobin, as well as erythrocyte counts, appear to reflect biochemical stress which affects energy reserves of *Barbus (Puntius) conchonius* .

The rapidity with which Hg affects hematological parameters in acute testing is more obvious upon consideration of the uptake of Hg by erythrocytes. Olson *et al.* (1973b) use rainbow trout (*Salmo gairdneri*) exposed 24 h to 0.275 ppb Hg as an inorganic (Hg^{+2}) or a organic (CH_3Hg^+). After brachial uptake, Hg is accumulated by the blood, equilibriating within 10 min. Almost 90% of Hg is bound to erythrocytes in 40 min. Leukocytes could be affected similarly. Analysis of uptake of both forms of Hg into the blood reveals a two-component system. A rapid component (half time: <1 min) probably coincides with initial binding of Hg to plasma protein and/or thiols. The second component (half time: about 11 min) may be a result of slower dissociation of bound Hg from "plasma sites" prior to erythrocyte uptake.

For erythrocytes, the ratio of bound to unbound Hg is constant, suggesting that saturation of trout erythrocytes with Hg involves sulfhydryl or other groups, which are responsible for plasma membrane integrity and are not saturated at the time of hemolysis (Olson, *et al.*, 1973b). Alternatively, disruption of the ionic balance and subsequent hemolysis could occur as a result of Hg levels below those required for total cell saturation. Sulfhydryl-containing compounds (albumin, cysteine, and reduced glutathione) fail to remove Hg previously bound to erythrocytes, suggesting that Hg binds erythrocytes tenaciously (Olson, *et al.*, 1973b). In chronic testing, fewer Hg-sulfhydryl interactions occur at the plasma membrane. Therefore, any altered patches of membrane might gradually be replaced over several weeks to restore normal function. Ultimately, the repair of abnormal patches of membrane could produce the transient effects which are a common feature of the hematological response of teleosts to Hg.

At least one environmental study shows that the hematological response to Hg exposures is long-lived. Reduced serum cortisol and alkaline phosphatase activity are measured in Hg-laden pike before and after a 1-yr depuration period. In their study, Lockhart and colleagues (1972) transfer four-hundred northern pike (*Esox lucius*) from Clay Lake, Ontario to an area relatively free of Hg contamination (Heming Lake, Manitoba) in 1970. Clay Lake is known to be heavily laden with Hg from use of mercurial slimicides in the pulp mill industry and emission of Hg from a chlorine plant (Fimreite and Reynolds, 1973). In this study, levels of Hg in skeletal muscle of walleye (*Stizostedion vitreum*) and burbot (*Lota lota*) are 16–22 ppm (Lockhart, *et al.*, 1972). Fish are assumed to have been exposed a life time since Hg releases began in 1962.

The tenacity with which Hg is bound to tissues of upper level predators in this environmental exposure is remarkable. A half-life of about 2 yr for Hg in pike is calculated based on levels in biopsy material taken from fish during 1971 (Lockhart, *et al.*, 1972). About 30% of Hg is lost in 1 yr—probably as a result of elimination processes and growth dilution. At the initiation of depuration, Clay Lake pike show <u>decreases</u> in serum glucose (59%), total serum protein (19%), and activity of alkaline phosphatase (45%). Reduced serum cortisol levels (74%) in Clay Lake pike could be a result of suppression of cortisol secretion or alterations in the rate of cortisol clearance. One year later, glucose levels are <u>elevated</u> (55%), total serum protein is almost normal, and serum alkaline phosphatase activities are still reduced (30%), as are cortisol levels (44%). The stress response (even 1 yr following transfer) might result in lower adaptability and lower survival of pike.

In summarizing hematological responses of fish to Hg, acute and chronic laboratory studies show a transient response of fish to Hg based on blood glucose levels, erythrocyte counts, and hemoglobin levels (Gill and Pant, 1981). However, chronic environmental exposures produce hematological changes which persist 1 yr following the initiation of depuration, in terms of activity of alkaline phosphatase and cortisol levels (Lockhart, *et al.*, 1972). Moreover, the tenacity of Hg binding to erythrocytes is exhibited by the failure of sulfhydryl-containing compounds (albumin, cysteine, and reduced glutathione) to remove Hg previously bound to erythrocytes (Olson, *et al.*, 1973b). Low-level, long-term tests would produce fewer Hg-sulfhydryl bonds at the plasma membrane of erythrocytes and other blood cells, allowing a slow replacement of the abnormal membrane constituents and probably leading to transient effects in the hematological response of fish to Hg. However, a life time exposure of upper level predators such as walleye and pike, produces a long-term stress response which is not easily reversed by depuration (Lockhart, *et al.*, 1972). Long-term effects are probably due to the marked elevations in the level of Hg which can accumulate in tissues of upper level predators.

Behavior: About 3 ppb inorganic Hg^{+2} is sufficient to alter behavior of previously conditioned goldfish (Weir and Hine, 1970) or the sensitivity of brook trout to a feeding disturbance (Drummond, *et al.*, 1974). Weir and Hine (1970) expose comet goldfish (*Carassius auratus*) 48 h to Hg^{+2} or to one of three other metals (or metalloids) following establishment of a conditioned avoidance response (23°C, pH 6.0–6.9, and 50 ppm hardness). The conditioned reflex paradigm involves a light stimulus (3 s), followed by light and a 14 V shock (2 s), then a dark, unelectrified interval (25 s). For a "correct response," goldfish swimming in that half of the aquarium where the light stimulus appears are required to move into the dark, unelectrified side within the 3-s light phase of the cycle. Groups of five fish are given twenty-five trials per day at 24 h and 48 h after initiation of exposures.

Of the metals studied (Hg, As, Se, and Pb), Hg is the most potent, producing statistically increased impairment of the conditioned reflex response at 0.003 ppm, a level 1/273 that of the 7-d LC50 for this species (0.82 ppm). For comparison, As, Se, and Pb impair the conditioned response at 0.10, 0.25, and 0.07 ppm in the water, respectively. These levels represent the 1/320, 1/48, and 1/1570 of the 7-d LC50 values for the goldfish (32, 12, and 110 ppm, respectively). Therefore, in goldfish, Hg^{+2} can induce central nervous system effects at lower elemental levels than As, Se, and Pb (Weir and Hine, 1970).

Drummond and colleagues (1974) also observe that Hg induces central nervous system effects at 3–12 ppb Hg (as CH_3Hg^+) in brook trout (*Salvelinus fontinalis*) at 9°C, 44–46 ppm hardness, and pH 6.0, 7.5, or 9.0. Apparently fish sensitivity to Hg is increased during the feeding period, as evaluated using cough frequency. Some brook trout are disturbed by the daily feeding period *per se*, and even though those individuals do not feed, cough frequency is increased. This increased irritability is noted at 3 ppb Hg, the same level of Hg which impairs the conditioned avoidance response in comet goldfish (Weir and Hine, 1970).

Behavioral changes of fish succumbing to Hg-poisoning suggest that the metal induces severe neurological abnormalities (Matida, *et al.*, 1972; MacLeod and Pessah, 1973). For example, MacLeod and Pessah (1973) expose rainbow trout fingerlings (*Salmo gairdneri*) to aqueous phenyl mercuric acetate (PMA, used as a slimicide) or inorganic Hg^{+2} at 5, 10, or 20°C (pH 7.5–7.8 and hardness 90 ppm, as $CaCO_3$).Visible signs of Hg-poisoning include loss of equilibrium, bursts of erratic swimming, a darkened coloration, and onset of inactivity just prior to death. As with studies previously mentioned, Hg probably induces neurological damage in rainbow trout fingerlings to cause behavioral and other changes.

Matida and co-workers (1972) observe symptoms of Minamata disease in rainbow trout (*Salmo gairdneri*) consuming toxic shellfish (*Horumomura mutabilis*) collected from Minamata Bay in 1960. Edible parts of the shellfish are ground into a fine powder to produce a fish food containing about 48 ppm Hg, with 33% as CH_3Hg^+. Fish suffering from Minamata disease swim rolling from side to side, maintain body position vertically by arching their bodies, lose the ability to turn in front of glass walls of aquaria or other barriers, and show an apparent decline in vision. In contrast, impaired vision is not found in *Barbus (Puntius) conchonius* (Gill and Pant, 1981). Apparently, in rainbow trout, accumulation of a specific level of CH_3Hg^+ may lead to the initiation of neurological abnormalities similar to those induced by Hg in man and other vertebrates (Friberg and Vostal, 1972).

Fish accumulating Hg from the water also develop Minamata disease. Half of fish develop symptoms of the disease when held 72 d at 100 ppm (i.e. eye hemorrhages, rolling swim, inability to change direction, and abnormal body position). In some exposures for a longer period of time (e.g. 189 d to 25 ppm) no hemorrhage occurs, but fish swim rolling from side to side and cannot stop in front of obstacles. Of several mercurial compounds tested, CH_3Hg^+ is most easily accumulated and localizes in the muscle (up to 13 ppm, wet weight) and in the brain (up to 16 ppm). High levels of Hg in the brain could initiate the behavioral changes observed by Matida and co-workers (1972), as well as other workers (Weir and Hine, 1970; MacLeod and Pessah, 1973; Drummond, *et al.*, 1974). In conclusion, Minamata disease occurs in fish as in humans, developing as a result of Hg uptake from dietary and/or aqueous sources. Fish show neurological damage at 3 ppb, based on behavioral changes. Elevated brain Hg levels could cause these changes, although some areas of the brain probably are more sensitive to low levels of Hg than are other areas.

Summary

Among the elements, Hg is one of the most hazardous to fish (and other animals)—primarily because of the tenacity with which Hg binds sulfhydryl groups. The presence of aqueous Hg stimulates mucous synthesis from the body surface of teleosts, mucus interacts with Hg to hold the element in an indefinitely stable condition, the Hg-mucus complex is sloughed, and the synthetic process is repeated. Up to 80% of the body burden of Hg can be trapped by the mucous coat. Although inorganic mercurial salts such as Hg^{+2} are efficiently held in this ion resin trap, mucus seems not to hinder the passage of the univalent alkylmercury salt CH_3Hg^+. The difference in the ion resin trapping capacity of mucus is one of the major reasons for the marked increase in CH_3Hg^+ uptake and subsequent toxicity of the organic mercurial, compared with the inorganic Hg^{+2}.

The liver plays a central role in the binding, storage, and redistribution of mercurials which enter peripheral circulation. Methylation of Hg^{+2} can occur in the livers of some fish species (e.g. albacore and yellowfin tuna, but not rainbow trout and mackerel). Vitamin B_{12} derivatives appear to be the methylating agents, producing CH_3Hg^+, which can bind metallothionein in some organs but not in others. Hepatic metallothionein appears to bind CH_3Hg^{+2} intact, in a least some species. With increasing levels of intracellular Hg, metallothionein is eventually saturated by the mercurial, during a time in which hepatic levels of Cu and Zn decrease. Upon saturation of the metal-binding protein, Hg is thought to spill over to the enzyme pool (that of high molecular weight proteins). At this time, pathological effects first appear. Possibly, some of the effects (e.g. 50% reductions in growth of chum salmon) are a consequence of the decrease in intracellular levels of essential elements Cu and Zn, rather than a direct toxic effect of Hg.

The tissue of greatest concern from the human standpoint is skeletal muscle. Mercury can concentrate in muscle to levels of 27.8 ppm. Methylation of up to 100% of total Hg in the muscle probably plays a major role in the lengthy half life of Hg in muscle—estimated at 2–3 yr. Many areas of the world have endemic teleost populations with Hg levels in edible tissues above 0.5 ppm, the so-called "action level." Acres of surface waters have been banned as fishing areas as a consequence. Even oceanic fauna (and flora) are being exposed to Hg, as it enters the upper 100 m at greater levels as a result of the anthropomorphic activities of man.

Presumably, over millions of years, vertebrates may have built an increased tolerance for Hg; so, it is remotely possible that life may now depend on Hg as a useful trace element. The key word is "trace" since the beneficial or harmful nature of Hg is determined by concentration, chemical form, and many other factors. One scientist states that "too little Hg in the environment might be as dangerous as too much" (Goldwater, 1971). However, to err on the side of "too little Hg" seems the better alternative, in view of the remarkable bioaccumulation of Hg in upper trophic level species, the intransigent nature of the element, and the potential for chromosomal damage. The journalistic outcry of the 1970's may have produced a brief public hysteria over the environmental Hg problem, but perhaps this has served to awaken some of the public to the severity of the effects of Hg. Today, although funding and enthusiasm have diminished, the basic problems of mercurial contamination remain and undoubtedly have become more detrimental to fish.

References

Abelson, P.H., Methyl mercury, Science, 169 (1970) 237.

Ackefors, H., III. Effects of particular pollutants: Mercury pollution in Sweden with special reference to conditions in water habitat, Proc. Roy. Soc. Lond. B., 177 (1971) 365.

Bache, C.A., W.H. Gutenmann, and D.J. Lisk, Residues of total mercury and methylmercuric salts in lake trout as a function of age, Science, 172 (1971) 951.

Barber, R.T., A. Vijayakumar, and F.A. Cross, Mercury concentrations in recent and ninety-year-old benthopelagic fish, Science, 178 (1972) 636.

Boulton, P., and L.J. Hetling, A statistical analysis of the mercury content of fresh-water fish in New York state, Department of Environmental Conservation, Report #19, Albany, New York, (1972) 1.

Bouquegneau, J.M., ATPase activity in mercury intoxicated eels, Experientia, 33 (1977) 941.

Brown, D.A., and T.R. Parsens, Relationship between cytoplasmic distribution of mercury and toxic effects to zooplankton and chum salmon (*Oncorhynchus keta*) exposed to mercury in a controlled environment, J. Fish. Res. Bd. Can., 35 (1978) 880.

Burrows, W.D., and P.A. Krenkel, Studies on uptake and loss of methylmercury-203 by bluegills (*Lepomis macrochirus* Raf.), Environ. Sci. Technol., 7 (1973) 1127.

Drummond, R.A., G.F. Olson, and A.R. Batterman, Cough response and uptake of mercury by brook trout, *Salvelinus fontinalis*, exposed to mercuric compounds at different hydrogen-ion concentrations, Trans. Amer. Fish. Soc., 103 (1974) 244.

Fimreite, N., and L.M. Reynolds, Mercury contamination of fish in northwestern Ontario, J. Wildlife Management, 37 (1973) 62.

Forrester, C.R., K.S. Ketchen, and C.C. Wong, Mercury content of spiny dogfish (*Squalus acanthias*) in the Strait of Georgia, British Columbia, J. Fish. Res. Bd. Can., 29 (1972) 1487.

Friberg, L., and J. Vostal, Mercury in the Environment: An Epidemiological and Toxicological Appraisal, CRC Press, Cleveland, Ohio, (1972) 1.

Gill, T.S., and J.C. Pant, Effects of sublethal concentrations of mercury in a teleost *Puntius conchonius*: Biochemical and haematological responses, Ind. J. Exp. Biol., 19 (1981) 571.

Gill, T.S., and J. Pant, Mercury-induced blood anomalies in the freshwater teleost *Barbus conchonius*, Wat. Air Soil Pollut., 24 (1985) 165.

Gillespie, D.C., and D.P. Scott, Mobilization of mercury sulfide from sediment into fish under aerobic conditions, J. Fish. Res. Bd. Can., 28 (1971) 1807.

Goldwater, L.J., Mercury in the environment, Sci. Am., 224 (1971) 15.

Goyer, R.A., Toxic Effects of metals: The Basic Science of Poisons, in Klaassen, C.D., M.O. Amdur, and J. Doull (Ed.), Casarett and Doull's Toxicology, 3rd ed., MacMillan Publ. Co., New York, New York, (1986) 582.

Hamilton, S.J., and T.A. Haines, Bone characteristics and metal concentration in white suckers (*Catostomus commersoni*) from one neutral and three acidified lakes in Maine, Can. J. Fish. Aq. Sci., 46 (1989) 440.

Hammond, A.L., Mercury in the environment: Natural and human factors, Science, 171 (1971) 788.

Heisinger, J.F., and W. Green, Mercuric chloride uptake by eggs of the ricefish and resulting teratogenic effects, Bull. Environ. Contamin. Toxicol., 14 (1975) 665.

Imura, N., S.-K. Pan, and T. Ukita, Methylation of inorganic mercury with liver homogenate of tuna fish, Chemosphere, 5 (1972) 197.

Jensen, S., and A. Jernelöv, Biological methylation of mercury in aquatic organisms, Nature, 223 (1969) 753.

Jernelöv, A., and H. Lann, Mercury accumulation in food chains, Oikos, 22 (1971) 403.

Johnels, A.G., T. Westermark, W. Berg, P.I. Persson, and B. Sjostrand, Pike (*Esox lucius* L.) and some other aquatic organisms in Sweden as indicators of mercury contamination in the environment, Oikos, 18 (1967) 323.

Kendall, M.W., Acute effects of methyl mercury toxicity in channel catfish (*Ictalurus punctatus*) kidney., Bull. Environ. Contam. Toxicol., 13 (1975) 570.

Kendall, M.W., Acute effects of methyl mercury toxicity in channel catfish (*Ictalurus punctatus*) liver, Bull. Environ. Contam. Toxicol., 18 (1977) 143.

Klaassen, C.D., Heavy metals and heavy-metal antagonists, in Gilman, A.G., L.S. Goodman, T.W. Rall, and F. Murad (Ed.), The Pharmacological Basis of Therapeutics, MacMillan Publ. Co., New York, New York, (1985) 1605.

Kleinert, S.J., and P.E. Degurse, Mercury levels in Wisconsin fish and wildlife, Department of Natural Resources, Madison Technical Bulletin No. 52, Madison, Wisconsin, (1972) 1.

Köhler, A., Regeneration of contaminant-induced liver lesions in flounder--Experimental studies towards identification of cause-effect relationships, Aquat. Toxicol., 14 (1989) 203.

Lock, R.A.C., P.M.J.M. Cruijsen, and A.P. van Overbeeke, Effects of mercuric chloride and methylmercuric chloride on the osmoregulatory function of the gills in rainbow trout, *Salmo gairdneri* Richardson, Comp. Biochem. Physiol., 68C (1981) 151.

Lockhart, W.L., J.F. Uthe, A.R. Kenney, and P.M. Mehrle, Methylmercury in northern pike (*Esox lucius*): Distribution, elimination, and some biochemical characteristics of contaminated fish, J. Fish. Res. Bd. Can., 29 (1972) 1519.

MacKay, M.J., M.N. Kazacos, R.J. Williams, and M.I. Leedow, Selenium and heavy metals in black marlin, Mar. Pollut. Bull., 6 (1975) 57.

MacLeod, J.C., and E. Pessah, Temperature effects on mercury accumulation, toxicity, and metabolic rate in rainbow trout (*Salmo gairdneri*), J. Fish. Res. Bd. Can., 30 (1973) 485.

Matida, Y., H. Kumuda, S. Kimura, Y. Saiga, T. Nose, M. Yokote, and H. Kawatsu, Toxicity of mercury compounds to aquatic organisms and accumulation of the compounds by the organisms, Bull. Freshwater Fish. Res. Lab., 21 (1972) 197.

McKim, J.M., G.F. Olson, G.W. Holcombe, and E.P. Hunt, Long term effects of methylmercuric chloride on three generations of brook trout (*Salvelinus fontinalis*): Toxicity, accumulation, distribution, and elimination, J. Fish. Res. Bd. Can., 33 (1976) 2726.

McKone, C.E., R.G. Young, C.A. Bache, and D.J. Lick, Rapid uptake of mercuric ion by goldfish, Environ. Sci. Technol., 5 (1971) 1138.

Miller, G.E., P.M. Grant, R. Kishore, F.J. Steinkruger, F.S. Rowland, and V.P. Guinn, Mercury concentrations in museum specimens of tuna and swordfish, Science, 175 (1972) 1121.

Murphy, P.G., and J.V. Murphy, Correlations between respiration and direct uptake of DDT in the mosquito fish *Gambusia affinis*, Bull. Environ. Contamin. Toxicol., 6 (1971) 581.

Olson, B.H., and R.C. Cooper, *In situ* methylation of mercury in estuarine sediment, Nature, 252 (1974) 682.

Olson, G.F., D.I. Mount, V.M. Snarski, and T.W. Thorslund, Mercury residues in fathead minnows, *Pimephales promelas* Rafinesque, chronically exposed to methylmercury in water, Bull. Environ. Contamin. Toxicol., 14 (1975) 129.

Olson, K.R., and P.O. Fromm, Mercury uptake and ion distribution in gills of rainbow trout (*Salmo gairdneri*): Tissue scans with an electron microprobe, J. Fish. Res. Bd. Can., 30 (1973a) 1575.

Olson, K.R., H.L. Bergman, and P.O. Fromm, Uptake of methyl mercuric chloride and mercuric chloride by trout: A study of uptake pathways into the whole animal and uptake by erythrocytes in vitro, J. Fish. Res. Bd. Can., 30 (1973b) 1293.

Olson, K.R., K.S. Squibb, and R.J. Cousins, Tissue uptake, subcellular distribution, and metabolism of $^{14}CH_3HgCl$ and $CH_3{}^{203}HgCl$ by rainbow trout, *Salmo gairdneri*, J. Fish. Res. Bd. Can., 35 (1978) 381.

Pan, S.K., N. Imura, and T. Ukita, Fractionation and characterization of mercury-methylation factor in tuna liver, Chemosphere, 6 (1973) 247.

Pentreath, R.J., The accumulation of inorganic mercury from sea water by the Plaice, *Pleuronectes platessa* L., J. Exp. Mar. Biol. Ecol., 24 (1976a) 103.

Pentreath, R.J., The accumulation of mercury from food by the plaice, *Pleuronectes platessa* L., J. Exp. Mar. Biol. Ecol., 25 (1976b) 51.

Phillips, G.A., and D.R. Buhler, The relative contributions of methylmercury from food or water to rainbow trout (*Salmo gairdneri*) in a controlled laboratory environment, Trans. Amer. Fish. Soc., 107 (1978) 853.

Rehwoldt, R., L.W. Menapace, B. Nerrie, and D. Alessandrello, The effect of increased temperature upon the acute toxicity of some heavy metal ions, Bull. Environ. Contamin. Toxicol., 8 (1972) 91.

Reinert, R.E., L.J. Stone, and W.A. Willford, Effect of temperature on accumulation of methylmercuric chloride and *p,p*'DDT by rainbow trout (*Salmo gairdneri*), J. Fish. Res. Bd. Can., 31 (1974) 1649.

Rodgers, D.W., and F.W.H. Beamish, Uptake of waterborne methylmercury by rainbow trout (*Salmo gairdneri*) in relation to oxygen consumption and methylmercury concentration, Can. J. Fish. Aq. Sci., 38 (1981) 1309.

Rottschafer, J.M., J.D. Jones, and H.B. Mark Jr., A simple, rapid method for determining trace mercury in fish *via* neutron activation analysis, Environ. Sci. Technol., 5 (1971) 336.

Rucker, R.R., and D.F. Amend, Absorption and retention of organic mercurials by rainbow trout and chinook and sockeye salmon, Prog. Fish. Cult., 31 (1969) 197.

Smith, F.A., R.P. Sharma, R.I. Lynn, and J.B. Low, Mercury and selected pesticide levels in fish and wildlife of Utah: I. Levels of mercury, DDT, DDE, dieldrin and PCB in fish, Bull. Environ. Contamin. Toxicol., 12 (1974) 218.

Trump, B.F., R.T. Jones, and S. Sahaphong, Cellular effects of mercury on fish kidney tubules, in Ribelin, W.E., and G. Migaki (Eds.), The Pathology of Fishes, The University of Wisconsin Press, Madison, Wisconsin, (1975) 585.

Varanasi, U., P.A. Robisch, and D.C. Malins, Structural alterations in fish epidermal mucus produced by waterborne lead and mercury, Nature, 258 (1975) 431.

Weir, P.A., and C.H. Hine, Effects of various metals on behavior of conditioned goldfish, Arch. Environ. Hlth., 20 (1970) 45.

Weis, J.S., and P. Weis, Effects of heavy metals on development of the killifish, *Fundulus heteroclitus*, J. Fish. Biol., 11 (1977) 49.

Weisbart, M., The distribution and tissue retention of mercury-203 in the goldfish (*Carassius auratus*), Can. J. Zool., 51 (1973) 143.

Westöö, G., Methylmercury as percentage of total mercury in fish and viscera of salmon and sea trout of various ages, Science, 181 (1973) 567.

Williams, D.M., and H.V. Weirs, Mercury in the marine environment: Concentration in sea water and in a pelagic food chain, J. Fish. Res. Bd. Can., 30 (1973) 293.

Winge, D.R., R. Premakuman, and K.V. Rajagopalar, Metal-induced formation of metallothionein in rat liver, Arch. Biochem. Biophys., 170 (1975) 242.

Wobeser, G., N.O. Nielsen, R.H. Dunlop, and F.M. Atton, Mercury concentrations in tissues of fish from the Saskatchewan River, J. Fish. Res. Bd. Can., 27 (1970) 830.

Wobeser, G., Acute toxicity of methyl mercury chloride and mercuric chloride for rainbow trout (*Salmo gairdneri*) fry and fingerlings, J. Fish. Res. Bd. Can., 32 (1975) 2005.

Wood, J.M., F.S. Kennedy, and C.G. Rosen, Synthesis of methyl-mercury compounds by extracts of a methanogenic bacterium, Nature, 220 (1968) 173.

CHAPTER IX
INTERACTIONS

Introduction

Few environmental contamination sites release only one toxic element into aquatic ecosystems. Coal-fueled electric generating stations, automobiles, mines, smelters, and other contaminate sources each release numerous elements. For example, particulate wastes from lignite-fired power plants release Se, Zn, Hg, Mn, Cr, Cu, and Ag into the water—so much so that fish kills result (Texas Department of Water Resources, 1978). In addition, As, Cd, Pb, and Ni can be discharged, providing a peerless mixture of a dozen or more elements in excess concentrations (Natusch, *et al.*, 1974).

Such elemental mixtures, even though similar, are often so unique that comparison between sites is difficult, if not impossible. Moreover, once internalized by vertebrates, each element within the mixture is a unique participant in a complex interaction (Fig. 1). And each is present in a specific chemical form and a specific ratio to other elements. Moreover, interactions among elements can be direct or indirect. In direct interactions, the elements compete for biological ligands, react chemically with one another, and/or alter metabolism by interacting with components of biochemical pathways (Sandstead, 1976). Synergistic and additive responses result. Indirect interactions at different steps in metabolic pathways ultimately influence some physiological process antagonistically. Needless to say, the impact of a given element on a fish species is determined by a multitude of parameters, even in a situations involving only a few elements in a mixture.

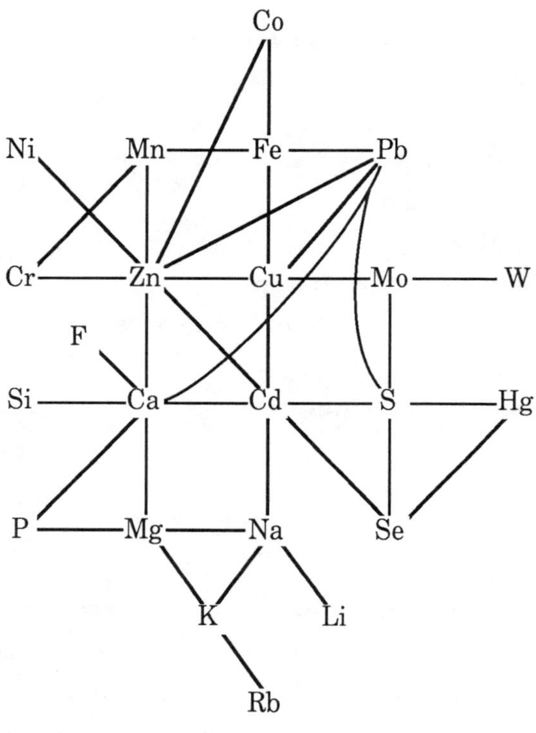

Figure 1. Solid lines interconnect elements to show interactions known to exist under experimental conditions in vertebrates. Although most reactions shown are antagonistic (competitive), synergistic and additive interactions are common. And just as Zn and Cu components exist for Cd-induced toxicity (shown), so other interactions stem from increased or decreased quantities of the elements involved. (An original diagram based on Sandstead, (1976); Hill, (1963); and others.)

In studies of feral fish, often the most abundant element is considered, without regard to moderate levels of other elements. Similarily, in laboratory experiments, reagent grade chemicals are added to solutions and trace contaminates of reagent grade chemicals or leachates from aquaria can go unnoticed and lead to a similar complexity of exposures. This chapter condenses a number of published studies which involve two or more elements present in the same exposure. As is the case for the studies upon which this section is based, information presented represents a simplification of a more complicated scheme in nature.

Evaluation of the Toxicity of a Mixture

Brown (1968) estimates the toxicity of a mixture of metals (A and B, for example) by dividing the concentration of each metal by the calculated 48-h LC50 for that metal when the metal is added to aqueous solutions. This provides the proportion of the lethal concentration as follows:

$$\frac{[\text{Metal A or B}]}{\text{48-h LC50}} = \text{Proportion of Lethal Concentration}$$

Values for each metal are totaled to give the proportion of the 48-h LC50 for the metals in the mixture. If the value of the sum is <1.0, less than half of fish usually die in 48-h. If the value is >1.0, then more than half of fish are likely to die in the same time period. As an example, suppose that the concentration of Zn and Cu in the water are 0.62 and 0.12 ppm, respectively, and the 48-h LC50 for Zn and Cu are 2.8 and 0.4 ppm, respectively. Zinc contributes $0.62/2.8 = 0.22$ of the lethal concentration; while, Cu contributes $0.12/0.4 = 0.30$ of the lethal concentration. The total of the two is 0.52; therefore, less than half of fish are predicted to die in 48 h.

Empirically, this method assumes that a metal is a stressor, producing a degree of shock with nonspecific results. Summation of stressors might seem reasonable were it not for the fact that metals often target specific systems and are capable of producing antagonistic and synergistic effects, as well as the additive effects upon which Brown's (1968) procedure are based. However, Brown does acknowledge the under-estimation of toxicity by summation of proportions and the existence of 50% mortalities at a sum of 0.7 rather than of 1.0.

Toxicity of Specific Mixtures of Elements

Copper and Zinc: Additive and synergistic relationships among pairs of metals are concentration-dependent. This is known to be true for Cu and Zn (Doudoroff and Katz, 1953; Lloyd, 1961; Sprague and Ramsay, 1965; Birge and Black, 1979). For example, if a 1:1 ratio of Cu:Zn is used on catfish, goldfish, or bass embryos; hatchability (a measure of embryonic survival) varies with the concentration of the metals involved (Fig. 2). Synergism occurs at high concentrations; whereas, additive effects are noted at concentrations of ≤0.5 ppm for bass, ≤1.0 ppm for catfish, and ≤10.0 ppm for goldfish.

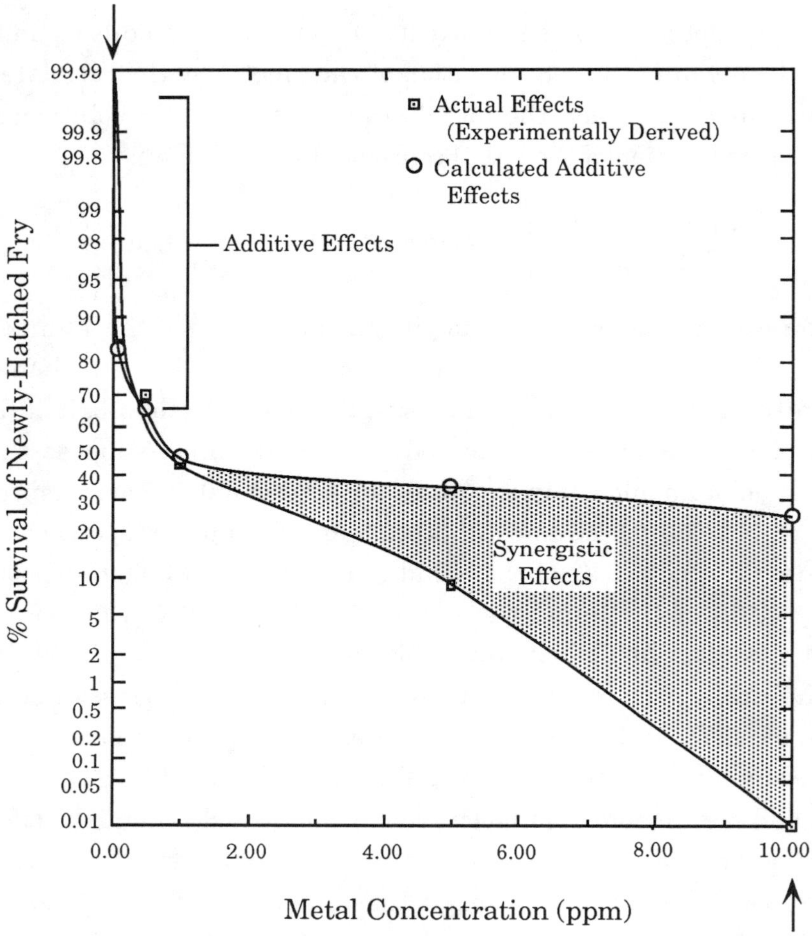

Figure 2. Synergism and additive effects on hatchability of channel catfish (*Ictalurus punctatus*) embryos exposed to equal concentrations of Cu and Zn. Copper and zinc are additive at lower concentrations, but synergistic at higher concentrations. (This figure is derived from numerical data presented by Birge and Black (1979).)

Elemental interactions are complex. Over 25 yr ago, Sprague and Ramsay (1965) wrote that no simple way can be used to express the effects of two or more pollutants when these occur together. More recent work corroborates this fact (Table 1) and shows some of the factors determining whether Cu-Zn interactions are synergistic or additive. These factors include concentration (Sprague and Ramsay, 1965; Birge and Black, 1979), solubility or water hardness (Lloyd, 1961), the parameter measured (e.g. ILL or LT50) (Sprague and Ramsay, 1965), and the species considered (Birge and Black, 1979). Brief discussion details results of some of these studies.

Sprague and Ramsay (1965) expose Atlantic salmon (*Salmo salar*) to Cu and Zn, alone or in combination at 17°C, pH 7.3, and 14 ppm total hardness as $CaCO_3$. When used singly, Cu has an ILL (incipient lethal level) of 32 ppb Cu. In comparison, the ILL for Zn is 420 ppb, making Zn approximately thirteen times less toxic than Cu under these conditions in Atlantic salmon. By setting 420 ppb Zn and 32 ppb Cu each equal to 1.0 toxic units, the ILL for Zn-Cu mixtures is simply additive (Table 1). But if LT50 (rather than ILL) is measured, Cu and Zn interact synergistically. Also, when mixtures are stronger (totaling two and five toxic units), fish die faster (synergism). Therefore, at higher Zn and Cu concentrations, survival time is potentiated (synergistic) in Atlantic salmon.

Lloyd (1961) expose rainbow trout (*Salmo gairdneri*) to Cu or Zn singly or to Cu-Zn mixtures (15.5–18.0°C, 320 or 15–20 ppm total hardness). The 3-d LC50 levels are 3.5 ppm Zn and 1.1 ppm Cu in hard water (320 ppm). Semi-logarithm plots of LT50 *versus* metal concentration suggest that both metals exert their toxic action in a similar way. A 6:1 ratio of a mixture of soluble Zn to soluble Cu causes additive toxicity in hard water (Table 1), but synergistic toxicity in soft water (15–20 ppm). The 7-d LC50 values are 0.56 ppm Zn and 0.044 ppm Cu in soft water when the two elements are tested singly. Curves for median survival *versus* metal concentration are similar to those in hard water for single metal exposures and for joint metal exposures (6:1 ratio of soluble Zn to soluble Cu) at low concentrations. However, at higher concentrations of mixtures of soluble forms of the two metals (Table 1), synergistic interactions are observed (i.e. survival times are shorter than predicted). This Zn-Cu synergism at high metal levels is reviewed by Doudoroff and Katz (1953).

Sellers *et al.* (1975) cannulate rainbow trout (*Salmo gairdneri*) to monitor opercular and buccal pressure changes (ventilation amplitude, ventilation rate, and coughing frequency) during exposure to Cu and Zn

Table 1. Copper-zinc interactions at various concentrations and elemental ratios. Copper and zinc levels are given as ppb; water hardness is provided as ppm.

Effect	Parameter	[Zn]	[Cu]	[Zn]:[Cu]	Hardness	Species	Reference
Synergistic	LT50[1]	420	32	13:1[4]	14	Atlantic salmon (*Salmo salar*)	Sprague & Ramsay, 1965
Additive	ILL[2]	420	32	13:1[4]	14	Atlantic salmon (*Salmo salar*)	Sprague & Ramsay, 1965
Synergistic	LT50	560	44	13:1[5]	15–20	Rainbow trout (*Salmo gairdneri*)	Lloyd, 1961
Synergistic	Ventilatory Activity[3]	690	70	10:1	51–68	Rainbow trout (*Salmo gairdneri*)	Sellers et al., 1975
Additive	Hatchability	100–500	100–500	1:1	100	Largemouth bass (*Micropterus salmoides*)	Bridge & Black, 1979
Additive	LT50	3500	1100	3:1[5]	320	Rainbow trout (*Salmo gairdneri*)	Lloyd, 1961
Synergistic	Hatchability	≥1000	≥1000	1:1	100	Largemouth bass (*Micropterus salmoides*)	Bridge & Black, 1979

[1]Median survival time or lethal time for 50% of the group of fish being tested.
[2]Incipient lethal level or concentration above which fish can no longer survive for an indefinate period of time.
[3]Ventilatory activity is evaluated in terms of ventilation amplitude, ventilation rate, and coughing frequency.
[4]Equilivant to the ratio of one toxic unit of Zn to one toxic unit of Cu.
[5]Equilivant to a 6:1 ratio of soluble Zn to soluble Cu.

singly and in combination. Intermediate concentrations of Cu or Zn (0.07 ppm Cu and 0.69 ppm Zn) elicit greater responses in ventilation amplitude, ventilation rate, and coughing frequency than higher concentrations. Higher Cu or Zn levels cause more variation in ventilation pressure changes than lower Zn levels. Singly or in combination, Zn causes increased coughing frequency; whereas, Cu does not. Therefore, combined exposure to Zn-Cu results in synergism at intermediate levels of the elements. In fact, coughing frequency at 0.04 ppm Cu and 0.66 ppm Zn combined is about one and one-half times greater than the sum of the responses to the single toxicants. The marked individual variability in ventilatory activity is noteworthy and is generally considered a result of irritation of buccal (and possibly gill) membranes. Amplitude may be low and ventilation rate may be high for one subject (or for one concentration) but not for a second subject (or concentration). Unfortunately, the authors do not report total respiratory activity for individual fish using ventilation rate and amplitude date (e.g. volume of water breathed per unit time). Perhaps the individual variation would be reduced in such comparisons.

Copper and Mercury: Evaluation of hatchability of trout embryos shows synergistic, additive, and antagonistic relationships between Cu and Hg (Fig. 3). As with Cu and Zn, synergistic interaction exists at high Cu and Hg concentrations in the water (Birge and Black, 1979). Additive effects are noted at an intermediate level of about 0.03 ppm of equal proportions of either element. Antagonism is noted at low levels ($\leq$0.01 ppm of equal proportions of each metal). Moreover, the complexity of elemental interactions is confirmed in such comparisons, although Cu-Hg interactions seem less complicated for channel catfish (*Ictalurus punctatus*) and goldfish (*Carassius auratus*) than for largemouth bass (*Micropterus salmoides*) and rainbow trout (*Salmo gairdneri*) (Birge and Black, 1979). The LC50 values show Hg to be twenty-five times more toxic than Cu to bass, trout, catfish, and goldfish under conditions of this series of studies.

Interactions between Cu and Hg at the epidermis of fish hint of the role of mucus in metal poisoning of fish. Epithelial mucus from plaice (*Pleuronectes platessa*) binds Cu and Zn at levels 100-fold and 20-fold greater, respectively, than levels in water during exposures of fish to low aqueous concentrations (Coombs, *et al.*, 1972). Dialysis of mucus against deionized water results in only a small decrease in the concentration of Cu and Zn bound to mucus. Glycoproteins low in sialic acid, aromatic and sulfur-containing amino acids, phosphate, and sulphate appear to be involved in binding of the two divalent cations. Moreover, Cu^{+2}, Zn^{+2}, and Hg^{+2} precipitate fresh plaice mucus in the order Cu > Zn > Hg (Jones, 1938). Mucus serves a protective function by binding excess aqueous metals as a precipitate.

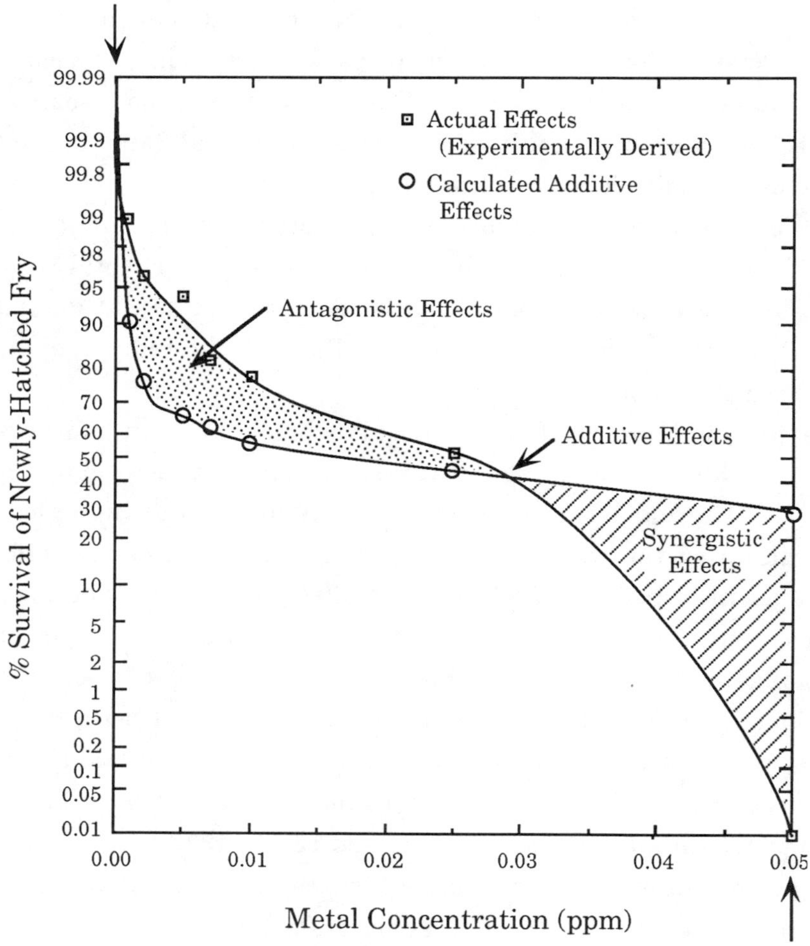

Figure 3. Synergism, additive, and antagonistic effects on hatchability of rainbow trout (*Salmo gairdneri*) embryos exposed to equal proportions of Cu and Hg. Copper and mercury are antagonistic at lower concentrations, additive at intermediate concentrations, and synergistic at higher concentrations. (This figure is derived from numerical data presented by Birge and Black (1979).)

Copper and Methylmercury: Roales and Perlmutter (1977) find that blue gouramis (*Trichogaster trichopterus*) respond to Cu or CH_3Hg^+ or both at 10% of the 96-h TLm (90 ppb) at 26–28°C. In this study, antigens are injected 1 and 3 wk following exposure of gouramis to Cu or Hg. Viral neutralization tests show that exposure of blue gouramis to 9 ppb Cu, Hg, or both causes appreciable decreases in antibody production. However, exposure to both Cu and Hg does not result in appreciable changes compared with single exposures to the same metals. The exact mechanism whereby these elements decrease antibody production is unknown, but could involve breakdown of antibodies, interference with complement production, alteration of protein synthesis, cytotoxic effects on the hemopoietic centers in fish (e.g. spleen), or hematologic changes. These effects are exerted at levels of Cu and Hg less than those affecting growth and reproduction.

Copper, Zinc, and Nickel: Brown and Dalton (1970) find that a mixture of Cu:Zn:Ni at a concentration ratio of 1:5:40 produces additive effects over the 0.14–0.48 ppm Cu concentration range in 1-yr-old rainbow trout (*Salmo gairdneri*) (15.3–18.4°C, pH 7.4–7.8, 240 ppm water hardness). These results agree with the Birge and Black (1979) study which shows additive effects for 1:1 mixtures of Cu:Zn in the 0.1–0.5 ppm concentration range for goldfish, catfish, and bass. Amazingly, the results of the embryonic (hatchability) studies of Birge and Black (1979) are reasonably close to those of studies involving 1-yr-old trout (Brown and Dalton, 1970) for two of three species tested. Unfortunately, of species tested, trout are not exposed to 1:1 ratios of Cu and Zn mixtures (Birge and Black, 1979); therefore, no comparison is possible between results of Birge and Brown and those of Brown and Dalton.

Selenium and Mercury: The presence of Se reduces the toxicity of (i.e., is antagonistic toward) Hg, Cd, As, and Ag in some studies (Frost and Lish, 1975; Turner and Rudd, 1983a; Turner and Swick, 1983b; Cuvin and Furness, 1988), but not in others (Huckabee and Griffith, 1974). As with interactions between Zn and Cu (Fig. 2) and between Cu and Hg (Fig. 3) (Birge and Black, 1979), interactions between Hg and Se are known to be concentration-dependent (Kim, *et al.*, 1977). Interactions between Hg and Se can be synergistic at low aqueous Hg levels (≤0.07 ppm) and antagonistic at high Hg levels (≥0.10 ppm) (Kim, *et al.*, 1977). Other studies show that Se and Hg interact synergistically (Huckabee and Griffith, 1974). Most workers, however, observe antagonistic interactions between Se and Hg (Frost and

Lish, 1975; Turner and Rudd, 1983a; Turner and Swick, 1983b; Cuvin and Furness, 1988).

Kim, Birks, and Heisinger (1977) find that Se (as selenite) is antagonistic to Hg^{+2} (0.10–0.19 ppm) in aqueous solutions for northern creek chubs (*Semotilus atromaculatus*) (24–25°C and 50 ppm water hardness). A 48-h pretreatment with 3 ppm Se reduced Hg^{+2}-induced mortality in chubs. Although ≥12 ppm Se causes heavy mucous secretion and death in this species, 3 ppm Se pretreatments prevent Hg^{+2}-induced poisoning.

Cuvin and Furness (1988) expose *Phoxinus phoxinus*, fresh-water minnows, to 0.01–0.30 ppm Hg^{+2} in the presence or absence of 0.05–0.60 ppm Se (as selenate). Selenium protects minnows against Hg^{+2}-induced toxicity; therefore, Se is antagonistic at 2.5:1 molar ratio of Hg:Se. However, a 1.3:1 molar ratio of Hg:Se causes increased mortality (synergism) compared with Hg only (at 0.3 ppm Hg). Therefore, the studies of Cuvin and Furness (1988) and Kim *et al.* (1977) indicate that antagonistic and synergistic toxic interactions between Se and Hg are a function of the concentration of the two elements and the molar ratio of one to the other.

High concentrations of Hg and Se are noted for being synergistic in carp embryos (Huckabee and Griffith, 1974). Huckabee and Griffith (1974) strip and fertilize eggs of carp (*Cyprinus carpio*) for exposure to 1–5 ppm Se (as selenite, since addition of SeO_2 to water forms SeO_3^{-2}) in the presence or absence of 1–5 ppm Hg (as $HgCl_2$) at 26°C and pH 7.3. In Table 2, only those embryonic exposures to 1–2 ppm Hg and 1–5 ppm Se are considered since gradual effects are produced in those concentration ranges. Normally, embryos hatch in 60–72 h with >96% hatchability. No embryos hatch in ≥4 ppm Hg^{+2} (administered singly), but even 5 ppm Se does not deter hatchability. A level of 3 ppm Hg^{+2} decreases hatchability to 12%. The presence of Se at increased aqueous levels decreases hatchability; therefore, Se potentiates Hg-induced toxicity in carp embryos. Possibly, the high affinity of Hg and Se for sulfhydryl groups on the outer embryo membrane causes osmotic imbalance. However, Hg and Se may be accumulated by internal embryonic tissues resulting in irreversible injury, rather than simply interacting with the outer membranes.

Since interactions between Se and Hg are known to be synergistic but have not been fully elucidated, it is not wise to attempt use of Se

Table 2. Mercury-selenium interactions at various concentrations and elemental ratios.

Effect	Parameter	[Hg] (ppm)	[Se] (ppm)	[Hg]:[Se] Concd. [1]	[Hg]:[Se] Molar [2]	Species	Reference
Antagonistic	% Mort. [3]	0.10-0.16	3.0	0.03:1 0.05:1	0.08- 0.13	Chubs *(Semotilus atromaculatus)*	Kim *et al.*, 1977
Synergistic	% Surv. [4]	0.30	0.60	0.50:1	1.25	Minnows *(Phoxinus phoxinus)*	Cuvin & Furness, 1988
Synergistic	Hatchab. [5]	1.00- 2.00	1.00- 5.00	1:1 0.4:1	1.00- 2.50	Carp *(Cyprinus carpio)*	Huckabee & Griffith, 1974

[1]Ratio of the concentration of Hg to Se.
[2]Molecular weight ratio of Hg to Se.
[3]Mercury-induced percent mortality during a 48-h exposure which followed a 48-h, 3-ppm Se pretreatment.
[4]Percent survival of minnows in a 10-d exposure to both Hg and Se in solution.
[5]Hatchability of carp embryos, a measure of embryonic survival

supplementation to decrease the severity of point-source Hg contamination, as has been suggested (Turner and Rudd, 1983a; Turner and Swick, 1983b). Synergistic interactions even on a single life stage, as that observed on carp embryos (Huckabee and Griffith, 1974), have the potential of eliminating an entire population of fish.

Accumulation of Elements from Mixtures

Copper, Zinc, Mercury, Iron, and Manganese: A few environmental studies address accumulation levels for mixtures of metals. Cross and workers (1973) catch fish at 2500 m deep near Cape Hatteras for analysis of levels of Hg in white muscle. Mercury levels increase with body weight ($p<0.001$) for bluefish (*Pomatomus saltatrix*) and morid (*Antimora rostrata*). Bluefish are epipelagic (living in the part of the ocean into which light penetrates) and morids are bathyl-demersal (living near the sea bottom in a biogeographic realm about 180–1800 m deep). Mercury accumulation is probably increased as a result of high lipid solubility, high electronegativity, and/or high affinity for sulfhydryl groups. Decreasing levels of all metals except Hg are noted for morids—an effect possibly due to growth dilution effects, compositional changes in muscle, and/or dietary changes in metal levels. In contrast to Hg levels, the concentrations of Mn, Fe, Cu, and Zn decrease or remain unchanged. In white muscle, the concentration factors (CF) for Hg, Mn, Cu, Zn, and Fe are 3700, 110, 200, 2100, and 2300, respectively. Obviously, metal accumulation patterns vary as a function of species, fish size, and metal analyzed.

Selenium and Mercury: Freeman *et al.* (1978) determine levels of Se and Hg in skeletal muscle from about fifty swordfish (*Xiphias gladius*) collected from the northwestern Atlantic Ocean. About 80% of fish have muscle levels of Hg above the World Health Organization limit of 0.5 ppm. In those swordfish having muscle levels up to 2.0 ppm Hg, Se levels exceed those of Hg. Therefore, a molar excess of Se to Hg is measured rather than the predicted molar relationship between the two elements. Since nearly 100% of Hg in swordfish muscle is likely to be present as CH_3Hg^+ (see Chapter VIII) and since the presence of Se has been shown to lower the toxicity of CH_3Hg^+, swordfish muscle might not be as toxic to humans as it would be without Se. However, the levels of up to 4.4 ppm Hg in swordfish muscle in this study remain a cause for concern. Consumption of such elevated levels of mercurials in the CH_3Hg^+ form could be hazardous to human health.

As indicated, Kim *et al.* (1977) observe antagonism between aqueous Se (as selenite) and Hg^{+2} in northern creek chubs (*Semotilus atromaculatus*) exposed at 24–25°C and 50 ppm water hardness. A 48-h pretreatment with 3 ppm Se decreases Hg^{+2}-induced mortality throughout the concentration range for Hg (0.07–0.19 ppm). However, with identical Se pretreatment, low aqueous Hg levels (0.01–0.07 ppm), favor elevated Hg uptake (synergism); whereas, higher Hg levels (0.10–0.16 ppm) favor reduced Hg uptake (antagonism). Therefore, toxicity and uptake data agree for higher Hg levels but not for lower Hg levels.

Cuvin and Furness (1988) expose *Phoxinus phoxinus*, fresh-water minnows, to 0.01–0.30 ppm Hg^{+2} in the presence or absence of 0.05–0.60 ppm Se (as selenate). As indicated, selenium protects minnows against Hg^{+2}-induced toxicity at a 2.5:1 (but not at a 1.3:1) molar ratio of Hg:Se. For this species, Se-induced potentiation of toxicity induced by Hg^{+2} may be due to increased Hg uptake in the presence of Se. The presence of Hg decreases Se elimination. Biological half-lives for Hg and Se (in single exposures) are 62 d and 16 d, respectively. However, the half-life of Se increases to 27 d during simultaneous exposure of minnows to 0.05 ppm stable Hg and 10μCi ^{75}Se (as selenite). Apparently, in this teleost species, identical results are obtained with both the selenite and the selenate. These results suggest that a Hg-Se complex is formed during simultaneous exposure of minnows to the two elements.

Opposite results are obtained for perch and pike from Clay Lake, a site receiving Hg contamination for at least 20 yr (Turner and Rudd, 1983a; Turner and Swick, 1983b). In this study, fish exposed for several generations to excess Hg (see Chapter VIII) are exposed to Se, resulting in concentration-dependent reductions in Hg levels. Therefore, fish containing excessive levels of Hg for a lifetime process Hg and Se in a different way than previously unexposed fish. That is, normally, levels of both elements increase simultaneously (Freeman, *et al.*, 1978; Boecker and Sorensen, 1988; Cuvin and Furness, 1988). However, in the Turner studies (Turner and Rudd, 1983a; Turner and Swick, 1983b), the presence of Se decreases Hg uptake. As with other elemental combinations, the concentration and molar ratios are of great importance in accumulation of Hg and Se in combined exposures of fish (Kim, *et al.*, 1977).

Koeman *et al.* (1973) conduct multielemental analyses of livers from the common seal (*Phoca vitulina*), dolphins (*Delphinus delphis*, *Tursiops truncatus, Lagenorhynchus obscurus*, and *Sotalia guianensis*), and porpoises (*Phocoena phocoena*). Levels of Hg and Se are strongly correlated, but no similar relationship exists between Se and Cd, As, Sb, Zn, or other elements. About 57% of Hg and Se are associated with nuclei and cell walls of a fresh liver homogenate of a seal. Moreover, only 2–14% of total Hg in seal liver is CH_3Hg^+. The 1:1 molar ratio for these marine mammals is not characteristic of other species, which can contain forty times more Se than Hg.

The Freeman *et al.* (1978) study parallels that of Ohi *et al.* (1975), who work with rodents. More Se is accumulated if CH_3Hg^+ is present. Male Wistar rats are divided into four groups and fed diets as follows: (1) 20 ppm CH_3Hg^+, (2) 20 ppm CH_3Hg^+ plus 3 ppm Se (as selenite, in drinking water), (3) 3 ppm selenite in water, and (4) basal diet (with 0.4 ppm organic Se). The presence of Se delays and mitigates the expression of CH_3Hg^+ intoxication, based on growth and morbidity. The CH_3Hg^+ and total Hg levels in the brain, liver, kidney, and blood increase in like fashion. However, Se accumulation in all organs (especially the kidney) is accelerated in rats treated with Se and Hg. In fact, an eight-fold increase in Se uptake is noted on day 41 for the kidney. In addition, the presence of 3 ppm Se in drinking water increases growth rate in rats consuming 20 ppm CH_3Hg^+ in the diet. Protective effects of Se are noted at 4 wk ($p<0.001$). Mercury causes renal enlargement and Se causes hepatic enlargement, but a mixture of the two elements does not produce a change in organ size. The exact mechanism of action for Se-induced protective effects in vertebrates poisoned with CH_3Hg^+ has not been elucidated.

Selenium, Mercury, Zinc, and Iron: In some areas in which environmental Se and Hg are elevated, however, Hg is not accumulated and other elements interact with Se. One such site is Martin Lake in east Texas site at the Texas Utilities Generating Company (TUGCO) electric generating station. Because of the manner in which Se-contaminated fly ash, scrubber sludge, and bottom ash are dispersed from ash ponds, fish kills result. A legal suit raised by the Texas Attorney General's Office (TAGO) against TUGCO has lead to numerous TAGO studies of sunfish inhabiting Martin

Lake (Sorensen, *et al.*, 1982a; 1982b; Sorensen and Bauer, 1983a; Sorensen, *et al.*, 1983b; 1983c; Sorensen and Bauer, 1984a; 1984b; Sorensen, 1986; Sorensen and Thomas, 1988a; Sorensen, *et al.*, 1988b).

Since the liver is considered one of the major accumulators of Se and Zn (Marafante, 1976; Matthiessen and Brafield, 1977; Sorensen, 1986; Sorensen and Thomas, 1988a; Sorensen, *et al.*, 1988b), multielemental analysis of hepatic tissue is used to establish whether the levels of Se, Zn, Fe, and Hg in contaminated fish differ from those of reference lake fish. Redear sunfish (*Lepomis microlophus*) show higher hepatic Se and Zn (356% and 132% increases, $p<0.001$) but lower hepatic Fe (54% decrease, $p<0.001$) (Table 3) (Sorensen and Thomas, 1988a; Sorensen, *et al.*, 1988b). No change is noted for Hg concentrations in the liver. Therefore, hepatic concentrations of Se and Zn are directly correlated, while Se and Fe are inversely correlated, and Hg levels are not altered.

Table 3. Comparison of the levels of Se, Zn, Fe, and Hg (in ppm, wet weight) in the livers of redear sunfish from Martin Lake from a July 1986 collection (Sorensen and Thomas, 1988a; Sorensen, 1988b). Values are mean ± one standard error of the mean.

Element	Lake Tyler[1]	Martin Lake[2]	Change (%)[3]
Se	2.081 ±0.185	7.398 ± 0.387***	356% Increase
Zn	18.793 ± 0.812	24.858 ± 0.844***	132% Increase
Fe	446.345 ± 72.285	239.370 ± 20.914***	54% Decrease
Hg	0.142 ± 0.026	0.174 ± 0.021	No Change

***Significantly different from the reference lake value by $p \leq 0.001$.

[1]Reference lake

[2]Contaminated lake.

[3]Percent change in Martin Lake fish compared with reference fish.

Therefore, and in reference to Table 3, despite the passage of eight years since the initial discharges of Se into the lake and despite an alleged reduction in the annual discharge of Se, hepatic levels of Se in redears are about four times higher than those of reference fish from Lake Tyler in this study (Sorensen and Thomas, 1988a; Sorensen, *et al.*, 1988b). The source of this contamination probably can be ascribed to excess sedimentary levels of Se because it has been established that elevated water levels of Se would not have to be maintained continuously if periodic pulses could sufficiently load the food web (Turner and Swick, 1983b). Periodic pulses would elevate sedimentary Se levels, resulting in loading of the food web. In July 1986, hepatic Se in Martin Lake redears had decreased about 25% from the July 1981 values of 10.0 ± 1.3 ppm (Sorensen, *et al.*, 1982b). Because redears consume sediments, along with benthic organisms (e.g. snails), this species remains in intimate contact with Se-contaminated sediments. Slow reduction in hepatic pools of Se (25% in 5 yr) is expected under conditions in which up to 75% of the redear gut contains Se-contaminated sediment.

Tests of water from East Ash Pond and West Ash Pond at Martin Lake show water levels of 2200–2700 ppm Se, 260–1200 ppm Zn, and 0.2–0.3 ppm Hg (Texas Department of Water Resources, 1978). No aqueous level of Fe is presented. Despite the fact that ash pond water contains higher elemental concentrations than does water in the main basin of Martin Lake, elevated levels of Se and Zn probably account for the elevated hepatic levels of Se and Zn in redears. Normally, aqueous levels of Se, Zn, and Hg are ~6, <20, and <0.2 ppb, respectively, for surface waters in east Texas (Texas Department of Water Resources, 1978).

To simulate environmental exposures of this same species to Se at electric generating stations, a laboratory exposure of redear sunfish was conducted (Boecker and Sorensen, 1988). In this report, young adults are exposed for 7 d or 14 d to measured aqueous levels of Se at 0.0015, 0.8128, 7.7843, 33.973, or 71.8840 ppm Se. Mercury levels were maintained at 0.0025, 0.0175, 0.1520, 0.6700, or 1.4517 ppm to provide an exposure in which redears were exposed to both Se and Hg (Fig. 4). Levels of Zn and Fe are held constant at approximately 0.0395 ppm Zn and 0.6467 ppm Fe. This experiment tests whether hepatic levels of Se, Hg, Zn, and Fe are altered within a 1- and 2-wk period under conditions in which Se and Hg are elevated and Zn and Fe are stable.

Under these conditions, both Se and Hg are present in the water in elevated concentrations at a ratio of fifty to one (50 atoms of Se for every atom of Hg) (Fig. 4). Hepatic levels of both Se and Hg increase following 7-d (Fig. 5) and 14-d (not shown), although Hg uptake is more variable than Se uptake. Statistically significant levels of Se are present after 7-d exposures to 0.8128 ppm Se ($p<0.001$), compared with 0.152 ppm Hg ($p<0.01$) under these conditions. The 14-d exposure data reveals the same trend, although

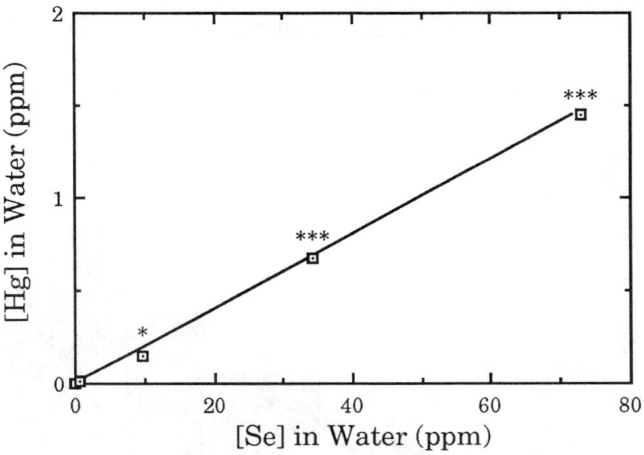

Figure 4. Correlation between aqueous levels of Se and Hg for 7-d and 14-d exposures of redear sunfish (*Lepomis microlophus*) to 0.0015–71.884 ppm Se and 0.0025–1.4517 ppm Hg in a simultaneous exposure. Significance levels are provided between control levels of Hg (0.0025 ppm)-Se (0.0015 ppm) and higher levels of the two elements ($p<0.001$, ***; $p<0.01$, **; and $p<0.05$, *). (This is an original figure derived from data collected for Sorensen and Thomas (1988a) and Sorensen, *et al.* (1988b) in collaboration with Dr. P. Thomas, Dr. T. Bauer, and Dr. F. Iskandar.)

the severity of the effect is more pronounced. Elevated levels of Hg ($p<0.05$) accumulate in redear livers at measured aqueous levels of 0.0175 ppm Hg after 14-d exposures.

Similarly, the 14-d concentration factors (CF) are statistically significant for Hg; whereas, 7-d CF data show no significant trends (Fig. 6). The 14-d Hg CF values are about 50.4, 9.4, and 2.2 for redears at aqueous levels of 0.0025, 0.0175, and 0.152 ppm Hg, respectively. In contrast, 14-d Se CF values are about 1211.3, 5.3, and 1.5 for redears at aqueous levels of 0.0015, 0.8128, and 7.7843 ppm aqueous Se (Fig. 6A). The 7-d CF values are about 1240.0, 4.2, 1.1, 0.4, and 0.3 for redears at measured aqueous levels of 0.0015, 0.8128, 7.7843, 33.973, and 71.884 ppm aqueous Se. These data agree with that of others who show that CF is inversely related to the exposure concentration of the element (Eisler, 1971; Eisler, 1974; Pascoe and Mattey, 1977). Ultimately, this relationship is probably due to a finite limit in the ability of fish to accumulate metals from the aquatic habitat (Eisler, 1974). However, other factors apparently alter this result since Saiki and

Lowe (1987a) publish data which does not show an inverse relationship between concentration factor and Se levels in the water (see Chapter II).

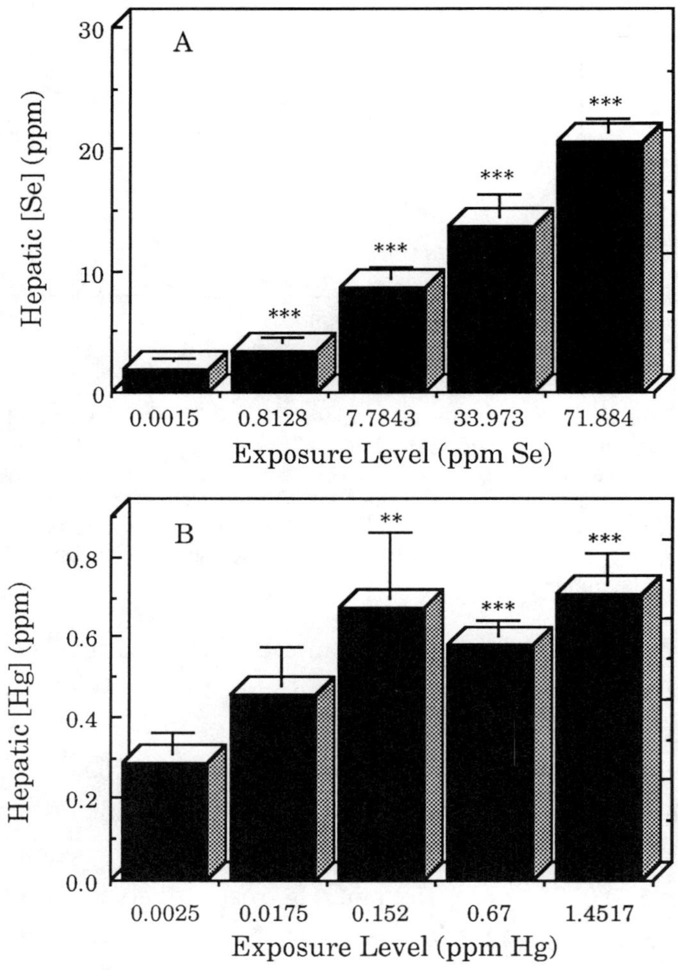

Figure 5. Accumulation of Se and Hg in livers of redear sunfish (*Lepomis microlophus*) following 7-d exposures to aqueous Se (A) and Hg (B) simultaneously at 26°C, pH 8.32, and 169 ppm total hardness. Averages and one standard error of the mean are presented. Asterisks provide significance levels between control fish (0.0025 ppm Hg and 0.0015 ppm Se) and fish exposed to higher levels of the two elements ($p<0.001$, *** and $p<0.01$, **). (This is an original figure derived from data collected for Sorensen and Thomas (1988a) and Sorensen, *et al.* (1988b) in collaboration with Dr. P. Thomas, Dr. T. Bauer, and Dr. F. Iskandar.)

Histopathological evaluation of field-exposed and laboratory-exposed redears shows similarity in effects (Boecker and Sorensen, 1988). Livers show focal necrosis, vacuolation, fatty infiltration, and architectural disorganization. Kidneys show proliferative glomerulonephritis, periglomerular fibrosis, and degeneration of proximal convoluted tubule cells.

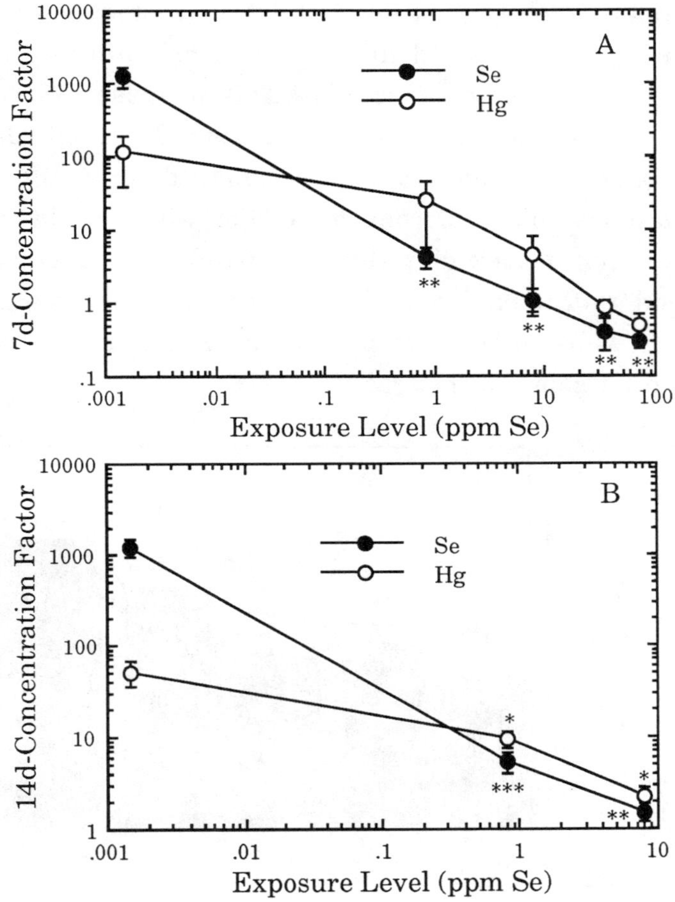

Figure 6. Hepatic concentration factors (CF) for 7-d (A) and 14-d (B) exposures of redears (*Lepomis microlophus*) to Se and Hg in the water (26°C, pH 8.32, and 169 ppm total hardness). Mean and one standard error of the mean are provided, as are significance levels of exposed *versus* control fish ($p<0.001$, ***; $p<0.01$, **; and $p<0.05$, *). Hepatic CF are inversely related to Se and Hg exposure levels in all cases except 7-d CF for Hg (A). (This is an original figure derived from data collected for Sorensen and Thomas (1988a) and Sorensen, *et al* (1988b) in collaboration with Dr. P. Thomas, Dr. T. Bauer, and Dr. F. Iskandar.)

The severity of these alterations increases with exposure to higher concentrations of Se. Changes are consistent with those observed following environmental exposures of this or congeneric species to Se in Martin Lake and Belews Lake in North Carolina (Sorensen, 1986).

Hepatocytes, as well as all cells of the vertebrate body, are protected from the action of toxic levels of ions such as Se, by a system of enzymes commonly referred to as the antioxidant defense system. At least one of these enzymes increases three-fold in activity (nonsignificant) with increased exposure to Se (Hilton, *et al.*, 1980). This increase in activity probably compensates for the oxidative stress induced by Se. The hydrogen peroxide and superoxide ion produced by Se-damaged or aging cells must be decomposed to prevent membrane destruction which would otherwise occur. Several such enzymes are glutathione-s-transferase, superoxide dismutase, catalase, and glutathione peroxidase. Recent preliminary studies indicate that a 7-d exposure to about 72 ppm selenate results in an increase in the activity of glutathione-s-transferase (Fig. 7).

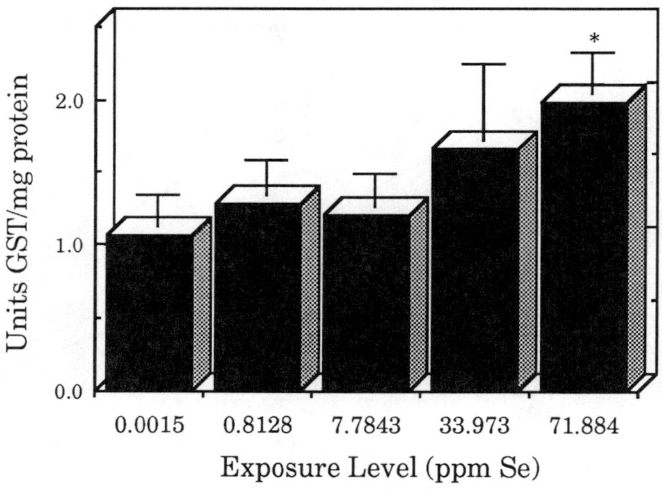

Figure 7. Activity of glutathione-s-transferase per mg protein in livers of redear sunfish (*Lepomis microlophus*) exposed for 7-d to the measured Se concentrations indicated (0.0015–71.884 ppm Se) in the water (26°C, pH 8.32, and 169 ppm total hardness). The GST activity is significantly elevated ($p<0.05$, *) after the 7-d exposure to 71.884 ppm Se in the water. (This is an original figure derived from data collected for Sorensen and Thomas (1988a) and Sorensen, *et al.* (1988b) in collaboration with Dr. P. Thomas, Dr. T. Bauer, Dr. F. Iskandar, and Dr. M. Murphy.)

Summary

In conclusion, numerous elements are released by a single contamination site, to make each elemental mixture unique. However, individual elements which are the most abundant generally are studied without reference to others present in a complex mixture. Moreover, numerous interactions have been documented for various laboratory animals; and synergistic, antagonistic, and additive biological effects are possible for fish exposed to mixtures. Categorization of elemental mixtures as synergistic, antagonistic, or additive depends on the concentration of the elements in the mixture, the solubility or water hardness, the molar and concentration ratios of elements, the parameter measured, the species considered, and other factors.

The ratio of the concentration or molecular weight of one element to the other is of primary importance. In catfish, a 1:1 ratio of toxic units for Zn:Cu produces additive effects at lower concentrations, but synergistic effects at higher concentrations (Birge and Black, 1979). In trout, a 6:1 ratio of a mixture of soluble Zn:soluble Cu produces additive toxicity in hard water, but synergistic toxicity in soft water (Lloyd, 1961). A 1:1 ratio of Cu:Hg produces synergistic effects at high concentrations, additive effects at intermediate levels, and antagonism at low levels in trout (Birge and Black, 1979). Also in trout, a 1:5:40 ratio of the concentrations of Cu:Zn:Ni and a Cu level of 0.14–0.48 ppm Cu, produces additive effects (Brown and Dalton, 1970). Most studies show that Hg and Se are antagonistic (Frost and Lish, 1975; Turner and Rudd, 1983a; Turner and Swick, 1983b), but other studies show either synergistic (Huckabee and Griffith, 1974) or synergistic and antagonistic effects (Kim, *et al.*, 1977; Cuvin and Furness, 1988). A 2.5:1 molar ratio of Hg:Se protects minnows from Hg-induced toxicity, but a 1.3:1 molar ratio causes increased mortality (synergism). In contrast, Se potentiates Hg-induced toxicity in carp embryos (Huckabee and Griffith, 1974). Therefore, toxicity studies evidence the complex nature of elemental interactions in fish.

Studies of accumulation are correlated with those of toxicity at high, but not low, aqueous concentrations of Hg in a Hg-Se mixture (Kim, *et al.*, 1977). The presence of Se can protect minnows against Hg-induced toxicity; whereas, the presence of Hg can decrease Se elimination and prolong the half-life of Se (Freeman, *et al.*, 1978). Higher hepatic Se and Zn and lower levels of hepatic Fe have been found in fish exposed environmentally to Se

(Sorensen, 1986; Sorensen, *et al.*, 1988b). When aqueous Se:Hg concentrations are 50:1, hepatic concentrations of the two elements depend upon exposure time and exposure concentration (Sorensen and Thomas, 1988a; Sorensen, *et al.*, 1988b). Needless to say, the intricate interactions among various elements in teleosts are deserving of more attention in future scientific studies.

References

Birge, W.J., and J.A. Black, Effects of copper on embryonic and juvenile stages of aquatic animals, in Nriagu, J.O. (Ed.), Copper in the Environment. Part II: Health Effects, John Wiley and Sons, New York, (1979) 373.

Boecker, C.S., and E.M.B. Sorensen, Histopathological alterations in teleosts following selenate exposure, Toxicologist, 8 (1988) 109.

Brown, V.M., The calculation of the acute toxicity of mixtures of poisons to rainbow trout, Water Res., 2 (1968) 723.

Brown, V.M., and R.A. Dalton, The acute lethal toxicity to rainbow trout of mixtures of copper, phenol, zinc, and nickel, J. Fish. Biol., 2 (1970) 211.

Coombs, T.L., T.C. Fletcher, and A. White, Interaction of metal ions with mucus from the plaice (*Pleuronectes platessa* L.), Biochem. J., 128 (1972) 128.

Cross, F.A., L.H. Hardy, N.Y. Jones, and R.T. Barber, Relation between total body weight and concentrations of manganese, iron, copper, zinc, and mercury in white muscle of bluefish (*Pomatomus saltatrix*) and a bathyl-demersal fish *Antimora rostrata*, J. Fish. Res. Bd. Can., 30 (1973) 1287.

Cuvin, M.L.A., and R.W. Furness, Uptake and elimination of inorganic mercury and selenium by minnows *Phoxinus phoxinus*, Aq. Toxicol., 13 (1988) 205.

Doudoroff, P., and M. Katz, Critical review of literature on the toxicity of industrial wastes and their components to fish. II. The metals, as salts, Sew. Indust. Wastes, 25 (1953) 802.

Eisler, R., Cadmium poisoning in *Fundulus heteroclitus* (Pisces: Cyprinodontidae) and other marine organisms, J. Fish. Res. Bd. Can., 28 (1971) 1225.

Eisler, R., Radiocadmium exchange with seawater by *Fundulus heteroclitus* (L.) (Pisces: Cyprinodontidae), J. Fish. Biol., 6 (1974) 601.

Freeman, H.C., G. Shum, and J.F. Uthe, The selenium content in swordfish (*Xiphias gladius*) in relation to total mercury content, J. Environ. Sci. Hlth., A13 (1978) 235.

Frost, D.V., and P.M. Lish, Selenium in biology, Ann. Rev. Pharmacol., 15 (1975) 259.

Hill, C.H., G. Matrone, W.L. Payne, and C.W. Barber, *In vivo* interactions of cadmium with copper, zinc, and iron, J. Nutr., 80 (1963) 227.

Hilton, J.W., P.V. Hodson, and S.J. Slinger, The requirement and toxicity of selenium in rainbow trout (*Salmo gairdneri*), J. Nutr., 110 (1980) 2527.

Huckabee, J.W., and N.A. Griffith, Toxicity of mercury and selenium to the eggs of carp (*Cyprinus carpio*), Trans. Am. Fish. Soc., 103 (1974) 822.

Kim, J.H., E. Birks, and J.F. Heisinger, Protective action of selenium against mercury in northern creek chubs, Bull. Environ. Contam. Toxicol., 17 (1977) 132.

Koeman, J.H., W.H.M. Peeters, C.H.M. Koudstaal-Hol, P.S. Tjioe, and J.J.M. De Goeij, Mercury-selenium correlations in marine mammals, Nature, 245 (1973) 385.

Lloyd, R., The toxicity of mixtures of zinc and copper sulfates to rainbow trout (*Salmo gairdneri* Richardson), Ann. Appl. Biol., 49 (1961) 535.

Marafante, E., Binding of mercury and zinc to cadmium-binding protein in liver and kidney of goldfish (*Carassius auratus* L.), Experientia, 15 (1976) 149.

Matthiessen, P., and A.E. Brafield, Uptake and loss of dissolved zinc by the stickleback *Gasterosteus aculeatus* L., J. Fish Biol., 10 (1977) 399.

Natusch, D.F.S., J.R. Wallace, and C.A. Evans Jr., Toxic trace elements: Preferential concentration in respirable particles, Science, 183 (1974) 202.

Ohi, G., S. Nishigaki, H. Seki, Y. Tamura, T. Maki, H. Maeda, S. Ochiai, H. Yamada, Y. Shimamura, and H. Yagyu, Interaction of dietary methylmercury and selenium on accumulation and retention of these substances in rat organs, Toxicol. Appl. Pharmacol., 32 (1975) 527.

Pascoe, D., and D.L. Mattey, Studies on the toxicity of cadmium to the three-spined stickleback *Gasterosteus aculeatus* L., J. Fish Biol., 11 (1977) 207.

Roales, R.R., and A. Perlmutter, The effects of sub-lethal doses of methylmercury and copper, applied singly and jointly, on the immune response of the blue gourami (*Trichogaster trichopterus*) to viral and bacterial antigens, Arch. Environ. Contam. Toxicol., 5 (1977) 325.

Saiki, M.K., and T.P. Lowe, Selenium in aquatic organisms from subsurface agricultrual drainage water, San Joaquin Valley, California, Arch. Envir. Contam. Toxicol., 16 (1987a) 657.

Sandstead, H.H., Interactions of cadmium and lead with essential minerals, in Nordberg, G.F. (Ed.), Effects and Dose-Response Relationships of Toxic Metals, Elsevier Sci. Publ. Co., Amsterdam, (1976) 511.

Sellers, C.M., A.G. Heath, and M.L. Bass, The effect of sublethal concentrations of copper and zinc on ventilatory activity, blood oxygen and pH in rainbow trout (*Salmo gairdneri*), Water Res., 9 (1975) 401.

Sorensen, E.M.B., C. Harlan W., and J. Bell S., Renal changes in selenium-exposed fish, Amer. J. Forensic Med. Pathol., 3 (1982a) 123.

Sorensen, E.M.B., T.L. Bauer, J.S. Bell, and C.W. Harlan, Selenium accumulation and cytotoxicity in teleosts following chronic, environmental exposure, Bull. Environ. Contam. Toxicol., 29 (1982b) 688.

Sorensen, E.M.B., and T.L. Bauer, Hematological dyscrasia in teleosts chronically exposed to selenium-laden effluent, Arch. Environ. Contam. Toxicol., 12 (1983a) 135.

Sorensen, E.M.B., J.S. Bell, and C.W. Harlan, Hisptopathological changes in selenium-exposed fish, Amer. J. Forensic Med. Pathol., 4 (1983b) 111.

Sorensen, E.,M,.B., C. Harlan W., J. Bell S., T.L. Bauer, and A.H. Pradzynski, Hepatocyte changes following selenium accumulation in a freshwater teleost, Amer. J. Forensic Med. Pathol., 4 (1983c) 25.

Sorensen, E.M.B., and T.L. Bauer, Planimetric analysis of redear sunfish (*Lepomis microlophus*) hepatopancreas following selenium exposure, Environ. Toxicol. Chem., 3 (1984a) 159.

Sorensen, E.M.B., and T.L. Bauer, A correlation between selenium accummulation in sunfish and changes in condition factor and organ weight, Environ Pollut., 34 (1984b) 357.

Sorensen, E.M.B., The effects of selenium on freshwater teleosts, in Hodgson, E. (Ed.), Reviews in Environmental Toxicology, Elsevier Science Publishers B.V., Amsterdam, (1986) 59.

Sorensen, E.M.B., and P. Thomas, Selenium accumulation, reproductive status, and histopathological changes in environmentally exposed redear sunfish, Arch. Toxicol., 61 (1988a) 324.

Sorensen, E.M.B., T.L. Bauer, and M.G. Krause, Multielemental analysis of the hepatopancreas of selenium-exposed sunfish, Toxicologist, 8 (1988b) 108.

Sprague, J.B., and B.A. Ramsay, Lethal levels of mixed copper-zinc solutions for juvenile salmon, J. Fish. Res. Bd. Can., 22 (1965) 425.

Texas Department of Water Resources, Fish kill investigation of Lake Martin Creek, Unpublished data, (1978) 1.

Turner, M.A., and J.W.M. Rudd, The English-Wabigoon River System: III. Selenium in lake enclosures: Its geochemistry, bioaccumulation, and ability to reduce mercury bioaccumulation, Can. J. Fish. Aq. Sci., 40 (1983a) 2228.

Turner, M.A., and A.L. Swick, The English-Wabigoon River System: IV. Interaction between mercury and selenium accumulated from waterborne and dietary sources by northern pike (*Esox lucius*), Can. J. Fish. Aq. Sci., 40 (1983b) 2241.

APPENDIX 1

APPENDIX 1. EPA Water Quality Criteria Summary for Priority Pollutants[1].

		L.O.E.L. (ppb)[2]				
		Freshwater		Saltwater		
Element	Domestic H_2O Std (ppm)	Acute	Chronic	Acute	Chronic	Date[3]
Se	0.001	260	35	410	54	1980
As(+3)	-	44	40	69	36	1985
Pb	0.05	8.2	3.2	140	5.6	1985
Zn	5.00	320	47	170	58	1980
Cd	0.01	3.9	1.1	43	9.3	1985
Cu	1.00	18	12	2.9	-	1985
Hg	0.002	2.4	0.012	2.1	0.025	1985

[1]Criteria are compiled by Christopher S. Zarba, U.S.EPA, 401 M. Street S.W. (WH-585), Washington, DC 20460. This table is a modification of one published in <u>USEPA Quality Criteria for Water 1986</u>, EPA 440/5-86-001.

[2]L.O.E.L. = Lowest Observed Effect Level for an average of 24 h.

[3]Data of reference by the Federal Register.

INDEX

A

B

C

D

E

F

G

H

N

O

Q

R

S